TRANSISTOR CIRCUITS
AND
APPLICATIONS

PRENTICE-HALL SERIES IN ELECTRONIC TECHNOLOGY

Dr. Irving L. Kosow, EDITOR

Charles M. Thomson, Joseph J. Gershon, and Joseph A. Labok,
consulting editors

Plate 1. *At the Bell Telephone Laboratories, some of the first investigations leading to the invention of the transistor were made with this apparatus. Seated is Dr. William Shockley, who initiated and directed the Laboratories' transistor research program. Standing are Dr. John Bardeen, left, and Dr. Walter H. Brattain, key scientists in bringing the transistor to reality. The trio received the 1956 Nobel Physics award for their invention of the transistor, which was announced in 1948. Photo courtesy of Bell Telephone Laboratories, Incorporated.*

TRANSISTOR CIRCUITS
AND
APPLICATIONS

SECOND EDITION

LAURENCE G. COWLES

Senior Electronic Design Engineer
The Superior Oil Company
Houston, Texas

PRENTICE-HALL, INC., Englewood Cliffs, New Jersey

Library of Congress Cataloging in Publication Data

COWLES, LAURENCE G
  Transistor circuits and applications.

  Bibliography: p.
  1. Transistor circuits. I. Title.
TK7871.9.C69 1974   621.3815'3'0422   73-15576
ISBN 0-13-930073-2

Printed in the United States of America

10 9 8 7 6 5 4 3 2 1

Prentice-Hall International, Inc., *London*
Prentice-Hall of Australia, Pty. Ltd., *Sydney*
Prentice-Hall of Canada, Ltd., *Toronto*
Prentice-Hall of India Private Limited, *New Delhi*
Prentice-Hall of Japan, Inc., *Tokyo*

Grau, teurer Freund, ist alle Theorie
Und grün des Lebens goldner Baum.
(All theory, dear friend, is dull gray,
And only the golden tree of life is green.)

GOETHE

I often say that when you can measure what you
are speaking about, and express it in numbers, you
know something about it; but when you cannot express
it in numbers, your knowledge is of a meager and unsat-
isfactory kind; it may be the beginning of knowledge,
but you have scarcely, in your thoughts, advanced to
the stage of science, whatever the matter may be.

LORD KELVIN

Contents

TWO

THREE

FOUR

FIVE

SIX

SEVEN

EIGHT

NINE

TEN

ELEVEN

TWELVE

THIRTEEN

WAVE SHAPING AND NONSINUSOIDAL WAVES   240

FOURTEEN

HIGH-FREQUENCY CIRCUITS AND UHF APPLICATIONS   263

FIFTEEN

SIXTEEN

Preface
to the First Edition

This volume presents a practically oriented description of transistor circuits for technicians, junior engineers, and practicing electronic engineers. It brings to beginning electronic courses at the sophomore college level an introduction to the most commonly used transistor and field-effect circuits. The volume is also useful as a source of material for applications, as a laboratory manual, or for individual study. The circuits have been carefully designed as examples of good engineering practice using inexpensive components and should appeal to the experimenter.

The book is intended to provide a practical course on transistor circuits or to supplement and update basic treatments of electronics. Its purpose is to introduce the reader to transistor circuits and applications as simply as possible, using soundly based principles that open avenues of learning. The practical objectives of this book are achieved by a number of interesting departures from former methods of analysis. The transistor circuit calculations are simplified by using only those transistor parameters and circuit components that are truly of practical importance. Circuit theory and the algebra of the gain-impedance relations in transistor amplifiers are avoided by examining a circuit to find the stage current gains and applying simple rules and calculations. Concern for transistor temperature problems is minimized by using circuits that employ improved devices and exhibit the advantages of planar transistors and the latest types of field-effect devices.

An important aim of this book is to explain feedback biasing in practical terms and to show how easily feedback circuits can be analyzed and understood. Textbook methods that are primarily concerned with circuit calculations usually overemphasize the importance of transistor parameters and thereby complicate the understanding of practical circuits and application problems. On the other hand, the universal use of either local or overall feedback makes the transistor subordinate to the resistors of the feedback circuits. In fact, the very purpose of feedback is to insure that the circuit performance is determined by the resistors.

Today's transistor devices are so much improved that they are best used in direct-coupled pairs or in closely coupled arrays, as in integrated circuits. The characteristics of such amplifiers can be found either by calculating the open-loop gain and using the classical feedback equation or by analyzing the closed-loop gain-impedance relations. For applications studies we are primarily concerned with the closed-loop overall gain-impedance relations, so that the circuit approach is satisfactory for the purposes of this book. Whether the amount of feedback is adequate for the proposed application is easily and best determined by a measurement. The simplicity of the method of analysis employed in this text encourages practical people to use gain and impedance calculations as a quantitative guide in their daily transistor work.

The first half of the book presents a practical introduction to transistors and transistor amplifiers; the remainder describes common applications of solid state devices and illustrates the many uses for semiconductors. The student is prepared for circuit analysis by the introduction of basic material on semiconductor rectifiers, power supplies, regulators, and special purpose devices. Throughout the book it is assumed that the author's *Analysis and Design of Transistor Circuits* will be used for more detailed information concerning circuits, design, and as a source for references.

Laboratory skills are as important as technical understanding in practical work. The last chapter on experimental techniques is written to help readers in need of practical training and experience in transistor application. In the Appendix selected laboratory experiments and problems have been included for school use or to help the individual reader who learns best by putting theory into practice. The experiments are planned to demonstrate laboratory techniques and to show how experimental results give meaning to theoretical relations. The experiments are graded to advance the student quite rapidly because the circuits illustrated throughout the book, all carefully tested, provide material for many additional experiments. Most of the circuits indicate a bias resistor that may need a slight adjustment in the laboratory, although a typical value for the resistor is usually shown.

The design of transistor circuits is always a complex problem requiring experience and an understanding of many technical details—some beyond the purposes of this volume. Even a single-stage transistor circuit design may become more involved than is generally appreciated. The advanced principles of transistor circuit design are covered in the author's previous work to which reference is made above. The advanced book has references, design data, and

*D. Van Nostrand Co., Inc., Princeton, N.J., 1966.

extensive application of the methods presented in this volume. Together, these two books present a closely integrated introduction to transistor circuit application and design.

I wish to acknowledge my indebtedness to The Superior Oil Company for many hours of support while I developed an understanding of transistors, and to my colleagues for their help and keen interest in this book. The interest of friends and family is quite rewarding, and to them I dedicate this work. Finally, and with much pleasure, I thank my wife, Alice, for her many helpful suggestions and the long hours she gave to the improvement of the manuscript.

Preface
to the Second Edition

The widespread acceptance and use of TRANSISTOR CIRCUITS AND APPLICATIONS has justified a careful revision and improvement of the first edition. I have tried to make the book easier to read and understand by expanding explanations, particularly in the introductory chapters and in sections concerned with feedback. The principal additions to update the book include new material on single-stage transistor and FET amplifiers, feedback, feedback stability, high-frequency circuits, and high-frequency diodes. My thanks go to readers who have written concerning the book, and, wherever possible, their suggestions for improvements have been used in the revision. I am particularly grateful to William O. Wottlin for many helpful suggestions that have come from his use of the first edition as a Seismic Instrument Instructor at the Exploration and Training Center of the Shell Oil Company in Houston.

Readers desiring material primarily concerned with design are referred to my most recent book entitled TRANSISTOR CIRCUIT DESIGN, Prentice-Hall, Inc., 1972.

LAURENCE G. COWLES

List of Symbols

B_i Current loss factor
B_v Voltage loss factor

C_{be} Base-to-emitter capacitance
C_{gd} Gate-to-drain capacitance
C_{gs} Gate-to-source capacitance
C_I Input capacitance
C_M Miller effect capacitance
C_N Neutralizing capacitance
C_{ob} Collector-to-base capacitance

e_g AC generator or source voltage
e_I AC input voltage
e_O AC output voltage
e_p Peak voltage or peak-to-peak voltage
e_s AC signal voltage or switching voltage

F_B Feedback factor
f_β Beta cutoff frequency
f_c Cutoff frequency (half power, or 3 db)
f_h High frequency cutoff
f_l Low frequency cutoff
f_T Current gain-bandwidth product

G_i Current gain
g_m Transconductance
g_0 Zero gate voltage transconductance

G_p	Power gain
G_v	Voltage gain
h	CB short-circuit ac input resistance, h_{ib}
h'	Effective circuit h with external series resistors included
h_{ib}	Hybrid parameter, see h
i_b	AC base current
I_B	DC base current
i_c	AC collector current
I_C	DC collector current
I_{CO}	Collector cutoff current, temperature sensitive
I_{DSS}	Zero bias drain current,
i_e	AC emitter current
I_E	DC emitter current
i_I	AC input current
I_L	DC load current
i_O	AC output current
i_s	AC current in short-circuiting load
L_P	Primary inductance
L_S	Secondary inductance
n_I	Input winding turns
n_O	Output winding turns
P_{dc}	DC power
P_I	AC input power
P_O	AC output power
R_A	Bias resistor (usually adjustable)
R_B	Base resistor
r_b	Base resistance, internal
r_c	Collector resistance, internal
R_C	Collector resistor
r_e	Emitter resistance, internal
R_E	Emitter resistor
R_f	Feedback resistor
R_g	Generator or source resistor
R_I	Input resistance
R_L	Load resistance
R_M	Miller equivalent input resistance
R_N	Negative resistance
R_S	Source resistor

S	Usually R_B/R_E or R_f/R_L; approximately the dc current gain
S_c	Corrected S-factor (with collector feedback)
t	Time
V_B	DC base voltage (to ground)
V_{BB}	DC base supply voltage
V_C	DC collector voltage (to ground)
V_{CC}	DC collector supply voltage
V_D	DC drain voltage
V_{DD}	DC drain supply voltage
V_{DS}	DC drain-to-source voltage
V_E	DC emitter voltage (to ground)
V_{GS}	DC gate-to-source voltage
V_P	FET pinchoff voltage
V_R	Reverse voltage
V_Z	DC zener diode voltage
X_E	Reactance of emitter capacitor
X_L	Reactance of inductance L
Z_f	Feedback impedance
Z_{in}	Input impedance
Z_M	Miller equivalent input impedance
α	CB short-circuit current gain, i.e.,$-h_{fb}$; approximately 1
β	CE short-circuit current gain, h_{fe}; approximately 50
β'	Equivalent to $(\beta + 1)$; can be read as β with negligible error
ω	Frequency in radians per second $(2\pi f)$

Abbreviations

A	Ampere
B	Base
BW	Band width
C	Collector
CB	Common base
CC	Common collector
CE	Common emitter
D	Drain
dB	Decibel (see Appendix)
dBm	Decibel referred to 1 mW
E	Emitter
G	Gate
GHz	Gigahertz
H	Hertz
IC	Integrated circuit
kHz	Kilohertz
kΩ	Kilohm
mA	Milliampere
mH	Millihenry
mV	Millivolt

pF	Picofarad
p-p	Peak-to-peak
Q-point	Quiescent point
S	Source (FET)
SC	Short-circuited
μA	Microampere
μF	Microfarad
μH	Microhenry

TRANSISTOR CIRCUITS
AND
APPLICATIONS

CHAPTER ONE

Transistor DC Relations

Transistors are small, versatile devices that can perform an amazing variety of control functions in electronic equipment. In response to low power command signals, they control the flow of electricity at higher power levels. Transistors are used as rectifiers, amplifiers, oscillators, switches, and modulators.

Transistors replace vacuum tubes but have important advantages over and are very different from vacuum tubes: they are very small, and they require very little operating power. In properly designed circuits they are more reliable than and do not age like vacuum tubes. Transistors are available that operate with current flow in either direction and do not require heater circuits and heater power. Hence, the designer has much greater flexibility when designing circuits using transistors rather than vacuum tubes.

This chapter considers the static characteristics of transistors and shows how their performance curves are sometimes used to find the dc currents and voltages in an operating amplifier. The curves show that transistor amplifiers without feedback are impractical and, therefore, that transistor circuits require feedback circuits and methods of analysis quite different from those used for vacuum tubes. Because the static curves reflect the diode nature of transistors, we begin with a study of semiconductor diodes.

1.1 TRANSISTORS ARE COUPLED DIODES (Ref. 1)

Transistors are fabricated from semiconductor materials containing minute quantities of impurity atoms which control the electrical properties of the semiconductors. The control of the impurity distribution in manufacture requires a precise technology because the distribution of the impurities determines the characteristics of the transistor. While knowledge of the physical theory of semiconductors and of transistors is not essential for an elementary understanding of transistor circuits, the reader may enjoy and find it helpful to review the physicists' explanation of diode rectification found in most books on electronics. In this book we accept diodes and transistors as circuit elements with known electrical characteristics. Generally, we are not concerned with the internal construction or with the physicist's theory of semiconductors.

A diode has an interface between two different semiconductor materials —one having an excess of "holes" and the other having an excess of electrons. A voltage applied across the interface makes the holes and electrons move in opposite directions, depending on the voltage polarity. The resistance of the diode changes by many orders of magnitude, depending on whether the holes and electrons are caused to move together or to separate.

A transistor is a *pair of back-to-back semiconductor diodes* that are formed with the junctions so close that current in one diode affects the current in the other. In fact, *a transistor may be described accurately as a pair of coupled diodes*. For this reason we begin the study of transistors with a brief review of the electrical characteristics of a semiconductor diode.

1.2 DIODE CHARACTERISTICS (Refs. 1,12)

A diode (Fig. 1.1) is formed as a junction of two semiconductor materials of slightly different composition. Wire leads are connected to each semiconductor, each material by itself being a fairly good conductor. One of the semiconductors is called *p-type* material, and the other is called *n-type* material. The junction formed between the two semiconductors has quite remarkable and useful properties, and the entire device is known as a *junction diode*. As shown in Fig. 1.2(a), a diode conducts very well if connected in a circuit with the *p*-material side positive and the *n*-material side negative. The diode conducts only negligible current if the polarity is reversed [Fig. 1.2(c)]. When the diode is polarized to conduct, we say it is *forward biased*; when polarized to resist conduction, *reverse biased*. The dc voltage provided to make the diode a conductor or a nonconductor is sometimes called the *bias voltage*.

The current-voltage relations of a low-power diode are represented in Fig. 1.3. The first quadrant shows a current of 0.5 A (amperes) when the diode is forward biased by only 1.0 V (volts). The abrupt increase of the forward diode current at 0.5 V comes from the fact that the exponential current-voltage relation is conventionally represented with a linear voltage scale. The current increases exponentially with small increases of voltage and, if not limited by a series resistor in the circuit, would destroy the diode by overheating. The third quadrant shows that a reverse voltage of 100 V produces a current of only 0.03 μA. This current is constant and independent of voltage

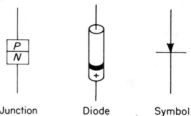

Figure 1.1. *A semiconductor diode.*   Junction   Diode   Symbol

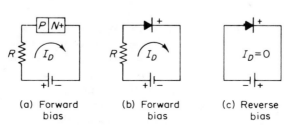

(a) Forward
bias

(b) Forward
bias

(c) Reverse
bias

Figure 1.2. *Diode polarity and current flow. Note: R is
required where shown to limit current.*

Figure 1.3. *Silicon diode current-voltage characteristics.*

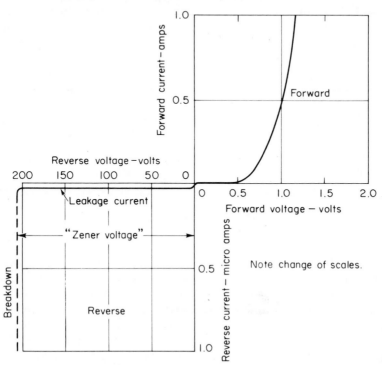

3

up to about 200 V. Above this voltage the diode can no longer prevent conduction and is said to **break down**. The diode will be destroyed by overheating unless the breakdown current is limited by an appropriately chosen resistor. Zener diodes are manufactured to make use of the sharp **breakdown characteristic** of the third quadrant for voltage regulation and control. They are available with reverse voltage ratings from about 3 V upward. High voltage ratings are easily obtained by connecting diodes in series.

Following electron tube convention, the *n*-type end of the diode is referred to as the **cathode** and is designated by a heavy line or by a + sign, as shown in Fig. 1.1. The + is a useful indication of the terminal that can be safely connected to the positive side of a voltage supply **without** a limiting resistor.

Diodes perform a number of functions but are better known for their ability to convert an ac voltage or signal to a series of half sine wave pulses of one polarity. The ac voltage is called the **signal**, the pulses are called the **rectified** output, and the diode may be referred to as a **rectifier**. The average or dc component of the series of half sine waves is 45 percent of the ac input rms value. When used as a rectifier, the plus terminal of the rectifier indicates the **positive** side of the rectified dc. We must remember that direct current cannot be obtained from a rectifier unless it is supplied from a diode loop that is capable of conducting dc. This means that the ac must be supplied either by a transformer with a **series diode** or by a coupling capacitor with a **shunt diode** as shown in Fig. 1.4(a) and (b).

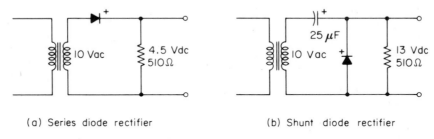

(a) Series diode rectifier   (b) Shunt diode rectifier

Figure 1.4. *Diode rectifiers.*

The rectifier without a capacitor produces a dc voltage that is 45 per cent of the rms ac voltage, whereas the rectifier with a capacitor produces a dc voltage nearly 140 per cent of the ac voltage. The lower dc voltage is the average value of the rectified voltage, whereas the higher dc voltage is approximately the peak value of the rectified wave. A large capacitor connected across the load in Fig. 1.4(a) increases the dc voltage to about 1 V less than the peak voltage of the rectified wave, as in Fig. 1.4(b).

1.3   TRANSISTOR CHARACTERISTICS (Refs. 3,7)

Transistors are fabricated by forming two semiconductor junctions in close proximity. As shown in Fig. 1.5, a *junction transistor* may be made as a *p-n-p* series of semiconductor materials with a wire lead connected to each. The *p-n* junction on one side forms one diode, and the *n-p* junction on the other side forms the other diode. However, when the center layer of *n*-type material is so thin that the two diodes are closely associated, we have a most unexpected result: if we connect to either of the *pn* junctions *with the lead attached to the other p-material open,* we find that each diode shows the expected rectification characteristics of an ordinary *pn* diode; however, when current is caused to flow in both diodes simultaneously, they interact in a manner explainable only by defining a new kind of circuit behavior—*transistor action.*

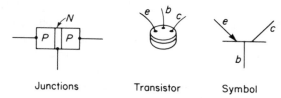

Junctions   Transistor   Symbol

Figure 1.5. *Junction Transistor.*

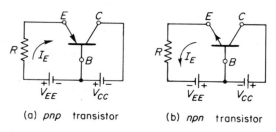

(a) *pnp* transistor   (b) *npn* transistor

Figure 1.6. *Transistor circuits showing bias polarities. Note: R limits the emitter current I_E.*

Figure 1.6(a) shows the circuit representation of a transistor connected to two bias batteries. The symbol for the transistor represents a pair of closely associated back-to-back diodes, with an arrow representing the emitter *p*-region and a single line representing the *n*-type layer that is shared by the diodes. The batteries are connected so that the diode on the left is forward biased and is conducting, while the diode on the right is reverse biased. The resistor R limits the emitter current, shown in Fig. 1.6 by an arrow, to a safe value. The diode on the right is shown with the arrow missing as a reminder that it is usually reverse biased. Because the diode on the left emits current into the transistor, it is called the **emitter**. The middle region is called the **base** and the right-hand p-region is called the **collector**. The entire device is called a **pnp transistor**. A transistor constructed as a series of *n-p-n* regions is called an **npn transistor**. The major difference between the two kinds of transistors is that the bias voltages are reversed for the *npn* transistor and the dc current directions are therefore reversed as shown in Fig. 1.6(b). The circuit symbol for the *npn* transistor shows the emitter arrow reversed in direction.

It should be noticed that both *pnp* and *npn* transistors are symmetrical with respect to the base. That is, the emitter and the collector of a given transistor are made of the same kind of material and are associated with the base in similar ways. In symmetrical transistors it makes little difference which region is used as the emitter and which is used as the collector. In most transistors, however, the emitter and the collector are not manufactured symmetrically. These devices exhibit a limited transistor action when connected in a circuit with emitter and collector leads interchanged. Early grown junction transistors were manufactured symmetrically, but more useful devices are now produced by making the emitter and collector different.

The characteristics of transistor action are described in Fig. 1.7, which is similar to Fig. 1.6 except for the addition of arrows to indicate the emitter, base, and collector currents found in an active transistor. If the transistor were simply a pair of diodes, we should expect to find, as in Fig. 1.6, a large current in the forward biased emitter diode and only a very small leakage current in the reverse biased collector diode. Because the transistor diodes are coupled, we find a large emitter current and an almost equally large collector current, their magnitudes represented by the length of the arrows in Fig. 1.7. The base current is small, being the difference of two nearly equal currents.

The current ratio obtained by dividing collector current by emitter current is a characteristic constant called the **current gain α**. The ratio of the collector current divided by the base current is called the **current gain β**. Although α is slightly less than 1, β is usually between 20 and 200. The current gain β is the most important transistor parameter, but the wide range over which β varies makes β anything but a constant. However, by good design transistor circuits are made to perform almost independently of β.

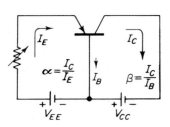

Figure 1.7. *Active transistor circuit.*

Transistors are *amplifiers* because the collector current varies with a change of current in the low resistance emitter diode. Because the collector current change occurs in a high impedance, reverse biased diode, high impedances can be used in the collector circuit, and the power change there may be many times greater than the power change in the emitter circuit. The power required for current flow in the collector circuit is furnished by the collector circuit battery, but the current is controlled by the emitter current. We call this process *power amplification* or *power gain*.

A simple picture of transistor action may be obtained by imagining that the current in the base is able to destroy the ability of the reverse biased collector diode to block current flow. For circuit studies, the collector current is shown as if from a constant current source, or generator connected across the collector diode. The current supplied by this generator is assumed to be proportional to—i.e., controlled by—the emitter or base current. Regardless of the representation, the transfer of energy from the collector battery to the collector load is controlled by the current in the emitter diode. In other words the collector circuit behaves as if it is coupled to the emitter circuit.

It is of interest to note that a transistor can be represented with surprising accuracy by a pair of equations known as the *Ebers* and *Moll relations* (Ref. 1). One equation represents the emitter diode; the other, the collector diode. Each equation has a coupling term: thus, a transistor is truly a pair of coupled diodes.

1.4 TRANSISTORS VS. VACUUM TUBES

Transistor characteristics are so different from those of vacuum tubes that a meaningful general comparison between transistors and vacuum tubes is quite impossible. The vacuum tube is easily understood as a voltage amplifier and many beginning students, familiar with vacuum tubes, hope for a useful comparison or a dual relationship. For those persons it may be helpful to explain the differences that make it impractical to compare the two devices or to use analogies.

A vacuum tube is usually operated with an external grid resistor that makes the grid insensitive to the grid leakage current in the tube and independent of the condition of the tube itself. In amplifier design and analysis it is simpler to consider the vacuum tube as a voltage amplifier and to neglect the current demands of both the grid resistor and the tube. This simplification of vacuum tube design is only one of several that are commonly used; it means that a vacuum tube voltage amplifier is not designed to achieve maximum voltage and power gain. Even so the design or analysis of a simple vacuum tube amplifier may be a formidable problem. Fortunately, the power gain of a vacuum tube is so high that the approximations made for analysis and design are usually negligible.

A transistor exhibits a linear relation between output and input current, but a nonlinear relation between output voltage (or current) and input voltage. For this reason the transistor may be described simply as a *current amplification device*. Moreover, compared with a vacuum tube, its power gain is low. As conservation of power gain is usually desirable, the effect of the transistor input impedance on a prior stage or on a driving source must be considered. This means that a transistor cannot be viewed merely as a current amplifier—the dual of a vacuum tube. Usually, we need to know both the current and voltage gain and, because the transistor is a low-voltage low-impedance device, the practical considerations of circuit design are very different from those of a vacuum tube.

Hence, the many differences between transistors and vacuum tubes cause practical considerations in their application to be very different; therefore, accepting these differences, we must approach transistors with an open mind, avoiding comparisons and analogies. Transistors have unique advantages, and offer a flexibility of design never possible with vacuum tubes. Some semiconductor devices are very much like their vacuum tube counterparts, but, because they are often directly coupled to transistors, we must understand these devices better than by the use of analogies.

1.5 DIRECT CURRENT RELATIONS IN A TRANSISTOR AMPLIFIER

A transistor by itself does not make a particularly useful amplifier. The dc bias currents or voltages must be established so that the transistor will respond without distortion to both positive and negative signal excursions from a given *quiescent* operating point (*Q-point*). We must also provide isolating transformers or capacitors so that the Q-point will not be disturbed when the amplifier is connected as part of a larger circuit. In this section we consider the dc relations and the problem of biasing a single transistor stage. By a *stage* we mean a single transistor with its bias and auxiliary circuit components. An amplifier is usually a series of stages and may include filters, power supplies, and various auxiliary circuits. When there is no input signal, the dc currents and voltages normally existing at the terminals of the transistor are referred to variously as the quiescent point conditions, bias values, or simply as the *Q*-point. Transistors used in *small signal amplifiers* (those in which the ac signal is small compared with the dc bias currents and voltages) are normally biased at emitter currents between 0.1 and 10 mA and collector voltages between 3 and 30 V. Bias currents and voltages below this range may cause distortion problems, while values above this range may cause excessive power dissipation in the transistor.

A *transistor stage* is shown in Fig. 1.8. The transistor emitter terminal is used by both the emitter circuit and the collector circuit. A circuit of this form

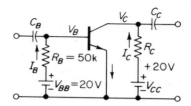

Figure 1.8. *Single-stage CE amplifier.*

is called a **common emitter**, or **CE**, connection. The battery V_{BB} on the left is connected through a current limiting resistor R_B and supplies a base current that forward biases the emitter diode. A silicon *npn* transistor is usually operated with a base-to-emitter voltage drop of about 0.6 V; a germanium transistor, with about 0.25 V. Transistors are operated with a 0.6 V (or 0.25 V) forward bias because this junction voltage produces a current density and temperature rise in the semiconductor which can be safely recommended for general purpose circuit designs. In either case the base current may be calculated by Ohm's law. For the silicon transistor the base current is given by:

$$I_B = \frac{V_{BB} - 0.6}{R_B} \qquad (1.1)$$

However, the base supply voltage is usually obtained from the collector supply, which is so much larger than the diode voltage drop that the diode voltage can be neglected. This means that the base current may be calculated *as if the emitter diode were a simple short circuit*, and Eq. 1.1 may be simplified to:

$$I_B = \frac{V_{BB}}{R_B} \qquad (1.1a)$$

For the example illustrated by Fig. 1.8 we substitute $V_{BB} = 20$ V, $R_B = 50$ kΩ (kilohms), and find that the base current is 0.4 mA.

The collector supply is polarized to reverse bias the collector diode, and a current limiting resistor is not required so long as the base current is properly limited. In a linear amplifier the base-to-emitter voltage is usually so small that the total collector-to-emitter voltage may be assumed to be the voltage across the collector diode. In practical circuit studies the **collector voltage** is generally measured from the collector to the common (ground) side of the circuit. The term **collector-to-emitter voltage** is used whenever the voltage across the transistor must be distinguished from the measured collector voltage.

In practical circuit studies the positive directions of current flow are usually chosen to agree with the actual directions, as shown in Fig. 1.7. The arrow on the emitter shows the direction of current in the emitter and, by its extension, the current direction in the base and the collector. The emitter current I_E is the sum of the base current I_B and the collector current I_C. This statement is expressed algebraically by the equation:

$$I_E = I_B + I_C \tag{1.2}$$

and the use of positive current directions makes the signs of the currents in Eq. (1.2) always conveniently positive.

A transistor stage such as that shown in Fig. 1.8. requires not only the biasing components, but also must have some means of introducing signal power in the input side and some means of transferring power changes in the collector circuit to other devices or stages of an amplifier. The amplifier in Fig. 1.8 has an input capacitor for inserting an ac input signal. For the present, we assume that the bias battery and resistor can be made large enough so that most of the signal flows into the transistor, and the portion lost in R_B is negligible. In the collector circuit a series resistor R_C has been included so that the collector current changes produce voltage changes that can be usefully employed. The output capacitor connected to the collector is a means of transmitting collector voltage changes to another stage or to a load. Obviously, if we want a large output voltage, we should make R_C large. However, in selecting this resistor we must understand its effect in reducing the collector voltage and know the magnitude of the collector Q-point current. In other words, we must know the *current-voltage characteristics of the transistor that is to be used in this stage*. The transistor characteristics used to obtain this information are known as the **static collector characteristics**.

1.6 THE STATIC COLLECTOR CHARACTERISTICS

A transistor, like an electron tube, has a nonlinear current-voltage characteristic that is most easily described graphically. The static collector characteristics represent the collector dc current-voltage relations. As shown in Fig. 1.9 for a low power transistor, such curves show how the collector current of a transistor depends on both the collector voltage and the base bias current. When the base is open, $I_B = 0$, the collector is like a reverse biased diode, and the collector current is the small reverse leakage current of a reverse biased diode. This current is practically independent of the collector voltage, as shown by line AB in Fig. 1.9.

As the base current is increased in 0.2 mA steps, the collector current increases in 5 mA steps, shown by the successively higher (nearly horizontal) lines. The fact that the lines are almost horizontal means that the collector current for a fixed base bias is practically independent of the collector voltage. Because the collector current is independent of the collector voltage, the current seems to come from a high resistance source and can be represented as a constant current.

From the static characteristics we may infer that the series resistor R_C in

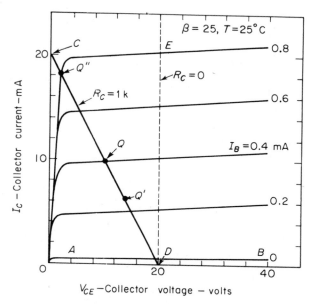

Figure 1.9. *CE collector curves with load lines.*

Fig. 1.8 lowers the collector Q-point voltage without appreciably changing the collector current and that the collector current is controlled by the base current only.

1.7 LOAD LINES

The combined effect of the base current and the collector load resistor may be examined by superimposing a load line on the static collector characteristics. For example, the voltage relations existing in the collector circuit shown in Fig. 1.8 may be represented by the linear equation:

$$V_C = V_{CC} - I_C R_C \qquad (1.3)$$

The line representing Eq. (1.3) has intercepts with the axes that are found by making V_C and I_C separately equal to zero. Short-circuiting the collector makes $V_C = 0$, and open-circuiting the collector makes $I_C = 0$. A line drawn on the collector characteristics between the short-circuit and open-circuit intercepts is called the *load line* for the resistor R_C because the line represents all possible collector voltages and currents when the load resistor is R_C. At high collector currents the collector-to-emitter voltage approaches 0 V. At low collector currents the collector voltage approaches the collector supply voltage. For small signal applications the amplifier shown in Fig. 1.8 is usually biased to make the collector voltage V_C about one-half the supply voltage V_{CC}.

Suppose now that the resistor R_C in Fig. 1.8 is 1000 Ω (ohms) and that the load line for R_C is to be drawn on Fig. 1.9, which represents the CE collector curves for the transistor used in Fig. 1.8. We know that with the collector short-circuited the collector current is 20 V $\div$ 1000 Ω, or 20 mA. This calculation gives the 20 mA intercept at C in Fig. 1.9. When the transistor is open-circuited the collector voltage is 20 V, the collector supply voltage. The intercept at D in Fig. 1.9 is therefore at 20 V. The line CD is the 1000 Ω load line; similarly, the line DE is the zero resistance load line. Each point on a load line represents a possible collector Q-point, and the line can be identified by using any convenient pair of Q-points. The base current at a Q-point on the line is the base current that will establish that Q-point. For example, the base current for the amplifier shown in Fig. 1.8 was found in Sec. 1.5 to be 0.4 mA; therefore, the Q-point of the amplifier falls at Q on line CD.

The load resistance R_C and the Q-point used for an amplifier design will usually depend on a number of related, yet conflicting, requirements. For this reason the designer's final selection of the circuit resistor values may require a difficult decision that can be, within limits, arbitrary. The collector supply voltage V_{CC} is usually a value that is conveniently available in power supplies or is conventionally used. The product of the Q-point current and voltage must also be within the power rating of the transistor, as this product represents power that must be dissipated by the transistor. Manufacturers' data specify limits of collector current and voltage as ratings that must be considered at the end points of the load line.

The last step in the design of an amplifier stage is usually the selection of a bias resistor that will supply the necessary base current. In a simple amplifier that must have a linear output vs. input response, we expect to find the Q-point somewhere near the center of the load line because this will permit approximately equal peak signal excursions. For the amplifier represented by the load line in Fig. 1.9 the Q-point Q requires a base current of 0.4 mA. If a 20 V collector supply is used to furnish the base current, as shown in Fig. 1.8, and the base-emitter diode is nearly a short circuit, as explained earlier, then the base current is 0.4 mA when the bias resistor is 50,000 Ω. However, if the amplifier is constructed and biased by a fixed resistor supplying the correct base current, we usually find that the Q-point does not fall, as predicted, near the center of the load line.

The real difficulty with this design will soon be recognized when several transistors, even of the same type, are used, or when the junction temperature of the transistor is changed. The difficulty is mainly that the Q-point is sensitive to a change in the transistor dc current gain. The static curves used in the design imply that the transistor current gain β is about 25. Actually, the current gain may vary from one unit to another by as much as ten to one.

The CE current gain, as defined in Sec. 1.8, is the ratio of the collector current change divided by the corresponding base current change, both measured at a fixed collector voltage. In Fig. 1.9 the line $R_C = 0$ represents a series of base currents and corresponding collector currents, all at the same collector voltage. When the base current is changed by 0.2 mA along this line, the collector current changes by 5 mA; hence, the short-circuit current gain is $5 \div 0.2 = 25$. However, as stated above, the current gain for this transistor may vary from as low as 15 to as high as 150, and a base current of 0.4 mA may correspond to a collector current of from 6 to 60 mA. As shown below, such a wide variation of collector current is usually intolerable in practical applications.

Many transistors are sold with only a specified minimum β; within a given type, β may range from 15 to 200 or more. If the transistor selected has a β of 15, the Q-point will fall at Q' instead of Q. Any unit having a β greater than 50 will cause the Q-point to fall at Q''. (All curves for $\beta > 50$ pass very near the point Q''.) Obviously, such a wide range of Q-points leaves much to be desired. Even if the transistors are carefully selected or the bias current is suitably adjusted, the β change with junction temperature changes can be enough to cause an unacceptably large variation of the Q-point. We see that the difficulty in this design is not that the emitter voltage drop is neglected in calculating the bias resistor, but that the static collector characteristics conceal the *essentially variable nature* of β. For this reason the static collector curves are useless for circuit design.

We conclude that the load line approach to transistor amplifier design (as used for vacuum tubes) is not useful except as a visual aid in power stage designs. The difficulty caused by the variability of β is resolved in Chap. 3 by using feedback to reduce and stabilize the stage current gain. An elementary amplifier has been discussed to illustrate an unsatisfactory design. However, in Chapter 2 the gain and impedance characteristics of transistors without feedback are used to explain how feedback improves a single-stage amplifier.

1.8   THE TRANSISTOR CURRENT GAINS—α (ALPHA) AND β (BETA)

The current gain of a transistor is an important characteristic parameter that describes the ratio of collector current to transistor input current. Because the collector current is under the control of current in the emitter loop, and may be considered as being proportional to either base current or emitter current, there exist two current gain parameters, α and β.

When the transistor is used with the emitter as the common terminal, the base terminal is called the input terminal, and the collector current I_C is expressed as being proportional to the base current I_B by writing:

$$I_C = \beta I_B \qquad (1.4)$$

When the resistance in the external collector circuit is small compared with the emitter diode resistance, the factor β is called the short-circuit CE *current gain.* A small external collector resistance is said to short-circuit the collector.

When the transistor is used with the base as the common terminal as shown in Fig. 1.7, the collector current is expressed as being proportional to the emitter current by writing:

$$I_C = \alpha I_E \tag{1.5}$$

Again, if the external load resistance is negligible, the factor α is called the short-circuit CB *current gain.*

For dc circuit analysis both current gains are defined as the ratio of the collector current divided by the corresponding input current I_B or I_E. For ac circuit analysis the current gain β (and similarly α) is defined as the ratio of the collector current change to the base current change. In the linear region of the static characteristics the dc and ac gains are nearly equal and may be measured either way. Moreover, for reasons explained in Sec. 1.10, the transistor current gain β is so variable that nothing is gained by making a distinction between dc and ac current gains.

For a good transistor, β is between 20 and 200. This, by Eq. (1.4), makes the base current from 5 to 0.5 percent of the collector current. For these values of base current Eq. (1.2) shows that the collector current will be 95–99.5 percent of the emitter current. Hence, by Eq. (1.5) α will be between 0.95 and 0.995. We observe that a 10 to 1 variation of β is equivalent to a 5 percent change of α. The variability of β must be taken into consideration in transistor work, but α can for most practical purposes be assumed to be 1.

1.9 SHORT-CIRCUITING LOADS (Ref. 1)

Except at very high frequencies, the external resistances in the collector circuit may be assumed to comprise a short-circuiting load. This means only that practical collector load resistances are small enough so that the α and β used in a calculation are the short-circuit α and β given in the commercial transistor data sheets (see Appendix). The assumption that a practical load short-circuits the transistor collector provides a useful simplification for transistor circuit analysis. This assumption simplifies the algebra of circuit calculations, making it easier to understand transistor circuits in general and easier to predict circuit performance when β varies.

With short-circuiting loads, Eqs. (1.4) and (1.5) may be combined as a double equality:

$$I_C = \beta I_B = \alpha I_E \tag{1.6}$$

Equation (1.6) is useful because it means that all three currents in a transistor may be written down if one of the current gains and one of the terminal currents are given. The double equality simply means that a transistor circuit problem having two circuit loops may be solved by using one equation and one terminal current.

1.10 CURRENT GAIN RELATIONS AND β'
(BETA PRIME)

Equations (1.2), (1.4), and (1.5) may be combined to eliminate the three transistor currents and leave a relation in terms of only α and β. The reader should easily find that:

$$\beta = \frac{\alpha}{1 - \alpha} \tag{1.7}$$

or:

$$\alpha = \frac{\beta}{\beta + 1} \tag{1.8}$$

It so happens that $(\beta + 1)$ is more useful in transistor circuit calculations than β, so it is convenient to give $(\beta + 1)$ the special symbol β'. From Eqs. (1.7) and (1.8) one finds that:

$$\beta' \equiv \beta + 1 = \frac{\beta}{\alpha} = \frac{1}{1 - \alpha} \tag{1.9}$$

As an example of the use of these equations, let us assume that β is given as 30. Equation (1.8) shows that α is $30 \div 31 = 0.97$. Equation (1.9) shows that β' is 31. Because β is usually 30 or larger, β' will be less than 3 percent larger than β, and we can use β' and β almost interchangeably. The current gain of a transistor is so variable from one unit to another that it is meaningless to think of β as having a precise value except as a convenience in making a calculation. That β is over 30 times more variable than α is shown by using α, which is in error by about 0.2 percent, to recalculate β, using Eq. (1.7). We find that $\beta = 32$, which is in error by about 7 percent. α and β are not constants by any means, and we must learn to recognize whether or not a particular circuit will be sensitive to the changes of β. Any circuit calculation that requires an exact value of β probably indicates that the circuit is impractically sensitive to the transistor and its environment.

It is helpful to review the difficulty that we encountered with the design of the amplifier in Fig. 1.8 by reference to Eq. (1.4). The problem is simply that with a fixed base bias current I_B, the collector current varies with β. If β can vary by a factor of 10 depending on the transistor, then the Q-point current I_C will vary by the same factor or until the short-circuit current is reached. If β can change by a factor of 2 because the junction temperature rises after the amplifier is turned on, then we must expect the Q-point current to change by a factor of 2. Unfortunately, we are usually faced with additional temperature-drift problems, so the design must be made relatively independent of β.

So far we have been discussing dc relations exclusively, so that the current gains α and β are dc ratios. In most work with transistors we are interested primarily in ac current gains, and it is convenient to use the same symbols α and β to represent the corresponding ac current gains. Because it is meaningless to imagine α or β as representing a precise number, we may reasonably let α and β represent **both dc and ac gains** and assume the reader is able to distinguish what is meant or what is important in a particular instance. Moreover, it should be recognized that β is always indefinite and that the variable nature of β is one of the principal problems in the application of transistors. A transistor circuit is practical only when there is enough feedback to make the gain and Q-point of a stage independent of the transistor β. To ensure adequate feedback we must know that β is high, but it is misleading and illogical to assign β a fixed value. Feedback will be discussed in detail in Chaps. 3 and 4.

1.11 HYBRID PARAMETERS

The hybrid parameters were developed to represent the early point-contact transistors. Although and because they lead to complicated circuit relations, there has been an overemphasis on their value for analysis and design. Transistor improvements made two of the four parameters useless in low-frequency analysis, and for high-frequency studies the hybrid parameters have been replaced by the more meaningful y and s parameters. Thus, the hybrid parameters and the hybrid equivalent circuits are essentially obsolete. Moreover, gains and impedances calculated with the hybrid parameters are seldom of practical value. The effect of a Q-point change is either neglected or lost in the formulas, and the calculations assume a particular value for the transistor current gain, which may vary more than 10 to 1 from one transistor to another.

For all practical purposes a transistor amplifier must have feedback. If an amplifier has adequate feedback, the designer can easily calculate the gain and impedance characteristics without knowing the transistor parameters. Thus, there is little reason to struggle with hybrid parameters and with the complicated open-loop, no-feedback, gain. The most useful transistor parameters are the current gain β, which is a figure of merit always given in the manufacturer's data sheet, and the short-circuit input impedance h, which is easily calculated without the data sheet. For these reasons the hybrid parameters are not used per se in this book or in the author's *Transistor Circuit Design.*

SUMMARY

A transistor is a pair of closely coupled semiconductor diodes that provide power gain when current in the high-impedance collector circuit is controlled by current in the low-impedance emitter circuit. The diode character of the transistor makes an understanding of diodes the key to an appreciation of transistors and transistor applications.

The dc current-voltage relations of semiconductor diodes and a coupling term, αI_E or βI_B, provide a correct description of transistor static characteristics. A transistor is made operable, or biased, by supplying a fixed current that forward biases the emitter diode and by supplying a voltage that reverse biases the collector diode. In a CE connection the collector current is β times the base current, and β is a highly variable factor.

The symbol β', defined by $\beta' \equiv \beta + 1 = \beta/\alpha$, is introduced to simplify circuit relations because β' and β are both large in practical applications and are essentially equal.

Practical circuits must be designed to make the Q-points and the gains independent of β. Therefore, practical circuits always use stabilizing feedback, and a load line study of transistor performance is no longer applicable. By abandoning load line methods, the analysis of transistor circuits is greatly simplified because we need consider only the feedback resistors that determine the impedance and gain characteristics of the amplifier and the bias resistors that fix the individual stage Q-points.

PROBLEMS

1-1. In Fig. 1.8 suppose that the collector supply voltage $V_{CC} = 20$ V, $R_C = 1000\,\Omega$, and the dc voltage from the collector to emitter is 15 V. (a) Find the emitter current, α, and β. (b) Find the value of R_B that will make the collector voltage 8 V.

1-2. Use Fig. 1.8. Assume $\alpha = 0.98$ and that the base current is 1 mA. Calculate the corresponding collector and emitter currents.

1-3. Derive Eqs. (1.7) and (1.8), as suggested in the text.

1-4. Refer to Fig. 1.9 and assume that β varies linearly with temperature and is double the 25°C value at 125°C. If $I_B = 0.4$ mA and is independent of temperature, at what temperature, approximately, will Q be at Q', and at Q''?

1-5. The transistor in Fig. P-1.5 has a CE current gain of 50 and a base-emitter voltage drop of 0.5 V. What is the value of R_A?

Figure P-1.5.

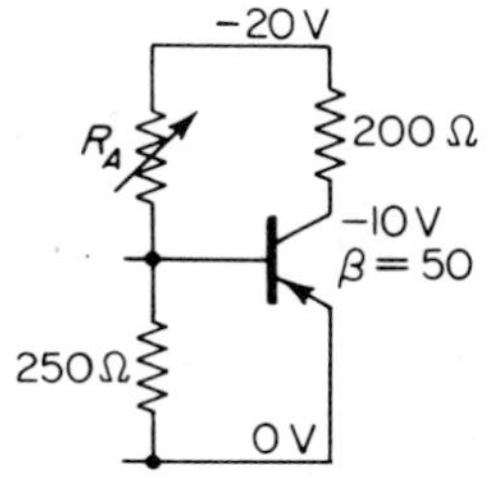

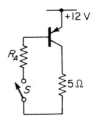

Figure P-1.6.

1-6. The 5 Ω heater in Fig. P-1.6 is to be turned on by closing the low current thermostat switch S. (a) What is the largest resistance R_A that you should specify? (b) What is the current in the thermostat?

1-7. Both transistors in the circuit of Fig. P-1.7 have a current gain of 100. (a) With S closed, what is the current in each resistor? (b) If the first transistor has a current gain of 10 and the second a gain of 100, what are the collector voltages?

1-8. Specify minimum current gains for each transistor in the circuit of Fig. P-1.7 that will ensure there is at least 3 A in the 5 Ω resistor. Neglect the collector-to-emitter voltage drops when possible.

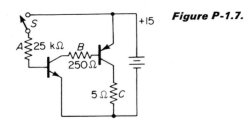

Figure P-1.7.

1-9. By a load line on Fig. 1.9 show what values of R_C may be used without appreciably changing the collector current as stated in Sec. 1.6 Ans: 0–1.8 kΩ.

1-10. (a) Repeat problem 1-9 assuming the bias current is reduced to 0.2 mA. (b) What are the load resistor limits if the transistor β doubles with an increased ambient temperature?

CHAPTER TWO

Transistor AC Gain-Impedance Relations

Transistor amplifiers are understood when we can predict their performance by analysis of a circuit. Thus, we must be able to calculate the input impedance and either the voltage gain or current gain (amplification) or both. This chapter introduces and explains fundamental relations that characterize transistor circuits. We will derive the *transistor gain-impedance relation*, which is a basic theorem that can be used for most gain calculations, and we will discuss the meaning and importance of the input impedance of a transistor. This chapter also presents examples which will be used to illustrate the impedance and gain characteristics of simple transistor amplifiers. We will begin by explaining what is meant by the term *impedance* and by defining *voltage gain* and *current gain*.

2.1 THE MEANING OF IMPEDANCE AND GAIN

The reader is presumed to understand the meaning of the terms *resistance*, *effective resistance*, and *impedance*. In circuit analysis the term *impedance* is used as a general term. In practical studies, however, the impedance is generally considered to be the effective ac resistance. Frequently, a resistor in the collector circuit is referred to as the *load impedance*, even though the circuit may be used later with an additional external load coupled by a capacitor. In such cases the gains calculated without the external resistor are corrected easily when a load is connected. In a similar way the gains calculated by assuming a zero impedance source may be corrected when a source having a significant impedance is connected.

The *input impedance* R_I of an amplifier is the ratio of the input voltage e_I to the input current i_I:

$$R_I = \frac{e_I}{i_I}$$

The internal output impedance of a transistor is nearly always assumed infinite.

Voltage gain G_v is defined as the ratio of output voltage e_O to the input voltage of the amplifier:

$$G_v = \frac{e_O}{e_I}$$

Current gain G_i is the ratio of the output current i_O to the input current i_I:

$$G_i = \frac{i_O}{i_I}$$

In this book the **power gain G_p** is always the ratio of the power output to the load, divided by the power input to the amplifier:

$$G_p = G_v G_i$$

2.2 TRANSISTORS AS AMPLIFIERS WITHOUT LOCAL FEEDBACK

This chapter deals with the characteristics of the simplest possible transistor amplifiers—those using transistors without feedback. Such amplifiers are rarely of practical importance because better performance characteristics can be secured by introducing feedback (see Chap. 3). However, one purpose of this chapter is to lay the groundwork for an understanding of feedback by discussing the characteristics of transistor amplifiers that make feedback desirable.

A transistor is a 3-terminal device: two terminals are required for the input and two for the output; therefore, it is necessary to make at least one of the terminals common to both input and output. The amplifiers are designated as CB, CE, or CC, depending upon which terminal is common. The three basic types of amplifiers are illustrated in Fig. 2.1. When a resistor is inserted between the common side of the transistor and the point used in common by the input and the output of the amplifier, it is useful to refer to the latter point as the ground, or occasionally as the power supply return. (A CE amplifier which uses the ground as common to the input and output circuits is shown in Fig. 3.2).

The characteristics of an amplifier differ considerably depending on which terminal is used in common, as will be explained in this chapter (Ref. 1, and Appendix). The CE amplifier gives the highest power gain and, having both voltage and current gain, is therefore the most generally adaptable. The CC amplifier cannot have a voltage gain greater than 1 but does offer a simple way to use a transistor for current gain and for coupling high-impedance sources to low-impedance loads. The CB amplifier offers an intermediate power gain, but because it has a current gain of 1 it cannot give a voltage gain except when the load impedance is greater than the input impedance. The CB

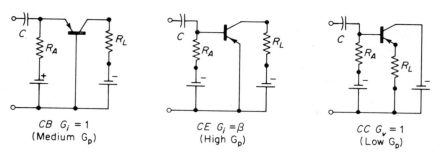

CB $G_i = 1$
(Medium G_p)

CE $G_i = \beta$
(High G_p)

CC $G_v = 1$
(Low G_p)

Figure 2.1. *Transistor amplifier types.*

amplifier is used much less frequently than the CE and CC amplifiers, but the input impedance of the CB amplifier is of basic importance in transistor circuit analysis.

Amplifiers that produce considerable amplification are usually operable only at small signal levels, and the gain of an amplifier is a measure of its effectiveness at small signal levels. The early stages of an amplifier use low-power transistors, and the dc bias currents and the terminal supply voltages are relatively small. Because the design of such amplifiers is simpler and quite different from that of high-level or power amplifiers, it is customary to call the early stages *small signal amplifiers*. The signal of such amplifiers is small compared with the static Q-point currents and voltages; therefore, the amplitude distortion of the signal is not an important consideration.

2.3 THE TRANSISTOR GAIN-IMPEDANCE RELATION (Ref. 1)

The gains and impedances of a transistor are closely related, and their relationship can be expressed in a simple formula. Once understood, this formula makes it possible for one to see at a glance just how a given amplifier can be expected to operate. The formula applies to almost any transistor amplifier, including amplifiers with feedback, so that it is of great value and utility. We will call this formula the *Transistor Gain-Impedance Relation* (TG-IR) and use the word *transistor* as a reminder that the relation is not useful for circuits that include vacuum tubes or field-effect devices.

The derivation of the Transistor Gain-Impedance Relation is simple. The amplifier represented in Fig. 2.2 has a pair of input and a pair of output

Figure 2.2. *Amplifier showing terms of the Transistor Gain-Impedance Relation.*

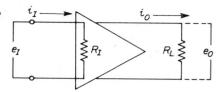

terminals, respectively. The ac signal input voltage is designated by e_I and the ac signal output voltage by e_O. We represent the input impedance of the amplifier by the resistor R_I and the output load resistor by R_L. Also, we designate the input current by i_I and the output current by i_O. By application of Ohm's law we know that the input voltage is equal to the product of the input current and the input resistance:

$$e_I = i_I R_I \tag{2.1}$$

The output voltage is likewise the product of the output current and the load resistance:

$$e_O = i_O R_L \tag{2.2}$$

Dividing Eq. (2.2) by Eq. (2.1), we have:

$$\frac{e_O}{e_I} = \frac{i_O R_L}{i_I R_I} \tag{2.3}$$

Now, we recognize the voltage ratio e_O/e_I as the voltage gain of the amplifier and the current ratio i_O/i_I as the current gain of the amplifier. Calling the voltage gain G_v and the current gain G_i, we can write Eq. (2.3) in the form:

$$G_v = G_i \frac{R_L}{R_I} \tag{2.4}$$

Equation (2.4) is the Transistor Gain-Impedance Relation, which we call TG-IR for short.

　　This relation tells us at once that if we wish to know the voltage gain of a transistor stage and we know the current gain, we must also know both the load resistance and the input impedance of the stage. Usually we can identify the load impedance R_L. The current gain can be determined by inspecting the circuit. The input impedance of a transistor or of an amplifier can be easily determined by applying a few simple rules. Any of the standard gain formulas require that one be able to identify and find these same quantities. The TG-IR requires nothing new and unifies the understanding of many transistor circuits.

　　As an example of the use of the TG-IR, let us consider an amplifier comprised of a series of *identical* R-C coupled or direct-coupled stages. The input impedance of each stage is the load impedance of the previous stage. Hence $R_I = R_L$, R_L/R_I in the TG-IR is unity, and the stage voltage gain is numerically equal to the stage current gain.

　　In practical amplifiers it is usually necessary to limit the current gain to about 20 per stage by local feedback or by overall feedback. Hence, one does not expect the voltage gain of transistor amplifiers to exceed about 20 per stage. As the TG-IR shows, voltage gain in excess of the tolerable current gain must be obtained by using an R_L/R_I ratio of more than 1. If the impedance ratio is less than 1, as is often the case, then one must expect voltage gains correspondingly lower than 20.

As a second example consider a CC stage (Sec. 2.10) which cannot have a voltage gain in excess of 1. If the current gain is limited to about 20 as the practical upper limit, then it follows that the input impedance R_I cannot be greater than 20 times the load impedance R_L. Where higher ratios of input-to-load impedance are required, one must expect to find additional CC stages or step-down transformers or less than unity voltage gain.

Before we can understand how to calculate the input impedance of an amplifier, we must examine the input impedance of a simple transistor. Also, to put the input impedance in a form that is useful for circuit calculations, we must use the T-equivalent circuit of the CB transistor.

2.4 THE CB TRANSISTOR T-EQUIVALENT

As indicated in the first chapter, a simple transistor by itself usually does not make a useful amplifier; but by examining the transistor by itself, we will learn how to calculate the gains and impedances of more complicated amplifiers. For circuit calculations it is convenient to represent a transistor as a T-network of resistances with an associated ac collector current source. The T-equivalent represents the ac resistances of a transistor at a particular operating point. These resistances may be obtained by ac measurements or derived graphically from the static characteristics.

We consider first the T-equivalent of the CB transistor because the intrinsic (or internal) input impedance h of the CB transistor is a fundamental parameter that must be used in many circuit calculations. Because the hybrid parameters are seldom used, we represent the CB short-circuit input impedance by h without subscripts.

A simplified form of the T-equivalent is shown in Fig. 2.3. It is called the short-circuited (SC) T-equivalent because the collector terminal is shorted to ground. Practical load resistances are usually small compared with the internal output resistance of the collector, so the T-equivalent is effectively short-circuited, even though a load resistance is used. The T-equivalent has an emitter resistance r_e, a base resistance r_b, and an arrow representing the collector ac current source. All parts of the T-equivalent except the load represent intrinsic, or internal, elements of the transistor.

The intrinsic resistances are ac resistances that depend on the emitter current and have only a negligible component of ohmic resistance. The ac components represent the dynamic resistance of the nonlinear junction; the

Figure 2.3. CB short-circuited T-equivalent. SC load means $R_L \ll r_c$.

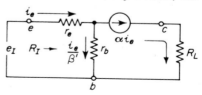

ohmic components, the resistances of the linear semiconductors. The dc circuit relations (Sec. 1.5) are represented separately as a nearly constant dc voltage drop across the junction. In small high-frequency transistors the base may be so thin that the base resistance is predominantly an ohmic "base-spreading resistance," but it is a safe assumption that most transistors used in small-signal low-frequency applications do not have an important ohmic component of resistance.

In the CB T-equivalent shown in Fig. 2.3 the input ac signal current is the emitter current i_e. Current in the collector circuit has as its source the constant current generator, represented by the arrow enclosed in a circle. This symbol represents the *active current source* and is a coupled current, αi_e or βi_b, depending on whether the emitter or the base current is considered to be the input. In the CB T-equivalent the active current is written as αi_e. Because small-signal ac currents represent small changes of dc currents and because the dc currents [Eq. (1.6)] are linear from the origin of the characteristic curves, we would expect the ac coupled currents αi_e and βi_b, to resemble the corresponding dc coupled currents. For these reasons the ac branch currents of the SC T-equivalent are related in the same way that the dc branch currents are related in Eq. (1.6). Using i to represent ac currents, we have:

$$i_c = \alpha i_e = \beta i_b \qquad (2.5)$$

It will be observed that the base current in the T-equivalent has been shown as i_e/β' by solving Eq. (2.5) for the base current:

$$i_b = \frac{\alpha}{\beta} i_e = \frac{1}{\beta'} i_e \qquad (2.6)$$

The load resistance is short-circuiting so that R_L is considered in a circuit calculation only to calculate an output voltage. The T-equivalent is now used to calculate the input impedance and the voltage gain of the CB amplifier.

2.5 THE CB TRANSISTOR INPUT IMPEDANCE AND VOLTAGE GAIN

The input impedance of the CB transistor is either measured or calculated by dividing the ac voltage e_I, applied between the input leads of the transistor, by the signal current i_e supplied to sustain the input circuit voltage drops. Equating the signal voltage to the voltage drops shown in Fig. 2.3, we have:

$$e_I = i_e r_e + \frac{i_e}{\beta'} r_b \qquad (2.7)$$

Dividing both sides of Eq. (2.7) by i_e, we obtain the input impedance:

$$R_I = \frac{e_I}{i_e} = r_e + \frac{r_b}{\beta'} \qquad (2.8)$$

Because the calculation of R_I assumes a short-circuiting load, Eq. (2.8) shows that the SC input impedance of a CB transistor is

$$h = r_e + \frac{r_b}{\beta'} \qquad (2.9)$$

This equation shows that the input impedance of the CB transistor is the sum of the emitter resistance and the base resistance divided by the current gain, β'. We note that the base resistance r_b can be quite large before it increases the input impedance, because the base current is relatively small. Many transistors are so constructed that the input impedance of the CB transistor is divided almost equally between the first and second terms of Eq. (2.9). The input impedance for a CB transistor is commonly referred to as the *SC input impedance* because it is measured with the collector connected to the common terminal; i.e., the external ac collector circuit is short-circuited.

The voltage gain of a CB transistor may be calculated by substituting values in the TG-IR, Eq. (2.4). Replacing G_i by the CB current gain α and the input impedance R_i by h, we obtain

$$G_v = \alpha \frac{R_L}{h} \qquad (2.10)$$

Since the current in R_L is from a current source, Eq. (2.10) implies that the transconductance of a transistor is α/h, or approximately $1/h$.

2.6 SHOCKLEY'S RELATION FOR THE CB INPUT IMPEDANCE (Refs. 1,3)

The input side of a CB or CE transistor is simply a forward biased diode. Thus, the input impedance of a transistor can be expected to vary with the diode current in the same way that the resistance of a forward biased diode changes with the diode current. We know that a forward biased diode has a low dc resistance and that the forward current increases rapidly for a small increase of the applied voltage. We would expect that the ac input impedance would be low and that it would decrease rapidly as the diode current is increased.

The dc current-voltage curve of a semiconductor diode is shown in Fig. 2.4, except that the current and voltage axes have been interchanged to put

Figure 2.4. *Emitter diode dc and ac relations.*

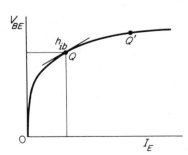

the diode current on the abscissa. This change is made because in transistor circuits we usually are interested in the base-emitter voltage when the emitter current is given (i.e., fixed by the bias circuit). Suppose that an emitter diode is biased at the point Q, so that the dc emitter current and the dc emitter voltage are coordinates of the point. If the emitter current is changed by small signal excursions above and below the emitter current I_E, the voltage will change by small excursions along the curve. For small changes, the curve is a segment of the tangent drawn at Q. The ac impedance of the diode is the ratio of the small-signal ac voltage divided by the ac current. Therefore, the ac resistance is proportional to the slope of the tangent at Q. At a higher emitter current, as at Q', the tangent is lower in value, making the ac resistance lower.

Shockley, one of the discoverers of the transistor, has shown that the input ac resistance, expressed in ohms, equals the constant 26 divided by the dc emitter current I_E in mA. Shockley's relation, written in algebraic form, is simply:

$$h = \frac{26}{I_E} \tag{2.11}$$

The h on the left is the well-known short-circuit input impedance of the common base transistor. This input impedance is important in circuit work because the reciprocal $1/h$ is the transconductance of a transistor. Shockley's relation equates h to the dc emitter current.

It will be observed that when the emitter current is 1 mA, the input impedance of the CB emitter diode is 26 Ω. This value is familiar to anyone who has studied transistor data sheets. Shockley's relation holds for any size transistor except power transistors operated at high emitter currents.

The importance of Shockley's relation can be appreciated by re-examining the equation for the CB voltage gain. Replacing h in Eq. (2.10) by $26/I_E$ from Shockley's relation, we obtain:

$$G_v = \alpha \frac{I_E R_L}{26} \text{ (Units: mA, } \Omega\text{)} \tag{2.12}$$

This reveals that the voltage gain of a simple CB transistor amplifier will be determined by the product of the load resistance and the emitter current. Equation (2.12) suggests that, within limits, the emitter current is an important parameter for the circuit designer, that high gain is not obtained at very low emitter currents, and that the emitter current is a means of adjusting the voltage gain of a transistor amplifier.

Shockley's relation accurately represents one of the two important transistor parameters and gives the input impedance of a CB transistor without using a data sheet. The fact that h is always given in transistor data as 26 Ω when $I_E = 1$ mA actually is a curious leftover from the earliest transistor data sheets.

2.7 A PRACTICAL CB AMPLIFIER

Shockley's relation combined with the TG-IR provides the means of finding the gain impedance characteristics of the practical CB amplifier shown in Fig. 2.5. We begin by determining the dc Q-point conditions of the transistor. Because the emitter diode voltage is small compared with 10 V, the emitter current is approximately the short-circuit current in the 1 kΩ resistor, 10 mA dc. The current gain α of a CB transistor is between 0.96 and 0.99 so the collector current is a little less than 10 mA. The dc IR drop in the collector supply resistor is 5 V, and the collector dc voltage is $10 - 5 = 5$ V. We note that the collector circuit dc load is the 510 Ω resistor, but the transistor's ac load is the resistance of 510 Ω and 600 Ω in parallel, which is 270 Ω.

Applying Shockley's relation, we find the input impedance of the amplifier is 2.6 Ω. The shunting effect of the 1 kΩ resistor that supplies emitter bias obviously can be neglected. We calculate the voltage gain of the amplifier by the TG-IR and find it is 270/2.6, or approximately 100. The Fig. 2.5 amplifier is not often useful because of the low-value input impedance and nonlinear character of the emitter diode which is across the input. Both difficulties can be alleviated by interposing a 23 Ω resistor in series with the capacitor. The resistor "swamps out" the small resistance changes of the diode and increases the input impedance of the amplifier to $23 + 2.6 = 26$ Ω. The TG-IR now shows that the voltage gain is reduced to 10, the amplifier gain is less dependent on the diode resistance, and the input current is more nearly proportional to the input voltage—that is, more linear. With the series emitter resistor, this amplifier makes a very stable wide-band high-frequency amplifier that should be useful up to the alpha cutoff frequency of the transistor.

The common base amplifier has illustrated one method of obtaining an increased input impedance—by exchanging voltage gain for input impedance. For applications that do not require the high frequency characteristics of the CB amplifier, a CE amplifier with the same voltage gain has about 50 times higher input impedance. The input impedance is higher because a CE transistor with the same emitter current and a current gain of β has an input current that is 20 to 200 times lower.

Figure 2.5. CB amplifier. Note: R_L is 510 Ω ‖ 600 Ω = 270 Ω.

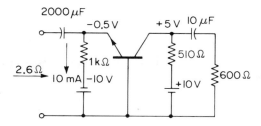

2.8 THE CE TRANSISTOR INPUT IMPEDANCE AND VOLTAGE GAIN

The input impedance of the CE transistor can be calculated in the same way as for the CB transistor. In the equivalent circuit of the CE transistor, shown in Fig. 2.6, the base current i_b is the (amplifier) stage input current;

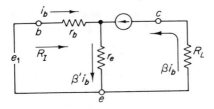

Figure 2.6. *CE SC T-equivalent;* $R_L \ll r_c/\beta$.

hence we use Eq. (2.5) to express the emitter and collector currents as multiples of the base current. Summing the voltage drops of the input circuit, we find:

$$e_I = i_b r_b + i_b \beta' r_e \tag{2.13}$$

and the input impedance becomes:

$$R_I = \beta' r_e + r_b \tag{2.14}$$

or:

$$R_I = \beta' h \tag{2.15}$$

Comparing Eqs. (2.15) and (2.8), we see that the input impedance of the CE transistor is β' times larger than the input impedance of the CB transistor. Transistor data sheets sometimes give the CE transistor input impedance h_{ie}, although Eq. (2.15) shows that h_{ie} will vary with β. For this reason we shall express the CE input impedance as βh so that the influence of β remains in evidence. As shown by Shockley's relation, h has a remarkably precise value that is determined by the emitter current, but β may vary by as much as ten to one. To understand the problems of transistor circuits, we must be careful not to conceal a basic difficulty by using h_{ie} as a fundamental parameter in the same way as h is used.

The voltage gain of the CE transistor can be found by inserting the current gain β and the input impedance [Eq. (2.15)] in the TG-IR. Hence, the CE voltage gain is:

$$G_v = \beta \frac{R_L}{\beta' h} \tag{2.16}$$

Because the betas cancel, the CE voltage gain is the same as in the CB voltage gain [Eq. (2.10)]. This means that the higher current gain of the CE transistor produces a higher input impedance and a higher power gain. It should be remembered, however, that the input impedance is sensitive to changes of β and is quite variable.

2.9 A CE AMPLIFIER GAIN AND IMPEDANCE CALCULATION

The amplifier shown in Fig. 2.7 is used as an example of a CE amplifier gain and impedance calculation. Because the transistor dc current gain is uncertain, it is necessary to assume that the bias resistor can be adjusted to

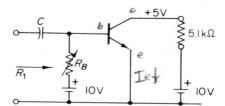

Figure 2.7. CE amplifier example for gain and impedance calculation.

set the Q-point as desired. For a Class A Q-point the collector voltage should be about one-half the supply voltage, 5 V, which requires a 1 mA collector current. By Shockley's relation, this current makes h about 13 Ω. Equation (2.15) shows that the CE amplifier input impedance depends on β and will be between 1000 Ω and 4000 Ω. Equation (2.16) shows that the CE voltage gain is independent of β, except for the variation of I_E with β, and will be approximately 200 (i.e., 5100/26).

A comparison of this amplifier (Fig. 2.7) with the CB amplifier shown in Fig. 2.5 is both interesting and instructive. It is evident that the CE amplifier has an input impedance βh that is about three orders of magnitude higher because β may be 100, and h is one order of magnitude higher because I_E is a factor of 10 lower. However, such dependence on β of both the input impedance and the Q-point cannot usually be tolerated in a practical amplifier. Because β may vary more than 10 to 1 from one device to another, the bias resistors must be individually adjusted for each CE stage, and the coupling capacitors must be individually selected if a particular frequency response is desired. These adjustment problems and the change of β with temperature generally make a CE stage impractical without feedback, even though its higher input impedance seems desirable.

The characteristics that make a simple CE stage so impractical can be corrected by adding enough emitter feedback to make a significant reduction of the gain. *Significant feedback* lowers the stage gain by a factor of 4 but makes an amplifier practical by making the gains relatively independent of β. With this improvement an amplifier with equal input and load impedances generally has stabilized voltage and current gains of about 20 per stage. Feedback for this purpose is usually provided by one of the feedback biasing circuits that are discussed in Chap. 3.

2.10 THE CC TRANSISTOR INPUT IMPEDANCE AND VOLTAGE GAIN

The circuit in Fig. 2.8 representing a CC transistor is similar to the circuit in Fig. 2.6 for the CE transistor, except that the emitter current has become the output current and the load resistance is in series with the emitter. The load impedance of a CB or CE transistor has a negligible effect on the input impedance because such loads are in series with the high-impedance collector. These loads do not affect the input current because they do not affect the branch currents. When the load is moved to emitter, as in Fig. 2.8, the base input current is reduced considerably because the emitter load carries the relatively high emitter current.

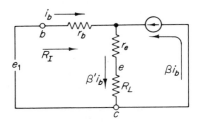

Figure 2.8. CC SC T-equivalent; $R_L \ll r_c / \beta$.

The input impedance of the CC transistor in Fig. 2.8 is calculated by adding the resistance R_L to the intrinsic emitter resistance r_e of Eq. (2.14). The input impedance of the CC transistor becomes:

$$R_I = r_b + \beta'(r_e + R_L) \qquad (2.17)$$

or:

$$R_I = \beta'(h + R_L) \qquad (2.18)$$

Whenever R_L is larger than h, the CC input impedance is several orders of magnitude larger than the transistor h, and Eq. (2.18) becomes:

$$R_I = \beta' R_L \qquad (2.19)$$

Thus, the input impedance is β times the load impedance.

The voltage gain of the CC transistor is obtained by substituting Eq. (2.18) into the TG-IR, so that:

$$G_v = \frac{R_L}{h + R_L} \qquad (2.20)$$

For load resistances, which are large compared with h, the voltage gain is:

$$G_v = 1 \qquad (2.21)$$

This means that the emitter voltage is the same as, or follows, the input voltage.

Consider the practical form of the CC transistor amplifier shown in

Fig. 2.9. Because the input resistance of a CC transistor, given by Eq. (2.19), varies with β, it is usually desirable to use a bias resistor R_b that is small compared to βR_L. This choice of R_B makes the input impedance constant and equal to R_B.

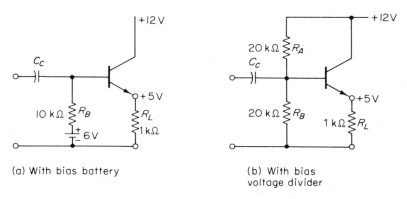

(a) With bias battery

(b) With bias
voltage divider

Figure 2.9. *CC amplifier example-a practical form.*

The current gain of the CC transistor amplifier is calculated by solving the TG-IR, [Eq. (2.4)], for G_i, thus:

$$G_i = G_v \frac{R_I}{R_L} \qquad (2.22)$$

Substituting G_v from Eq. (2.21) and R_B for the input impedance R_I, we obtain:

$$G_i = \frac{R_B}{R_L} \qquad (2.23)$$

Equations (2.21) and (2.23) show that the voltage gain and the current gain of a practical form of a CC amplifier are independent of the transistor and of the value of β. These two results show that a CC transistor amplifier can be designed to have gain characteristics that are substantially independent of the transistor. The voltage gain is independent of the transistor when h is small compared with the load R_L, and the current gain is independent of β if β is large compared to the resistor ratio R_B/R_L. Later, we shall find that this improvement of the amplifier is caused by *feedback*. The circuit (Fig. 2.9) has feedback because the load voltage and the input voltage are impressed on the transistor input in series opposition.

The principal application of the CC amplifier is in coupling high impedance circuits into low impedance loads. The amplifier transfers the input voltage to the load without requiring voltage step-down, as in a transformer. The practical form of the CC amplifier has a power gain and a current gain of about 10 and is used to replace transformers in many circuits.

2.11 A PRACTICAL CC AMPLIFIER (Refs. 1,2)

The principles of Sec. 2.10 are illustrated by the amplifier shown in Fig. 2.9(a). For simplicity the base bias is shown as supplied by a bias battery in series with the 10 kΩ base resistor R_B. If the base current is negligible, the base voltage is 6 V and the emitter follows the base. Allowing 0.5 V for the dc base-to-emitter voltage drop, the dc emitter voltage is 5.5 V. A quality transistor has a typical current gain of 100. Therefore, the indicated 5.5 mA emitter current implies a 55 μA base current, and the voltage drop in the base resistor is 0.55 V. We conclude that the emitter voltage is about 5 V and the emitter Q-point is satisfactory for high output voltage changes.

Because the emitter current is 5 mA, h is much smaller than R_L. By Eq. (2.20) we expect the ac voltage gain to be 1. The value of β for present day transistors is typically 100. Therefore, R_I, given by Eq. (2.18), is 100 kΩ. Equation (2.18) neglects the shunt effect of the resistor R_B. Hence, the input impedance of the CC stage is 10 kΩ in parallel with 100 kΩ, or approximately $R_B = 10k\Omega$.

Because the base resistor in a practical amplifier is usually small compared with βR_L, the input impedance of a CC stage is R_B and the current gain is R_B/R_L. For both amplifiers shown in Fig. 2.9 the input impedance is 10 kΩ, the current gain is 10, the voltage gain is 1, and the power gain is the product $G_v G_i$—again 10 (20 dB).

The circuit in Fig. 2.9(b) shows the more common way of biasing a CC amplifier stage from the collector voltage supply. The base resistor of Fig. 2.9(a) has been divided into two equal parts of 20 kΩ each in order to make a voltage divider. Thevenin's theorem applied to the voltage divider of Fig. 2.9(b) shows that the dc base voltage and the input impedances are the same for both amplifiers. For some purposes a slightly better Q-point can be obtained by moving the input capacitor and the base to a point about 5 kΩ higher on the voltage divider. This change is made to raise the dc emitter voltage to one-half the supply voltage, and does not appreciably alter the calculated input impedance, which is R_A in parallel with R_B.

A CE stage used to drive a CC amplifier might be expected to operate with the dc collector voltage at one-half the collector supply voltage. Because a change of several volts in the base voltage of a CC stage can be tolerated, a CE driver may easily be direct coupled to a CC output stage, especially when both stages use the same supply voltage.

The internal output impedance that the CC stage load "sees" looking back into the emitter is approximately the impedance seen by the base looking toward the signal source divided by β. For the practical amplifiers in Fig. 2.9 the load sees 10 kΩ/β, or about 100 Ω, which is 10 per cent of the load impedance. With a low-impedance source driving the CC stage, the internal output impedance is usually less than 1 per cent of R_L. A proof of this statement makes an interesting exercise for the more experienced reader.

SUMMARY

The performance characteristics of small-signal ac amplifiers are found by measuring or calculating ac gains and impedances. A very useful relation for the analysis of transistor amplifiers, the Transistor Gain-Impedance Relation (TG-IR), states that the voltage gain equals the current gain multiplied by the ratio of the load impedance to the input impedance. Generally the current gain and the load impedance are known; the input impedance is usually calculated.

Shockley's relation states that the short circuit input impedance of a transistor *h is inversely proportional to the emitter current*. Shockley's relation combined with the Transistor Gain-Impedance Relation reveals that the voltage gain of a stage without feedback varies with the emitter current. Because the emitter current and the input impedance both vary with β, a CE stage requires individual adjustments and is generally impractical without feedback.

A CC transistor amplifier with properly chosen base and emitter resistors is an example of an elementary feedback amplifier. The load voltage opposes the input voltage, thereby reducing the stage current gain from β to a value determined by the resistor ratio R_B/R_L. By making the stage current gain low compared with β, the CC amplifier is made substantially independent of the transistor.

Practical forms of the CC amplifier are used as step-down transformers that have a power gain of about 10. The CC amplifier is ordinarily biased to make the emitter voltage about one-half the collector supply voltage. These amplifiers are readily direct coupled to a CE driver stage.

PROBLEMS

2-1. An amplifier is required having a minimum voltage gain of 40, an input impedance of 1000 Ω, and an output load impedance of 100 Ω. (a) If the amplifier is built using transistors that have current gains between 20 and 100, what is the expected voltage gain variation? (b) The transistor's current gain with local feedback can be limited to vary between 10 and 20. What is the voltage gain variation to be expected with feedback?

2-2. Using the TG-IR, $I_E = 0.4$ mA, $\beta = 50$, and $R_L = 100$ Ω, show what voltage gain, power gain, and input impedance should be expected if a single stage transistor amplifier is constructed. Consider each type of amplifier: (a) CB, (b) CE, and (c) CC.

2-3. A 3-stage amplifier has a 3000 Ω input impedance and an overall voltage gain of 400. The output stage is a CC amplifier which has a current gain of 30. The load impedance is 10 Ω. If the first two stages are identical, what is the required current gain and what is the voltage gain per stage?

2-4. If a single-stage amplifier is required to have an input impedance of 2500 Ω, a 50 Ω load, and a voltage gain of 1 without the use of transformers, what emitter current is required if the amplifier is (a) a CB stage, (b) a CE stage, and (c) a CC stage?

2-5. (a) Assume β is 30 in the amplifier in Fig. P-2.5. Find the collector Q-point voltage, the input impedance, and the voltage gain. (b) What is the ac signal voltage on the collector if a 1 V signal is introduced at X? (c) Repeat part (a) assuming β = 100.

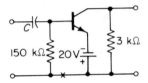

150 kΩ   20V   3 kΩ

Figure P-2.5.

2-6. A transistor (β = 50) is connected as a CB amplifier and the emitter current is adjusted to make the input impedance 520 Ω. If the load impedance is 5200 Ω, what is the voltage gain? (b) Repeat (a) assuming the transistor is operated as a CE amplifier. (c) What is the emitter current and the power gain of each amplifier?

2-7. A transistor (β = 50) connected as a CC impedance transformer has a 52 Ω load. (a) If the voltage gain must be at least 0.9, what is the lowest emitter current that can be tolerated? (b) What is the input impedance?

CHAPTER THREE

Single-Stage
Transistor Amplifiers

The single-stage small-signal amplifier is the key element of many electronic circuits. This chapter presents examples of the most common types of single-stage transistor amplifiers, but is primarily concerned with the CE stage and emitter feedback biasing circuits. A transistor is such a variable circuit element that a satisfactory amplifier cannot be designed without feedback. The current gain of transistors varies from one unit to another (even of the same type number), with temperature, and with the Q-point. Without feedback, the amplifier gain and the collector Q-point vary with β and change over a wide range. Without some way of limiting these variations, a single-stage transistor amplifier is impractical.

We show in this chapter that collector and emitter feedback biasing make the current and voltage gains of a stage relatively independent of the transistor β. Because the performance of a stage is then fixed by resistors, the transistor parameters are not required for circuit analysis, and the stage gain and impedance values can be found very simply. Without feedback, gain calculations are difficult and are practically meaningless. For these reasons the principles outlined in this chapter are essential for an understanding of well-designed amplifiers.

3.1 FEEDBACK PRINCIPLES (Ref. 2)

When a part of the output signal is combined with the input signal, an amplifier is said to have *feedback*. If the resulting input signal is reduced, the feedback is called *negative* or *inverse*. When the output signal opposes the input signal, the difference between these signals after amplification is phased so that the output is more like the input. Feedback limits the overall gain of the amplifier by reducing the amplifier input signal when the gain exceeds the limit predetermined by the feedback resistors. Feedback that reduces the gain of a stage brings many important advantages. In a single-stage amplifier, feedback biasing limits the dc current gain to a value fixed by the resistors and thereby reduces the amount the stage current gain can change with β.

The effectiveness of feedback in improving an amplifier is approximately proportional to the gain reduction, and the cost of the improved performance is reduced gain. High quality performance dictates low stage gain with feedback and makes the design expensive. However, a satisfactory compromise is usually secured by limiting the stage dc current gain to about one-fourth the transistor β.

Feedback in a single stage is called *local feedback*; over several stages, *overall feedback*. This chapter is concerned with single-stage local feedback, as distinguished from feedback applied over two or more stages. Feedback is introduced in a simple transistor amplifier by adding resistors, and the simplest way to explain the effect of local feedback is to show how these resistors affect the amplifier impedance and gain characteristics. The analysis of overall feedback tends to become a complex amplifier-feedback problem, whereas a single-stage amplifier may be analyzed more easily as a circuit problem. Either way, the final results are the same.

The degree of feedback used in a single-stage transistor amplifier may be set by following the practical rules set forth in Sec. 3.3. Because similar practical rules are not generally applicable to field-effect transistor (FET) amplifiers, a part of our study of feedback is deferred until Chap. 4.

3.2 EMITTER FEEDBACK AND THE CE INPUT IMPEDANCE

A circuit of a transistor CE amplifier having emitter feedback biasing is shown in Fig. 3.1, except that a bias resistor R_A has been omitted to simplify the present discussion. Feedback exists in this circuit because the load current in R_L flows to ground through the emitter resistor R_L and produces a feedback signal by varying the emitter voltage. The base is returned to the ground side of R_E via R_B, and the feedback is effective only because the ground side of the feedback signal also reaches the base.

The circuit relations required to establish a satisfactory amount of feedback can be understood by examining the input impedance of the amplifier shown in Fig. 3.1(a). To calculate the impedance looking into the amplifier, we must begin by finding the impedance looking into the base (with respect to ground) with the resistor R_B removed. Removing R_B leaves a common emitter transistor with a resistor R_E in series with the emitter as shown in Fig. 3.1(b). The input impedance of the CE transistor is given by Eq. (2.14) of the previous chapter as:

$$R_I = r_b + \beta' r_e \tag{3.1}$$

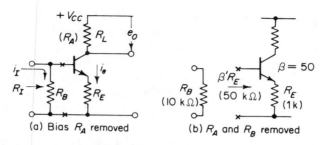

Figure 3.1. *CE amplifier illustrating emitter feedback.*

The effect of the emitter resistor is included by adding R_E in series with the intrinsic r_e; hence, Eq. (3.1) becomes:

$$R'_I = r_b + \beta'(r_e + R_E) \tag{3.2}$$

or:

$$R'_I = \beta'(h + R_E) \tag{3.3}$$

In a small-signal amplifier R_E will be between $100\ \Omega$ and $1000\ \Omega$, h is less than 5 percent of R_E and negligible, and β' will be at least 20. Suppose β' is 50 and R_E is $1000\ \Omega$; then the input impedance $\beta'R_E$ will be $50{,}000\ \Omega$.

Consider the effect of replacing the resistor R_B, as in Fig. 3.1(a). For a satisfactory design we must give R_B a value between $5R_E$ and $10R_E$, depending on the amount of feedback desired. This makes R_B $5000\ \Omega$ to $10{,}000\ \Omega$; hence, R_B in parallel with $\beta'R_E$ will be 10 to 20 percent less than R_B itself. This result shows that the input impedance of a stage with emitter feedback will be determined for all practical purposes by the resistor R_B. In fact, if a stage has enough feedback to ensure satisfactory performance, it automatically follows that the input impedance is very nearly the impedance of the base resistor, R_B. In practice, R_B is made small ($R_B = 5R_E$) if β is low (20); or, R_B is made larger if the minimum β is high.

3.3 SUMMARY OF EMITTER FEEDBACK REQUIREMENTS

The impedance relations that characterize an emitter feedback amplifier that has adequate ac feedback are:

1. The base resistor R_B of a practical stage is usually between $5R_E$ and $10R_E$.
2. The input impedance of the stage is nearly R_B.
3. The input impedance is at least one order of magnitude larger than βh, the input impedance of the transistor.

The higher input impedance (3) has been obtained at the cost of a reduction in the current and voltage gain below the gain obtainable when using the transistor without feedback. This means that an amplifier is made relatively independent of the transistor characteristics by improving the performance of each stage with negative feedback and by using more stages. Because the performance is almost independent of the transistor, we may expect that the current and voltage gains of a stage are fixed by the feedback resistors. We now will show how the current and voltage gains may be determined from the resistor values used in a feedback biased amplifier.

3.4 THE CE AMPLIFIER CURRENT GAIN

The current gain G_i of an amplifier stage is defined as the ratio of the collector ac load current to the ac signal input current. This current ratio is easily calculated. In Sec. 2.10, Eq. (2.21), we saw that as long as h is smaller valued than the series emitter resistor R_E, the emitter voltage follows the base voltage. We should expect, therefore, to find that in the amplifier of Fig. 3.1 both the dc and ac voltages of the base are transferred to the emitter resistor. Writing this statement as an equation, using i_I for the amplifier input current in R_B and i_e for the current in the emitter resistor, we have:

$$i_e R_E = i_I R_B \tag{3.4}$$

Dividing both sides by $i_I R_E$ gives:

$$\frac{i_e}{i_I} = \frac{R_B}{R_E} \tag{3.5}$$

The collector current is very nearly equal to the emitter current ($\alpha = 0.98$), so we may substitute i_c for i_e and write the last equation as:

$$G_i = \frac{i_c}{i_I} = \frac{R_B}{R_E} \tag{3.6}$$

Equation (3.6) asserts that the overall current gain of the amplifier is the ratio of the resistors: R_B/R_E. The feedback has reduced the current gain of the amplifier from β to a value between 5 and 10 fixed by the circuit resistors, as long as the current gain of the transistor is significantly higher than the resistor ratio. The effect of this resistance ratio is of such broad usefulness and importance that we give the ratio a special symbol S where:

$$S = \frac{R_B}{R_E} \tag{3.7}$$

and we call S the S-factor of the amplifier or, sometimes, the dc current gain. Low S-factors tend to make the amplifier performance independent of the transistor and reduce the Q-point shift caused by temperature changes. An S-factor of 10 to 20 is usually satisfactory when using present-day silicon transistors at room temperatures. A transistor power stage may have an S-factor of 5, of less.

3.5 THE CE AMPLIFIER VOLTAGE GAIN

The voltage gain of the CE amplifier (Fig. 3.1) is also easily determined from the values of the circuit resistors. Because the emitter voltage follows the amplifier input voltage, the stage gain G_v may be found by comparing the load voltage e_o with the emitter voltage e_I. If we assume that the base current is negligible, the collector and emitter currents are the same (for $\alpha > 0.97$ they are within 3 percent of each other). The voltage across the load e_o is:

$$e_o = i_c R_L \qquad (3.8)$$

and the input voltage is:

$$e_I = i_c R_E \qquad (3.9)$$

Hence, the voltage gain of the stage is:

$$G_v = \frac{e_o}{e_I} = \frac{R_L}{R_E} \qquad (3.10)$$

The dc voltage gain is independent of the transistor current gain only if the transistor current gain β is sufficiently high compared with the S-factor of the amplifier. The cost that must be paid for this stability of voltage gain is either low voltage gain (high R_E) or low input impedance (low R_B).

3.6 SUMMARY OF CE AMPLIFIER GAIN RELATIONS

For typically good designs the current and voltage gains of the CE transistor amplifier having local feedback are indicated by simple ratios of the resistors of the circuit, as follows:

1. Current gain:
$$S = \frac{R_B}{R_E} \qquad (3.7)$$

2. Voltage gain:
$$G_v = \frac{R_L}{R_E} \qquad (3.10)$$

3. Input impedance:
$$R_I = R_B \qquad (3.11)$$

A CE amplifier with emitter feedback and the form shown in Fig. 3.1(a) has identical dc and ac gains. If the emitter feedback requirements (Sec. 3.3) are met (adequate feedback), the transistor base current is negligible and the emitter follows the base.

3.7 CE AMPLIFIER BIASING AND DC FEEDBACK

The amplifier under discussion (Fig. 3.1) will not operate until the Q-point is properly set by adding a base current bias resistor, as shown in Fig. 3.2. The bias resistor R_A, connected from the collector supply voltage to the base, supplies the base Q-point current. This resistor is usually an order of magnitude larger than R_B, so its effect on the input impedance (and gains) can be neglected.

Most discussions of transistor biasing err in using the static characteristics to explain biasing and disregard the fact that a practical transistor amplifier must use feedback. With emitter feedback biasing, the performance of a given amplifier stage is almost independent of the transistor characteristics; thus, it should be possible to select the bias resistor by considering the resistors of the stage, giving only scant attention to the characteristics of the transistor itself. With adequate feedback, as specified in Sec. 3.3, the Q-point of an amplifier can be calculated by assuming the transistor has a very high current gain and negligible base current. However, if the transistor in use happens to have a low current gain, either we must expect to increase the bias current

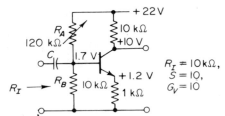

$R_I = 10\,k\Omega,$
$S = 10,$
$G_V = 10$

Figure 3.2. CE amplifier-example of bias calculation.

enough to supply the higher base current, or we must be able to tolerate a small Q-point shift from the calculated Q-point.

The simplest and best way of setting the bias of any amplifier is to adjust the bias resistor experimentally so that either the stage voltage gain or the peak output signal is maximized. Both criteria require about the same value of bias resistor. An amplifier for small-signal applications is usually biased to place the Q-point where the ac collector (output) voltage may be both increased and decreased from this Q-point. An amplifier is capable of producing a reasonably large peak output voltage only if the Q-point falls about midway between the upper and lower limits of the possible voltage excursions. One limit of the collector voltage is the voltage that exists with the collector open, and the other limit is the voltage that exists when the collector is connected to the emitter. The former condition means that the transistor is cut off, and the latter condition means that the transistor is turned on, like a switch.

For an example of the optimum Q-point condition consider the amplifier shown in Fig. 3.2. Observe that with the collector open the collector voltage is 22 V, the collector supply voltage, and with the collector connected to the emitter, the collector voltage is 1/11th of the supply voltage, or 2 V. The peak limits of the collector voltage are 2 V and 22 V, so a good Q-point design would make the collector voltage about 12 V. Actually, any value between 10 V and 14 V should prove a satisfactory Q-point for most purposes. In practice, the bias resistor R_A is adjusted to set the Q-point at the lower voltage

(10 V) using a high β transistor, so that the Q-point takes the higher value (14 V) if a low β transistor is used.

For a practical stability of the collector Q-point an emitter feedback stage must have adequate dc feedback, and the requirements for dc feedback are more severe than for ac feedback. If a collector voltage change of several volts is to be offset by feedback, the base-emitter voltage must change nearly 0.1 V, and adequate dc feedback exists only when the emitter voltage is reasonably larger, for example 0.6 V. This need for at least 0.6 V across R_E makes the emitter feedback stage relatively unsatisfactory unless the collector supply voltage exceeds 12 V. For the same reason it is a good practical rule that the collector supply voltage should exceed the dc voltage gain R_L/R_E (volts).

3.8 THE BIAS RESISTOR

A carefully **calculated** value for the bias resistor is of little worth because the resistor is always selected **experimentally** in actual practice. Only by experimental procedures is one brought into immediate contact with the total complex of the amplifier and the bias problem. There is value, however, in knowing that in emitter feedback designs when R_L is 10 or more times R_E the bias resistor R_A should be about 20 per cent larger than the value given by

$$R_A = R_B \frac{R_L}{R_E} \qquad (3.12)$$

For Fig. 3.2 Eq. (3.12) gives $R_A = 100$ kΩ, and the 120 kΩ value shown in Fig. 3.2 was obtained experimentally.

We may confirm that the 120 kΩ resistor is approximately correct by a simple calculation. Assuming the base current to be negligible, the bias resistors form a simple voltage divider, and the base voltage is:

$$V_B = \frac{R_B}{R_A + R_B} V_{CC} \qquad (3.13)$$

For the resistors in Fig. 3.2 Eq. (3.13) gives $V_B = 1.7$ V. Subtracting 0.6 V gives 1.1 V as the emitter voltage, and the collector voltage is 22 V minus 11 V, or 11 V. However, the only safe way to select a bias resistor is by an experimental evaluation of the amplifier performance. The calculations should be used as aids to understanding and as guides for experiments.

Given a desired Q-point, the value of the bias resistor may be found by reversing the procedure just outlined:

1. Select a (low) collector Q-point voltage.
2. Calculate the corresponding emitter current and voltage.
3. Find the base voltage—0.6 V above the emitter voltage.
4. Find R_A from the base voltage, but neglect the base current.
5. Adjust the bias resistor by making an experimental test.

3.9 SUMMARY OF EMITTER FEEDBACK BIASING (Ref. 2)

The operating characteristics of a well designed CE amplifier are easily recognized by examining the values of the resistors in the circuit. The observable characteristics, as indicated in Fig. 3.3, are:

1. The existence of adequate feedback (see Sec. 3.3)
2. The gain and impedance values (see Sec. 3.6)
3. The approximate value of the bias resistor (see Sec. 3.8)
4. The existence of adequate dc feedback for collector Q-point stability (see Sec. 3.7)

And finally, after adjusting the bias in the laboratory, the most important tests of performance are the measurements of the amplifier gain, frequency response, noise, large signal operation, and Q-point stability (with temperature and supply voltage changes).

The design of an amplifier requires a background of experience and a

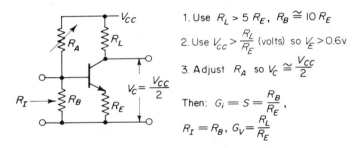

1. Use $R_L > 5 R_E$, $R_B \cong 10 R_E$

2. Use $V_{CC} > \dfrac{R_L}{R_E}$ (volts) so $V_E > 0.6\text{v}$

3. Adjust R_A so $V_C \cong \dfrac{V_{CC}}{2}$

Then: $G_i = S = \dfrac{R_B}{R_E}$,

$R_I = R_B$, $G_V = \dfrac{R_L}{R_E}$

Figure 3.3. *CE amplifier biasing relations.*

compromising of factors not easily reduced to a step-by-step procedure. However, the novice may adapt proven circuits to his purposes by scaling all components either larger or smaller, using a common scale factor. We examine now several variations of the CE amplifier.

3.10 THE EMITTER BYPASS CAPACITOR (Ref. 1)

In a CE amplifier, the emitter resistor reduces both the current and voltage gains of a stage, as shown by Eqs. (3.7) and (3.10). Reduced dc gains are required in order to stabilize the collector Q-point. However, for some purposes we need not reduce the ac gains. The addition of an emitter capacitor, as shown in Fig. 3.4, allows the ac emitter current to bypass the emitter resistor, and the ac current gain is restored to β while the ac voltage

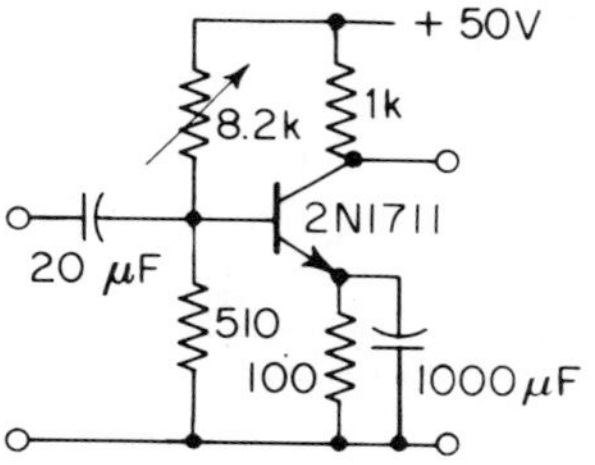

With C: $G_v = 400$, $G_i = \beta$, $R_I \cong 250\,\Omega$

Without C: $G_v = 10$, $G_i = 5$, $R_I = 500\,\Omega$

The gains and impedances shown on circuits in the author's books are generally measured values which may differ by 25 percent from a calculated value. The error is a price paid for simpler explanations and mathematical relations. A clear understanding of transitor circuits is of more value than a collection of complicated relations that may give higher accuracy.

Figure 3.4. *RC amplifier with bypassed emitter,*
$f_l = 70$ *Hz.*

gain is increased to the value given by Eq. (2.16), below. This technique is similar to the use of a cathode capacitor in vacuum tube amplifiers except that it is difficult to appreciate just how large the capacitor must be to restore the ac gain. Contrary to popular belief, the value of the capacitor is **not** determined by the resistance of the emitter resistor.

The reason an emitter bypass capacitor raises the gain of a CE amplifier may be explained by considering the voltage gain of a CE transistor as given in Sec. 2.8 by the equation:

$$G_v = \frac{R_L}{h} \qquad (2.16)$$

The effect of connecting a resistor R_E in the emitter lead is shown in Sec. 2.10 to change h in Eq. (2.16) to $h + R_E$ and the voltage gain of the CE amplifier reduces to:

$$G_v = \frac{R_L}{h + R_E} \qquad (3.14)$$

If the capacitor across R_E is to remove R_E from Eq. (3.14), the reactance of the capacitor X_E must be less than h. Therefore, the emitter is said to be bypassed to ground if:

$$X_E \ll h \qquad (3.15)$$

A bypass capacitor having a low reactance at the lowest expected operating frequency may have to be large and correspondingly expensive. When some reduction of the ac gain can be tolerated, a smaller capacitor is adequate, provided a resistor is used in series with the base or in series with the emitter. When the resistor is in series with the emitter bypass capacitor, as shown in Fig. 3.5, the reactance of the capacitor need only be smaller than the

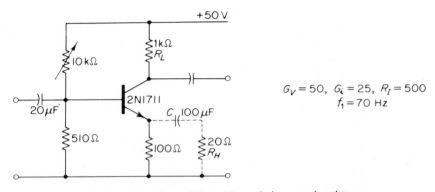

Figure 3.5. *RC amplifier with partly bypassed emitter.*

series resistance R_H, which in turn replaces R_E in Eq. (3.7) and (3.10) for the computation of the ac gains. We refer to this technique as a ***partially bypassed*** emitter.

As a practical example of an amplifier having the emitter resistor bypassed, consider the RC amplifier shown in Fig. 3.4. With the capacitor C removed we see by inspection that the voltage gain is 10, the current gain is 5, and the input impedance is 500 Ω. With the emitter capacitor in place, the ac current gain of the amplifier approaches the β value of the transistor, the voltage gain is 400, and the input impedance is indefinite. If β exceeds 100, the input impedance is about 500 Ω, but if β is 20 the input impedance may be only 50 Ω. Such a low input impedance is undesirable because large bypass and coupling capacitors are required, and the input impedance changes with β. The fully bypassed emitter should be avoided because the bypass makes the gain and impedances of a multistage amplifier very unpredictable.

The addition of a resistor in series with the emitter bypass capacitor improves the amplifier at the expense of lower gain, as shown in Fig. 3.5. With a 20 Ω resistor the current gain is stabilized by the resistors and lowered to 25, the voltage gain is 50, and the input impedance is 500 Ω. An important advantage in using the resistor R_H is that the capacitor C is a factor of 10 smaller than the capacitor required with the fully bypassed emitter. Both amplifiers have the same 70 Hz low-frequency cutoff. Because h for the transistor is about 2 Ω, the value of C is determined by the desired cutoff frequency and 20 Ω, the resistance of R_H.

3.11 COLLECTOR FEEDBACK BIASING (Ref. 2)

A transistor stage may be given feedback by connecting a resistor R_f between the collector and base, as shown in Fig. 3.6. This method of biasing a

$$S_c = \frac{S\beta}{S+\beta}$$

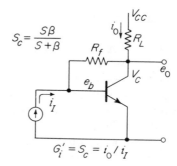

Figure 3.6. *Collector feedback stage with current source.*

$$G_i' = S_c = i_o/i_I$$

stage requires only one resistor instead of the three required for emitter feedback, and the adjustment of bias is much less critical than with emitter feedback.

The S-factor of a collector feedback stage is defined as the ratio

$$S \equiv \frac{R_f}{R_L} \tag{3.16}$$

and the current gain of the stage is given by the corrected S-factor,

$$S_c = \frac{\beta S}{\beta + S} \tag{3.17}$$

For significant feedback, which means the stage current gain is one-fourth the current gain without feedback,

$$S_c = \frac{\beta}{4} = \frac{3}{4}S \tag{3.18}$$

and

$$S = \frac{\beta}{3} \tag{3.19}$$

We observe by Eq. (3.17) that when β is very high compared with S, S_c is approximately equal to S, and with significant feedback S_c is only 25 per cent less than S. Thus, for simplicity we refer somewhat loosely to S as the current gain of the stage and use the corrected S_c only when a clear distinction is important, as in selecting resistors for a stage. If we assume β is about 150, we have significant feedback when $S = 50$ and the stage current gain is $S_c = 37$. If β drops to 100, the current gain decreases only to 33.

With collector feedback the collector voltage is proportional to the S_c/β ratio and is

$$V_C = \frac{S_c}{\beta} V_{CC} \tag{3.20}$$

Thus, with significant feedback the collector voltage is one-fourth the collector supply voltage, and for S small compared with β, the collector voltage is only a small fraction of the supply voltage. Because the collector voltage and the stage current gain cannot be independently adjusted, a collector feedback stage is usually designed with no more than significant feedback. A stage with a 20-V collector supply and significant feedback has only a 5-V collector voltage and may not be suitable when a large output signal is required. If the Q-point is set for a higher collector voltage by increasing S, the Q-point voltage is relatively more sensitive to a change of β, and the stage has the disadvantages of the CE transistor stage without feedback. (See Sec. 2.9.) When both a large signal output and significant feedback are required, a collector feedback stage should have a collector supply voltage of at least 50 V.

3.12 COLLECTOR FEEDBACK GAINS AND IMPEDANCES

When a collector feedback stage is driven by a current source (a high impedance), the stage has a current gain S and the input short-circuits the source. If the stage is driven by a voltage source (a low impedance) that is coupled by a large capacitor, as shown in Fig. 3.7, the ac voltage gain is independent of the current gain and is R_L/h, as given by Eq. (2.16). Replacing h by use of Shockley's relation, as in Sec. 2.6, we find that the intrinsic voltage gain of the CE stage is approximately 40 times the dc voltage drop in the collector load resistor. If the collector supply voltage is twice the dc voltage drop, the intrinsic voltage gain is 20 times the magnitude of the collector supply voltage. Thus, the practical upper limit of the intrinsic voltage gain of a CE stage is generally

$$G_v \cong [25\, V_{CC}] \tag{3.21}$$

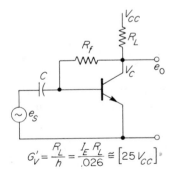

$$G_v' = \frac{R_L}{h} = \frac{I_E R_L}{.026} \cong [25\, V_{CC}]$$

Figure 3.7. Collector feedback stage with voltage source.

where the brackets indicate that Eq. (3.21) requires the numerical value of the supply voltage, because the voltage gain is dimensionless. For the stage shown in Fig. 3.7, if the collector supply voltage is 20 V, Eq. (3.21) indicates a base-to-collector voltage gain of approximately 500.

The input impedance of the CE stage shown in Fig. 3.7 is calculated by solving the *TG-IR* for R_I, and we find

$$R_I = S_c \frac{R_L}{G_v} \tag{3.22}$$

And, by combining Eq. (2.16) with Eq. (3.22), we obtain

$$R_I \cong Sh \tag{3.23}$$

Equation (3.23) shows that the input impedance of a CE stage is reduced by collector feedback from βh to Sh.

The input impedance of the stage illustrated in Fig. 3.7 may be found by substituting the known values of the current and voltage gains in Eq. (3.22). For $S = 50$ and $G_v = 500$, the input impedance is

$$R_I \cong \frac{R_L}{10} \tag{3.24}$$

Thus, the input impedance is low compared with the load impedance, and the high intrinsic voltage gain of a collector feedback stage cannot be developed in a series of identical, or iterated, stages. If the signal source has an impedance the same as the load, the source is effectively short-circuited by the amplifier, and the TG-IR shows that the iterated voltage gain is

$$G_v \cong S \tag{3.25}$$

The simplicity of the collector feedback stage makes it useful between equal source and load impedances, even though the iterated voltage gain is only one-tenth of the intrinsic gain. However, we must not forget that the input impedance is low and that a relatively large coupling capacitor may be required. The calculated gain and impedance characteristics of a collector feedback stage are given as answers to a problem at the end of this chapter.

A collector feedback stage has the disadvantage that the input impedance varies inversely with the ac load impedance, since S in Eq. (3.23) varies inversely with R_L. Because R_I varies inversely with R_L, a capacitance load appears at the input as an inductive reactance that may cause instability, even at relatively low frequencies. (An application of collector feedback that produces a tuned amplifier is described in Chapter 10.) Because of the close relation between the input impedance and the load impedance, the collector feedback stage cannot be used much above the audio frequencies. Thus, the collector feedback stage is generally used alone as an input stage or is placed between emitter feedback stages.

The ac feedback via R_f is reduced when a low-impedance load is coupled to the collector, as shown in Fig. 3.8. The additional load reduces the collector signal relative to the base signal and may increase the iterated voltage gain to β. Thus, with significant dc feedback a coupled load that is less than one-fourth the collector resistor R_L effectively bypasses the ac feedback. Similarly,

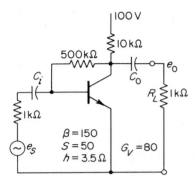

Figure 3.8. *Collector feedback stage with coupled load.*

a low-impedance source bypasses the feedback and may increase the voltage gain to the intrinsic gain given by Eq. (3.21).

3.13 COLLECTOR FEEDBACK SUMMARY

The collector feedback stage can be easily designed for room temperature applications and operates over a wide range of the supply voltage without requiring a carefully selected value of the bias resistor. Significant dc feedback exists when the bias resistor is adjusted to make the collector voltage one-fourth the supply voltage. Significant ac feedback exists also if R_S is greater than βh and the capacitor-coupled load impedance is greater than R_L.

The collector feedback stage has an iterated gain equal to S_c, which, with significant feedback, is three-fourths the S-factor. By reducing the feedback, a capacitor-coupled, low-impedance load tends to increase the input impedance and may increase the iterated gain to β. Similarly, a low-impedance source may increase the stage voltage gain to $[25 \, V_{CC}]$, the intrinsic CE voltage gain.

The collector feedback stage tends to reflect an inversion of the load impedance that can cause instability problems above the audio frequencies.

3.14 COMBINED COLLECTOR AND EMITTER FEEDBACK (Ref. 2)

A transistor stage with dc collector feedback and ac emitter feedback has important advantages. A stage with combined shunt and series feedback is illustrated in Fig. 3.9. The collector feedback is supplied through the resistor R_f, while ac signals are bypassed to ground by the capacitor C_f. The ac feedback is supplied by the emitter resistor R_E. The amplifier has the same number of components as an emitter feedback stage but has the advantage that the bypass capacitor can be a small mylar capacitor instead of a large

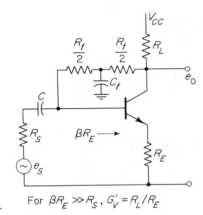

Figure 3.9. *Combined feedback stage.*

For $\beta R_E \gg R_S$, $G_V' = R_L/R_E$

electrolytic capacitor. The size of the capacitor is best found by trial and may be selected to provide a small low-frequency boost, as shown later in Sec. 10.3.

The bias may be adjusted by decreasing one or both halves of R_f to lower the collector voltage, or by connecting a resistor from the base to ground to raise the collector voltage. The base bias resistor is surprisingly effective in increasing the Q-point voltage while not changing appreciably the stage voltage gain or input impedance.

The procedure for designing a stage is relatively simple. The designer selects a suitable load resistor and makes the S-factor approximately $\beta/3$ if significant dc feedback is desired. If the transistor β is not known or measured, the S-factor should be made about one-half the specified minimum β, and the transistors can be selected at assembly by measuring the Q-point voltage. The emitter resistor is inserted to produce the desired amount of ac feedback. Usually, the emitter resistor does not change the Q-point, but the bias can be adjusted as described above.

With the collector feedback bypassed for ac signals and $R_E = 0$, the ac voltage gain from the source e_s to the collector is

$$G_v = \beta \frac{R_L}{R_S + \beta h} \tag{3.26}$$

For an iterated stage, which means that $R_S = R_L$, R_S is large compared with βh, and Eq. (3.26) reduces to

$$G_v = \beta \frac{R_L}{R_S} \tag{3.27}$$

An iterated stage is given significant feedback by inserting an emitter resistor that reduces the ac voltage gain by a factor of 4. Thus,

$$G_v' = \beta \frac{R_L}{4R_S} = \beta \frac{R_L}{R_S + \beta R_E} \tag{3.28}$$

Solving Eq. (3.28) for the input impedance of the stage gives

$$R_I = \beta R_E = 3R_S \qquad (3.29)$$

Equation (3.29) shows that significant feedback may be adjusted experimentally by increasing R_E until the base signal is three-fourths the no-load input signal e_S or by making

$$\frac{R_S}{R_E} = \frac{\beta}{3} \qquad (3.30)$$

With significant feedback the stage input impedance varies with β and is approximately $3R_S$, the current gain is β, and the iterated voltage gain is

$$G'_v = \frac{3}{4} \frac{R_L}{R_E} \qquad (3.31)$$

A combined feedback stage with practical component values is shown in Fig. 3.10. This stage with significant feedback gives a voltage gain of 40 between equal source and load impedances.

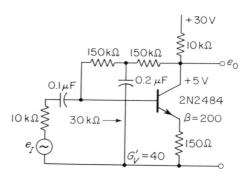

Figure 3.10. *Combined feedback stage with significant feedback.*

3.15 COMBINED FEEDBACK SUMMARY

A transistor stage with combined dc collector feedback and ac emitter feedback has the advantages of both the elementary collector feedback stage and the familiar emitter feedback stage. The separation of the dc and ac feedback allows greater flexibility in design, permits higher iterated gain, and eliminates the relatively large emitter bypass capacitor. Separation of the dc and ac feedback simplifies the circuit analysis, and by simple measurements we may know whether or not there is significant Q-point control or significant ac feedback.

Significant dc feedback exists when the S-factor R_f/R_L is approximately $\beta/3$, and significant ac feedback exists when R_S/R_E is approximately $\beta/3$. The voltage gain of a stage with significant ac feedback is approximately R_L/R_E, and the input impedance is approximately βR_E.

A combined feedback stage has the important advantage of not reflecting load changes back to the input, except as a relatively minor effect.

3.16 PRACTICAL AMPLIFIER VARIATIONS (Ref. 1)

It is interesting to examine some of the ways the CE amplifier may be modified to meet a variety of requirements. For example, the amplifier of Fig. 3.2 is not limited in its application to the 22 V supply voltage shown, although the selection of a collector voltage to carry through the Q-point computation is helpful. For small changes of the supply voltage the Q-point will remain near the center of the available collector voltage excursion. For large changes of the supply voltage (say, 30 V) the bias resistor should be changed. Generally, with minor adjustments a low-power amplifier will be satisfactory with supply voltages from 6 V up to the collector-emitter voltage rating of the transistors—about 50 V for audio transistors.

The impedance level of the amplifier may be increased or decreased by increasing the resistance of all the resistors by the same factor. This is a fairly safe procedure if all the resistors are increased ten times, or all are decreased ten times. For some applications we may scale all resistors as much as one hundred times larger. If we make the resistor values smaller, we must be careful not to exceed the power rating of the transistor.

For example, Fig. 3.4 shows the amplifier of Fig. 3.2 with the collector and emitter resistors scaled ten times smaller, but with bias resistors made twenty times smaller in order to lower the S-factor to 5. If we expect to use a 50 V collector supply, then the maximum power dissipation in the collector will occur when the collector-to-emitter voltage is just one-half the open circuit supply voltage. With the 1000 Ω collector load resistor and a 100 Ω emitter resistor the corresponding collector current will be 22.7 mA (25 V drop in 1100 Ω). The transistor collector will be dissipating 25 V at 22.7 mA, so the transistor must be able to dissipate 570 mW. If the collector supply voltage is 6 V and the collector load resistor is 100 Ω (again, all resistors are scaled ten times lower), the power to be dissipated will be only 81 mW. This shows that the power dissipation depends on the circuit and supply voltage, and that each case has to be considered after the resistors and the supply voltage are given.

We should remember in calculating the power dissipation that the maximum power transfer from the collector supply to the transistor always occurs when the dc voltage across the transistor is one-half the open-circuit voltage that would exist with the transistor removed.

3.17 TRANSFORMER-COUPLED AMPLIFIERS

Transformer-coupling of amplifier stages is sometimes used to increase the gain per stage or the peak output voltage. The more common use of

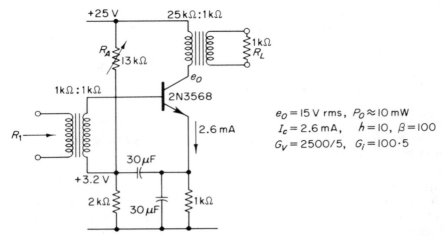

Figure 3.11. *Transformer-coupled amplifier.*

transformers is to obtain increased output power and to match into a fixed load impedance, especially low resistance loads. The circuit in Fig. 3.11 shows how the RC-coupled stage of Fig. 3.5 can be changed into a transformer-coupled stage in order to obtain increased voltage gain and improve the Q-point stability (i.e., low S-factor). The input transformer matches the input impedance of approximately 100 h, but is used mainly to provide a low dc resistance so that the S-factor is determined separately from the ac input impedance.

In a similar way, the output transformer provides a high output impedance and increased voltage gain without decreasing the collector dc voltage significantly below that of the 25 V supply. The transformer also gives a more efficient transfer of power to an external load. The transformer is bulky, heavy, and expensive, and it tends to be saturated by the collector Q-point current. Usually, we prefer to obtain the required voltage gain by using an extra RC-coupled stage and use transformers only for power stages or for unusual requirements of impedance transformation. The present example illustrates the interesting fact that *a series of iterated stages has a voltage gain equal to the product of the stage current gains and the transformer current ratio*. This fact is easily demonstrated by the Transistor Gain-Impedance Relation.

3.18 EMITTER-COUPLED AMPLIFIER

The amplifier shown in Fig. 3.12 is often found in integrated circuits because it provides a reasonably high gain without requiring an emitter capacitor. The amplifier consists of a pair of CE stages sharing a common

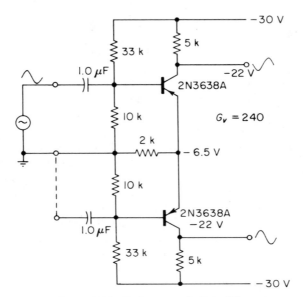

Figure 3.12. *Emitter-coupled amplifier.*

2 kΩ emitter resistor. Whenever the upper amplifier is driven by an input
signal and the lower amplifier is not driven, the emitter of the lower stage
presents a low dynamic resistance and the 2 kΩ resistor is effectively bypassed.
The emitter impedance is lowered still further when the lower base is bypassed
to ground. A base capacitor may be about β times smaller than an equivalent
emitter capacitor.

   If the emitter resistor is fully bypassed, as with a capacitor, the output
signal is on the upper collector only, and the voltage gain is R_L/h. Without
the emitter bypass the relatively high emitter resistor makes the emitter currents alike and causes the output signal to divide equally between the two
sides. Thus, the voltage gain to each collector is approximately $5000/2h =$
150, and the calculated gain for the full output is 300.

   Because the collectors offer equal out-of-phase signals, the amplifier is
useful also where the choice of an in-phase or an out-of-phase signal is desired.
The amplifier is known as a ***paraphase*** amplifier or a ***phase inverter*** when used
to couple a single-sided amplifier to a ***push-pull***, or a ***balanced***, power amplifier.
The emitter-coupled amplifier is called a ***long tail pair*** when the emitter
resistor is increased by including a voltage supply to offset the dc voltage drop.

3.19 DIFFERENTIAL DC AMPLIFIER (Ref. 1)

   The amplifier shown in Fig. 3.13 is a form of an emitter-coupled amplifier commonly found in oscilloscopes and recording instruments. The

amplifier, known as a **difference** or **differential amplifier**, converts a signal difference between the two inputs to equal but out-of-phase collector signals. When two equal but oppositely phased input signals are in series with the base resistors, the ac emitter signals cancel each other and the stage gain is calculated as if the emitter resistor is shorted.

If we assume the differential amplifier is driven by a stage that has a 10 kΩ internal impedance on each side, the effective voltage gain of the amplifier is calculated by using the internal (open circuit) voltage as the input signal. Because the base-to-base input impedance is small compared with the 20 kΩ source impedance, the differential voltage gain is approximately $\beta \cdot 10000/20000$, or about 50.

The differential stage shown in Fig. 3.13 is biased by making the emitter more positive than the point to which the base current returns. As with all direct-coupled stages, the bias current flows through the signal source and produces a small voltage drop. For an input stage the emitter and collector resistors are made about 100 times larger to increase the input impedance and reduce the base bias current. These changes increase the voltage gain to about 500 and the input impedance to at least 100 kΩ.

Differential amplifiers are particularly useful because signals common to both inputs are eliminated or reduced. A common mode signal like power line pickup drives both bases in phase with equal magnitude ac voltages, and the amplifier behaves as though the transistors were in parallel. The emitter resistor introduces emitter feedback, which reduces the common mode signal gain without reducing the differential signal gain.

Figure 3.13. Differential amplifier (single-stage).

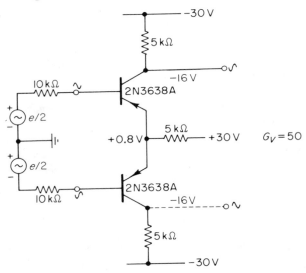

The effective emitter resistance is easily increased by replacing the emitter resistor by a transistor. Even with a low dc voltage drop the transistor provides a high internal resistance from collector to base. If the value of the effective emitter resistance is fifty times larger than that of the collector resistor, the stage has a common mode voltage loss of $\frac{1}{100}$. Because the common mode gain is low and the emitter easily follows the base, a differential amplifier will tolerate large common mode input signals without overloading and preventing the use of the amplifier for observing small signal differences.

Differential amplifiers are often used to observe a small signal difference between two points when these points also have large in-phase signals. For example, a pair of wires may transmit weak data signals, although the wires also have 60 Hz power line pickup. If the wires are connected to opposite inputs of a differential amplifier, the desired data signals appear amplified as a signal difference between the output collectors. If the power pickup is of the same phase and magnitude on both inputs, the interference will appear at the output collectors reduced in magnitude by the common mode signal loss.

Because the two sides of an amplifier are not exactly alike, a common mode signal produces a small difference signal that will interfere with the data signals and be amplified in succeeding stages. However, a well balanced differential amplifier transfers only a small part of the common mode signal into a difference signal. If the output difference signal is divided by the differential amplifier gain, the resultant is called an *equivalent differential input signal*. The ability of an amplifier to prevent conversion of a common mode signal into a difference signal is expressed by its common mode rejection ratio (CMR)—namely, the ratio of the common mode input signal divided by the equivalent differential input signal. An amplifier that has a differential gain of 30, a common mode gain of $\frac{1}{100}$, and a 1 percent voltage gain difference between the two sides will have a CMR advantage of $30 \times 100 \times 100 = 300,000$ (110 dB). A high quality differential amplifier will offer about 120 db CMR. However, an amplifier having a high CMR cannot reject a large pickup signal unless both inputs receive identical signals. For this reason we must often balance ac pickup by using an input signal balance adjustment.

Direct-coupled amplifiers are usually operated on a dc power supply which has both positive and negative voltages with respect to ground. The negative supply makes it possible to operate the base inputs at ground potential and to offset the dc voltage required by the common emitter resistor or its equivalent.

The output terminal Q-point voltage of a high gain dc amplifier changes slowly with time and temperature. These changes are referred to as Q-point drift, or simply as *drift*. Unless both sides of a differential amplifier are exactly alike, a part of the Q-point drift appears as a change in the differential dc output signal which cannot be distinguished from a change caused by the dc differential input signal. Differential drift is reduced in much the same way as common mode pickup signals are reduced: by local feedback and by carefully making both sides of the amplifier alike. However, temperature drifts arise from differences between the two sides of the amplifier, while the CMR ratio is increased by a factor of approximately 100 by common mode feedback. For this reason, the CMR ratio is much larger than the drift reduction that can be expected in a differential amplifier.

3.20   TEMPERATURE PROBLEMS (Ref. 1)

A difficulty common to transistor amplifiers is the shifting of the Q-point with temperature. If an amplifier is poorly designed (e.g., S-factor too high), or the transistor temperature is excessively high, the collector current increases with temperature until the amplifier becomes inoperative. If the operating temperature is too low, the transistor current gain decreases, and the collector current decreases until the stage may become inoperative. Germanium and the early silicon transistors were especially subject to a thermally sensitive collector-to-base current or leakage, *the I_{co} current, that doubles with every ten degree Centigrade increase in the junction temperature*. This exponential increase in the base current changes the stage bias with temperature and may be undesirable in equipment subjected to as little as a 20°C increase in temperature. The I_{co} drift can be kept under control by the use of low S-factors and low values of the collector circuit dc resistance. Because the planar silicon transistors have a negligibly small I_{co} current, they have since 1961 replaced almost all the early types of transistors except for germanium power transistors.

When planar transistors are used, the temperature effect is either the 2 mV/°C decrease of the base-emitter voltage drop with increased temperature or the approximately 1 %/°C increase of the current gain with temperature. Both these effects are more than an order of magnitude less troublesome than the earlier I_{co} problem. Both are linear, nonexponential effects, and cause approximately equal increases in the collector current.

The user of early transistors—germanium and grown junction silicon devices—was beset with severe temperature problems, even in single-stage amplifiers. The invention of the passivated planar construction in 1961 reduced the I_{co} leakage several orders of magnitude and made the two- and three-stage direct-coupled amplifiers a practical reality. We must remember when reading the transistor literature that most circuits described before 1965 and even later are either totally obsolete or else may be greatly improved by substituting modern silicon devices.

For room temperature applications the planar transistor may be considered almost independent of temperature, especially when used in single-stage or in two-stage amplifiers which have considerable feedback. We may summarize the simpler methods of alleviating moderate temperature effects as follows:

1. Use planar transistors or equivalent silicon devices.
2. Use low S-factors (low dc current gain) and low dc voltage gains.
3. Protect equipment from exposure to heat—as from vacuum tubes.
4. Use heat sinks where transistor case temperatures are high.
5. Avoid direct-coupled amplifiers except with dc feedback to limit the dc gain.

The control of temperature effects in high-power amplifiers and in amplifiers that are exposed to temperature extremes is a difficult problem, even for the experienced designer who understands the most effective use of each method.

SUMMARY

Feedback biasing makes the performance of a transistor CE amplifier almost independent of the transistor parameters. With emitter feedback the input impedance of the stage is R_B, the voltage gain is R_L/R_E, and the stage current gain S is R_B/R_E. The bias resistor R_A is best found by adjusting the resistor experimentally to set the collector Q-point voltage near $V_{CC}/2$. The collector current is approximately $(V_B - 0.6)/R_E$, and, with adequate feedback, the base bias voltage may be calculated by neglecting the base current.

A stage with collector feedback has a low input impedance and an iterated gain S, where S is approximately R_f/R_L. With significant feedback $S = \beta/3$, and the collector Q-point voltage is one fourth the supply voltage.

With combined dc collector feedback and ac emitter feedback a stage has a relatively high input impedance βR_E and an iterated voltage gain of nearly R_L/R_E. Significant ac feedback is ensured by making R_S/R_E equal to $\beta/3$.

An emitter bypass capacitor increases the voltage gain of a stage but decreases the input impedance and makes the ac characteristics of a stage vary with β. Emitter bypass capacitors are large and expensive and may be eliminated either by using more stages, combined feedback, or an emitter-coupled stage.

Interstage transformers are sometimes used for coupling a high impedance collector circuit to a low impedance base circuit when the operating power must be conserved or the available power is limited, as in power amplifiers. Generally an emitter follower, as described in Sec. 2.11, is a simpler and better means of coupling unequal circuit impedances.

Differential amplifiers are used where the amplifier must be insensitive to temperature, power supply variations, and common-mode pickup. A differential amplifier has a high degree of common-mode feedback and a minimum of local feedback. The common-mode feedback greatly reduces the Q-point drift and keeps the differential signal gain high. Q-point drift caused by differences between the two sides of the amplifier is controlled by using matched transistors mounted in close proximity, so that they experience similar temperature changes.

PROBLEMS

3-1. Assume the amplifier in Fig. P-3.1 is biased so that the collector-to-ground voltage is -10 V. (a) Find the input impedance with the 6000

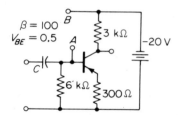

Figure P-3.1.

Ω resistor removed. (b) Find the input impedance with the 6000 Ω resistor replaced. (c) What is the percent error in assuming the input impedance to be 6000 Ω?

3-2. Repeat Prob. 3-1, assuming $\beta = 20$.

3-3. (a) What is the S-factor of the amplifier in Fig. P-3.1? (b) What is the voltage gain of the stage? (c) Show how these answers check in the TG-IR.

3-4. Three bias resistors are suggested for connection between A and B in Fig. P-3.1. (a) Should the resistor be 65 kΩ, 85 kΩ, or 100 kΩ? (b) Give the reasons for your choice.

3-5. (a) What is the collector Q-point voltage if the amplifier in Fig. P-3.1 is operated on a 40 V collector supply, using the 100 kΩ resistor between A and B? (b) What is the collector Q-point voltage if the supply voltage falls to 10 V?

3-6. If an emitter capacitor is used in series with 15 Ω to bypass the emitter resistor of the amplifier in Fig. P-3.1, (a) what is the stage input impedance, (b) the voltage gain, (c) the ac current gain, and (d) the dc S-factor?

3-7. Consider the amplifier in Fig. P-3.7. (a) What is the approximate S-factor, voltage gain, input impedance, and output impedance? (b) Find the base, emitter, and collector dc Q-point voltages.

3-8. For the transistor in Fig. P-3.7 calculate h and show how h affects the voltage gain and the input impedance of the stage.

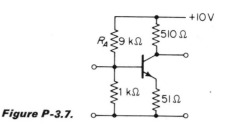

Figure P-3.7.

3-9. Refer to the amplifier in Fig. P-3.1 and assume that the collector resistor is interchanged with the base resistor. (a) Find an approximate value for the bias resistor and (b) describe the gain and impedance characteristics of the stage.

3-10. Refer to the emitter-coupled amplifier in Fig. 3.12. Assume that the lower emitter opens. Find the resulting stage gain and the Q-point voltages.

3-11. Why would a 2N3568 transistor be an unsatisfactory replacement for the 2N1711 shown in Fig. 3.5?

3-12. For the combined collector and emitter feedback amplifier shown in Fig. 3.10. (a) Calculate the collector and emitter Q-point voltages. (b) Calculate the expected iterated voltage gain. (c) Show that there is significant ac feedback. (d) Find the approximate low-frequency cutoff when each capacitor is considered alone.

3-13. Repeat Prob. 3-12 assuming $\beta = 70$.

3-14. Repeat Prob. 3-7 using a 20 V collector supply, and explain why the higher supply voltage makes a more practical stage.

3-15. For the collector feedback stage shown in Fig. 3.8, show that $h = 3.5\,\Omega$ and find the gain and impedance characteristics, with the resistor R_L removed, for: (a) $R_S = 0$, (b) $R_S = 1\ \text{k}\Omega$, and (c) $R_S = 10\ \text{k}\Omega$. Ans: (a) $G_V = [2500]$ or $2800 = 37(10000)/130$. (b) $330 = 37(10000)/1130$. (c) $36 = 37(10000)/10130$.

CHAPTER FOUR

Field-Effect Amplifiers and Feedback

A variety of field-effect semiconductor devices is rapidly replacing both vacuum tubes and transistors in applications requiring high impedances. Two kinds of field-effect devices are widely used—the junction field-effect or unipolar transistors (FET or J-FET) and the metal oxide semiconductors (MOS), known also as insulated gate (IG) devices. The characteristics of these devices are generally similar to the pentode vacuum tube and are different from the bipolar transistors of the previous chapters. Field-effect devices provide simple high-impedance circuits where needed, but usually produce low voltage gains when used alone. Hence, FET devices are often paired with transistors.

The application of FETs differs in an important way from the application of vacuum tubes. Although FETs were discovered along with the bipolar transistor in 1948, the difficulties of manufacturing a uniform product delayed the commercial availability of FETs for more than a decade, and even today the buyer must accept devices with a wide range of characteristics. The high impedance of FETs combined with the high g_m of transistors make the use of feedback an economical way of offsetting the inherent variability of FETs. This chapter includes a brief review of voltage gain calculations suitable for pentode-like devices and also resumes the study of feedback by explaining how a simple measurement may show whether a given circuit has a significant amount of feedback.

4.1 JUNCTION-TYPE FETs (Ref. 1)

The junction-type FETs are formed as a semiconductor filament of silicon called the *channel* and an associated diode called the *gate*. As shown in Fig. 4.1(a), the channel is reduced to a narrow gate-like region by forming the

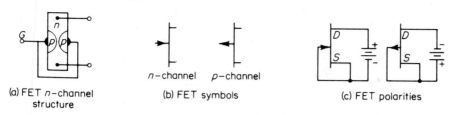

(a) FET n-channel structure

(b) FET symbols

(c) FET polarities

Figure 4.1. *The junction FETs.*

diode on opposite sides of the channel. Junction FETs are formed either as n-channel or p-channel devices and are represented by the symmetrical symbols shown in Fig. 4.1(b). A voltage that reverse biases the gate diode creates a field capable of controlling or *pinching off* current flow through the channel. Because the gate diode is reverse biased, the driving source sees a high input impedance. The FET can be operated without bias (zero bias) unless the peak input signal forward biases the gate by more than 0.5 V, the diode *threshold voltage*. In order to obtain high-voltage output signals, we must often let the input exceed 0.5 V and, therefore, we are forced to use bias and reduced voltage gain.

The junction FETs are readily available with a wide range of characteristics and in either p-channel or in similar n-channel types. Because the channel is a simple semiconductor filament, both ends are alike and the circuit symbols are symmetrical end to end. Whether an FET is used in a circuit as a follower or an amplifier is found by noting the channel polarities. The n-channel device is used with the battery connected as if for a vacuum tube. The channel terminal connected to the negative side of the battery, as shown in Fig 4.1(c), is called the *source* (S), and the terminal connected to the positive side is the *drain* (D). A p-channel device requires the opposite polarities, negative side to the drain. We can remember the drain polarity by observing that the arrow on the symbol points in to a positive drain or away from a negative drain. The source and the drain should always be indicated on circuit diagrams by the letters S and D.

An FET that has a low channel resistance may be used as a gate-controlled switch. Such devices usually have an ON resistance between 3 and 300 Ω and a gate leakage that is typically between 0.1 nA and 100 nA. The static characteristics of a switching type J-FET are shown in Fig. 4.2. If this device is operated with zero gate bias [$V_{GS} = 0$], a change of the gate voltage from $+0.5$ V to -0.5 V with a constant V_{DS} produces a drain current change of 3.0 mA. The zero bias mutual conductance g_0 is the ratio of the drain current change to the corresponding gate voltage change, which is 3000 μmho.

The static characteristics of an amplifier type J-FET are shown in Fig. 4.3. This device has a low zero bias drain current I_{DSS} and a mutual conduct-

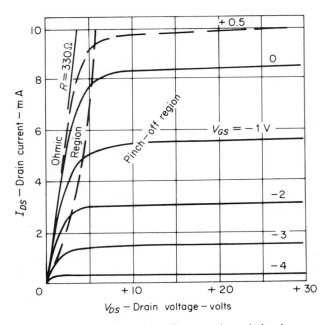

Figure 4.2. *Junction field-effect transistor drain charac-teristics, n-channel general purpose.*

Figure 4.3. *Junction field-effect transistor drain charac-teristics, n-channel amplifier.*

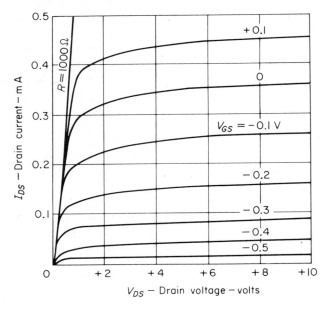

ance of 1000 μmho. The figure of merit g_m/I_{DSS} is 3.0 mho/amp for the amplifier type and is only 0.36 mho/amp for the switching type. The advantage of a high figure of merit is explained in Sec. 4.6. Observe that the low g_m device has the higher figure of merit.

The gate voltage required for channel current cutoff is called the pinchoff voltage V_p. The pinchoff voltage of FETs is usually between 2 V and 8 V. The switching type FET illustrated in Fig. 4.2 has a pinchoff voltage of 4 V and the amplifier-type pinchoff voltage is 0.6 V. For similar constructions, the figure of merit varies inversely with the pinchoff voltage.

The FET offers the advantages of simple biasing techniques, relative insensitivity to temperature, very low noise level, and useful power gain, even above 500 MHz. Because FETs have impedance and gain characteristics much like vacuum tube pentodes, most tube circuits are easily converted to use FETs. They are also useful in many applications as constant current resistors and as voltage-controlled relays (switches), choppers, and modulators.

4.2 INSULATED GATE FETs

The insulated gate semiconductors are similar to the junction FETs except that the gate is formed [see Fig. 4.4(a)] as a small high-quality capacitor, thus permitting the control of current flow in the channel without the disadvantages of the gate diode. For the control of the channel current by an electric field, the capacitor must have exceedingly small dimensions and the silicon dioxide dielectric must be such a thin layer that it has a low breakdown rating—about 30 V. The gate capacitance is so low, and the leakage resistance so high, that the gate is easily charged to a voltage above breakdown.

Gate damage is prevented only by using a grounded soldering iron and by protecting the gate from static discharges, as from one's body in dry weather. Insulated gate FETs should be picked up by the case rather than by the leads, and the leads should be shorted together during handling by foil or a wire loop. Some devices have built-in Zener diodes to protect the gate from low energy sources, but these devices have the leakage characteristics of a Zener diode, and reasonable precautions are still required in the application of insulated gate FETs.

Figure 4.4. *Enhancement MOS FET.*

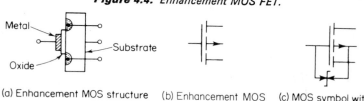

(a) Enhancement MOS structure   (b) Enhancement MOS symbol p-channel   (c) MOS symbol with Zener diode

The MOS devices are available in two types, characterized by two different modes of operation—the **depletion mode** and the **enhancement mode**. In the depletion type the channel is a conductor and current normally flows in the conducting channel. A voltage applied to the gate reduces the current and may cut off the current completely. In the enhancement type the channel is normally cut off and is turned on by applying a gate voltage or bias.

The enhancement MOS devices are presently constructed as shown in Fig. 4.4(a) with a p-channel substrate as one side of the gate capacitor. The symbol for these devices is shown in Fig. 4.4(b) with a broken line to suggest that the channel is open with zero bias. Some devices, as shown in Fig. 4.4(c), have a protecting Zener diode formed as a part of the structure. The p-channel enhancement MOS requires a negative drain voltage and a negative gate bias that allows the use of very simple bias techniques. As shown in Sec. 5.8, the enhancement MOS amplifier stages may be operated with the gate and the drain at the same potential, making it possible to direct-couple a series of stages, even though the stages operate from a common power supply.

The static characteristics of an enhancement mode MOS are shown in Fig. 4.5. This insulated gate device is operated with a negative voltage on both the gate and the drain. With a gate bias of -7 V, the MOS has a g_m of 3000 μmho with a 4 mA drain current. The channel is cut off for gate voltages below -5 V. Observe that the MOS can be operated with equal gate and drain voltages, as indicated by the dotted line in Fig. 4.5. This line is the locus of Q-points that may be used for a direct-coupled amplifier operating with equal gate and drain voltages.

Figure 4.5. *Transistor drain characteristics, p-channel enhancement type.*

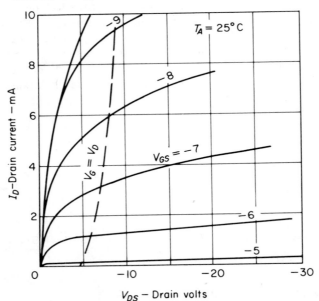

Depletion-type MOS devices are constructed by depositing a thin-film
n-channel on a semiconductor substrate and by depositing the oxide dielectric
and metal plate of the gate, as shown in Fig. 4.6(a). The symbol for the
n-channel MOS, given in Fig. 4.6(b), shows a substrate terminal which is
usually connected to the source. A depletion MOS operates with positive,
negative, or zero bias and is, therefore, approximately interchangeable with
the J-FETs. The insulated gate devices offer advantages at high frequencies
(200 MHz) and in integrated circuits, but IGs presently have higher noise
levels and are not as reliable as the J-FETs. Exceedingly high gate resistance
and low input capacitance make these devices attractive for high input
impedance applications, as in the input stages of electrometers.

The static characteristics of an *n*-channel depletion MOS are shown in
Fig. 4.7. This device may be operated with zero bias and both positive and
negative gate voltage excursions. The insulated gate prevents gate current.
Observe that the static characteristics show that the MOS device represented
in Fig. 4.7 is nearly interchangeable with the J-FET represented in Fig. 4.2.

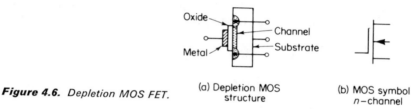

Figure 4.6. *Depletion MOS FET.*

(a) Depletion MOS
structure

(b) MOS symbol
n-channel

Figure 4.7. *MOS Transistor drain characteristics,*
n-channel depletion type.

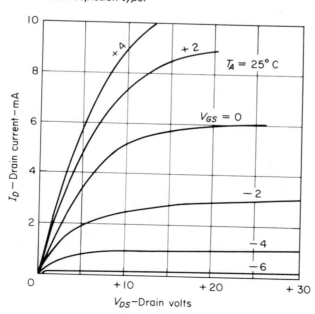

4.3 RESISTANCE-COUPLED FET AMPLIFIERS

A simple resistance-coupled (RC) n-channel FET amplifier is shown in Fig. 4.8. With a signal applied to the gate, as indicated by the sine wave, the source follows with the same signal polarity but at about $\frac{6}{10}$ the gate signal amplitude. Because the gate resistance is 300 times the source load resistance, the source follower has a power gain of $(0.6)^2$ (300), or about 100 times (20 db). The signal at the drain terminal is of opposite phase and is four times the input signal; hence, the power output is about seven hundred times the power input (28 db).

The source resistor R_S furnishes a negative gate bias that lowers the drain current and raises the drain voltage V_D to about one-half the drain supply voltage V_{DD}, thus permitting large ac output signals. The source resistor also causes negative feedback which stabilizes the Q-point and the ac gain, thus offsetting a part of the difference between one FET and another of the same type number. If used, a 5 μF capacitor connected across R_S reduces ac feedback and raises the voltage gain to 10.

Zero bias operation is effected by shorting R_S and lowering the drain resistor value enough to offset the effect of increased drain current and restore

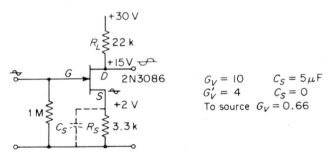

Figure 4.8. *FET amplifier for n-channel FET or depletion MOS.*

the drain Q-point voltage. For many applications the simplicity of zero bias operation justifies the selection of FETs or an adjustment of the load resistor. A 2N3086 with zero bias and a 10 kΩ resistor will give a voltage gain of about 7.

Amplifiers using depletion-type MOS devices may be constructed in the same way as FET amplifiers, i.e., as shown in Fig. 4.8. The characteristics of the two devices are similar enough to make them essentially interchangeable. The MOS, like the FET, gives greatest voltage gain when used with a bias and a high resistance load. With zero bias and a low value load resistor the gain is low, but the MOS amplifier tolerates larger input signals.

4.4 ENHANCEMENT MOS AMPLIFIERS

A single-stage amplifier using an enhancement MOS 2N4352 is shown in Fig. 4.9. This amplifier uses a p-channel device, so that the drain side of the battery is negative. Bias is supplied through a 20 MΩ drain-to-gate resistor, and a drain voltage of at least 20 V is required. Because the gate is practically an open circuit, the gate and drain voltages are equal (5 V) and about 1 V above the gate threshold voltage. Because the bias resistor R_f causes shunt feedback, the input impedance is (Sec. 4.13):

$$R_I = R_f/(G_v + 1) \tag{4.1}$$

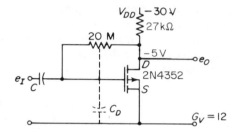

Figure 4.9. MOS amplifier; p-channel enhancement MOS.

where G_v is the gate-to-drain voltage gain. For this amplifier $G_v = 12$; hence, the input impedance is 1.5 MΩ. As long as the driving generator impedance is at least a factor of 5 smaller than the input impedance, the generator is not loaded and full voltage gain is obtained. This requirement is easily met in most amplifiers; thus, the effect of feedback in reducing the input impedance can be neglected except when selecting the interstage coupling capacitor. If the midpoint of R_f is bypassed to ground, the input impedance becomes 10 MΩ.

4.5 FET AND MOS AMPLIFIER GAIN CALCULATIONS

The field-effect devices are high input impedance and high internal output impedance devices, similar to a vacuum tube pentode. Amplifier gain calculations are greatly simplified by assuming that the input and output impedances are so high they can be neglected. The gain calculated under this assumption is sufficiently accurate for practical circuit analysis. However, a method that takes the internal output impedance into consideration is described also so that we may understand both the exact and the approximate formulas.

Field-effect devices, like small-signal pentodes, may be represented by the equivalent circuit shown in Fig. 4.10. The dc gate and drain supply voltages are not shown because the equivalent circuit is used only to calculate ac currents and voltages. The gate is represented simply as an open-circuited

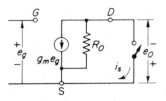

Figure 4.10. *Norton equivalent circuit.*

terminal having an ac voltage e_g. The internal output impedance of the device is represented as an ac resistance R_O. The internal generator (the arrow enclosed by a circle) is represented as a constant current source $g_m e_g$. This current is proportional to the gate voltage e_g, and the proportionality factor g_m is called the **transconductance** of the device. The equivalent circuit means simply that an ac input voltage e_g across the input resistor creates a constant ac current $g_m e_g$ through any connected load. When the drain is externally shorted (ac short) to the source, the entire generator current flows through the short and is:

$$i_s = g_m e_g \tag{4.2}$$

so that:

$$g_m = \frac{i_s}{e_g} \tag{4.3}$$

As Eq. (4.3) indicates, the mutual conductance is the short-circuit current per volt input.

When the drain is open, the generator current flows through R_O, and the drain ac voltage becomes:

$$e_O = g_m R_O e_g \tag{4.4}$$

By definition the **amplification factor** μ of a device is the open circuit voltage gain, e_O/e_g. Therefore, Eq. (4.4) gives Van der Bijl's well-known (vacuum tube) relation:

$$\mu = g_m R_O \tag{4.5}$$

Field-effect devices are sometimes represented by the Thevenin equivalent circuit of Fig. 4.11, where the generator is a constant voltage (rather than constant current) source μe_g and the internal output impedance of the device is in series with the generator. Both equivalent circuits lead to the same result, but with a high internal impedance the Norton equivalent is simpler to use. One should be able to use either equivalent, whichever is more convenient.

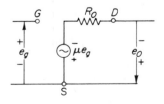

Figure 4.11. *Thevenin equivalent circuit;*
$\mu = g_m R_O$.

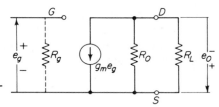

Figure 4.12. Norton equivalent of FET amplifier with load R_L.

Consider an amplifier represented by the Norton equivalent shown in Fig. 4.12 with a load resistor R_L connected externally from the drain to the source. The load R_L is in parallel with the internal impedance R_O, so instead of Eq. (4.4), we write:

$$e_O = g_m\left(\frac{R_O R_L}{R_O + R_L}\right)e_g \tag{4.6}$$

Field-effect devices usually have an internal impedance that is at least five times larger than the load, i.e.,

$$R_O = 5R_L \tag{4.7}$$

Combining Eqs. (4.7) and (4.6), and solving for the voltage gain, we find:

$$G_v = 0.83 g_m R_L \tag{4.8}$$

If it can be assumed that the internal output impedance is very large and negligible, then:

$$G_v = g_m R_L \tag{4.9}$$

A comparison of Eqs. (4.8) and (4.9) shows that a voltage gain calculation, neglecting the internal impedance of a field-effect device, will indicate a voltage gain that is 10 to 20 percent too high. This small error is not the real difficulty in a gain calculation. Rather, the difficulty is that the transconductance of field-effect devices is rarely given within a 2 or 3 to 1 limit by the manufacturer, and the Q-point conditions existing in a practical amplifier are rarely as favorable as those given on the data sheet. Equation (4.9) is accurate enough for practical gain calculations providing we are careful in selecting a value of g_m appropriate for the Q-point and the nontypical device in use. Let us see how the operating Q-point affects the transconductance of a field-effect device.

4.6   THE TRANSCONDUCTANCE OF FETs (Ref. 1)

The transconductance g_m of FETs in the pentode or pinchoff region is found experimentally to vary as the square root of the drain current I_D. This fact may be written as

$$g_m = g_O\sqrt{\frac{I_D}{I_{DSS}}} \tag{4.10}$$

where g_0 is the transconductance at the zero bias drain current I_{DSS}. The manufacturer usually gives typical values for g_O and I_{DSS}. The theory of FETs indicates that g_0 is related inversely to the gate voltage V_P required to pinch off the drain current, i.e.,

$$g_0 = \frac{2I_{DSS}}{V_P} \qquad (4.11)$$

From equations (4.9) and (4.11) the voltage gain of a zero biased amplifier is

$$G_v = g_0R_L = \frac{2I_{DSS}R_L}{V_P} \qquad (4.12)$$

In a resistance-coupled amplifier the dc voltage drop in R_L will be approximately one-half the drain supply V_{DD}; hence, Eq. (4.12) may be written as:

$$G_v \cong \frac{V_{DD}}{V_P} \qquad (4.13)$$

Equation (4.13) shows why high supply voltages are desirable and the advantage in using devices having low V_P values. The advantage is expressed commercially by calling g_0/I_{DSS} the *figure of merit* of an FET. Solving Eq. (4.11) for the figure of merit gives

$$\frac{g_0}{I_{DSS}} = \frac{2}{V_P} \qquad (4.14)$$

which shows that high figure of merit devices have low V_P values. Using Eq. (4.10), we can show that biasing an FET to reduce the drain current tends to improve the figure of merit. Hence, an amplifier will be satisfactory if biased to about one-half the zero bias drain current. The calculated voltage gain is usually found to be reasonably in agreement with the measured gain if the g_m of the device is assumed to be one-half to two-thirds the zero bias value.

MOS devices may not conform exactly to the relations given above, but Eq. (4.10) approximates actual operating conditions sufficiently for most gain calculations (and also for vacuum tube pentodes). We can assume, therefore, that the gain calculations outlined here for FET amplifiers are reasonably applicable also to insulated gate and pentode applications.

4.7 FET GAIN CALCULATIONS: AN EXAMPLE

The amplifier shown in Fig. 4.8 will be used as an example for a gain calculation. We begin by drawing the equivalent circuit showing the externally connected resistors. With the bypass capacitor C_s in place, the source resistor is ac short-circuited and the equivalent circuit is that of Fig. 4.12. The amplifier has a resistor R_g from gate to ground that is usually placed across the input in order to fix the input impedance at a known value, independent of temperature and humidity. Usually this resistor is so high it can be neglected as a load on the previous stage or generator. Hence, we assume that the gate is driven by a known signal voltage e_g.

The manufacturers give the zero drain g_0 as between 400 and 1200 μmho. The typical value is given as 800 μmho. However, the manufacturers give the typical zero bias drain current I_{DSS} as 1.5 mA whereas the amplifier is biased and the drain current is only 0.7 mA. Using the typical g_0 in Eq. (4.10), we find that g_m is about 560. Equation (4.9) then predicts a voltage gain of 12. For practical purposes, the calculated gain agrees with the measured gain as closely as we may expect. However, because g_0 may vary by a factor of 3, so may the voltage gain. When a more uniform voltage gain is required, we are forced either to select the FETs or to use gain stabilizing negative feedback.

4.8 SIGNIFICANT FEEDBACK

The variations in gain caused by the differences between FETs of the same type number are so great that local feedback must be used, as with transistors, to stabilize the gain of an amplifier. In a single-stage amplifier, as shown in Fig. 4.8, the voltage drop in the source resistor R_S causes a part of the output voltage to appear in series opposition to the input voltage. The feedback is called **series current feedback** because the voltage across the source resistor is in series with the input and is proportional to the load current.

The effectiveness of feedback in improving an amplifier is approximately proportional to the gain reduction produced by the feedback. If the distortion or the gain variation of an amplifier is to be reduced by a factor of 4, we must reduce the gain of the amplifier to one-fourth the gain without feedback. These statements imply that the gain without feedback is known or is measured. In actual practice, we need to know how to determine if a given stage is operating properly and that there is enough feedback to ensure a significant improvement in the gain stability of an amplifier—without knowing or measuring the gain without feedback. We shall show that the amount of local feedback in a stage may be easily estimated by measuring the ac input and feedback signals.

As an example of significant feedback consider the amplifier shown in Fig. 4.13, designed to have just enough feedback to reduce the voltage gain G_v by a factor of 4, 12 db of feedback. For present purposes we designate the 4 to 1 gain reduction as **significant feedback**, because a 4 to 1 improvement

Figure 4.13. *Feedback amplifier with minimum significant feedback, i.e.,* $e_S = 0.75e_I$ *and* $G_v' = 0.75\,(R_L/R_S)$.

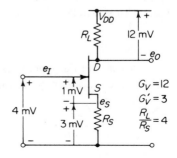

represents about a minimum for a significant improvement. The factor of 4 is selected as a practical value. (For some purposes we might use less feedback and for others more.) Expressed algebraically, significant feedback exists because the gain with feedback G'_v is just one-fourth the gain without feedback, i.e.,:

$$G'_v = \frac{G_v}{4} \qquad (4.15)$$

The ac voltages shown in Fig. 4.13 represent ac signals existing with a 4 mV input signal, this signal being chosen to make the gate-to-source voltage a normalized 1 mV. The 1 mV gate-to-source voltage produces a 12 mV signal across the load R_L; the resistor R_S was selected by the designer to establish 3 mV ac from the source to ground. To maintain these signals, we must supply a 4 mV gate-to-ground input signal; hence, with feedback the stage gain is 3. The 1 mV gate-to-source voltage represents the input signal required if the source feedback signal is bypassed; hence, the input signal is four times larger with feedback than without.

The important facts to be recognized are: (1) the gain is reduced because of the signal conditions existing in the input circuit; (2) the 4 to 1 relationship is shown by the input signals; and (3) the gain reduction is 4 because the feedback signal is three-fourths the input signal. Statement (3) says that *significant feedback exists if the ac source voltage is at least three-fourths the amplifier input signal*:

$$e_S = \frac{3}{4} e_I \qquad (4.16)$$

Feedback in the amplifier may be removed by using a capacitor to bypass the signal existing across the source resistor. With the capacitor, a signal input of 1 mV maintains a load voltage of 12 mV; therefore, the no feedback voltage gain is 12. Because the resistor ratio required to establish effective feedback varies with and is determined by the amount of voltage gain available without feedback, we often need to know how to set the resistance ratio when the voltage gain without feedback is known. The present example shows that significant feedback exists if the resistance ratio is one-third the gain without feedback; i.e.:

$$\frac{R_L}{R_S} = \frac{G_v}{3} \qquad (4.17)$$

Eliminating G_v from Eqs. (4.15) and (4.17) produces a useful relation for the gain with significant feedback in terms of the resistance ratio:

$$G'_v = \frac{3}{4} \frac{R_L}{R_S} \qquad (4.18)$$

An algebraic generalization of Eqs. (4.17) and (4.18) is obtained by writing the amplifier input signal e of Fig. 4.14 as

$$e = e_I - B_v e_o \qquad (4.19)$$

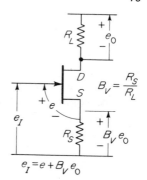

Figure 4.14. *J-FET feedback amplifier illustrating negative feedback.*

And the output signal is

$$e_O = G_v e \qquad (4.20)$$

Eliminating e from Eqs. (4.19) and (4.20), we obtain

$$G'_v = \frac{e_O}{e_I} = \frac{G_v}{1 + B_v G_v} \qquad (4.21)$$

The term $B_v G_v$ is the amplifier gain without feedback multiplied by the feedback voltage loss factor. The product $B_v G_v$ is the gain around the feedback loop and is called the **loop gain**. Eq. (4.21) states that the gain with feedback is the gain without feedback divided by 1 plus the loop gain. When the loop gain is large compared with 1, the gain with feedback is the reciprocal of the loss factor and is independent of G_v. The feedback is called **negative feedback** when the feedback signal opposes the input signal and $B_v G_v$ is positive.

The voltage loss factor for the amplifier shown in Fig. 4.13 is

$$B_v = \frac{R_S}{R_L} \qquad (4.22)$$

B_v is the reciprocal of the resistance ratio R_L/R_S. Our definition of significant feedback is equivalent to a requirement that the feedback loop gain $B_v G_v$ is 3, which is the meaning of Eq. (4.17).

We should observe that Eq. (4.21) is simply a mathematical representation of the voltage relations existing in the amplifier input circuit. Equation (4.17) tells the designer that significant feedback requires a loop gain of 3 or a resistance ratio that is one-third of the gain without feedback. Equation (4.18) tells the experimenter that significant feedback actually exists if the measured gain with feedback is three-fourths the resistance ratio. If the gain is lower than three-fourths the resistance ratio, the feedback is not of practical significance. Equation (4.16) is a similar test for significant feedback in terms of the input voltage and the feedback voltage. All three tests apply equally well to transistor amplifiers. when the base-to-emitter input impedance is large compared with the equivalent source impedance.

4.9 AN AMPLIFIER WITH ADEQUATE FEEDBACK (Ref. 1)

When feedback is applied over two or more stages, the reduction of the overall gain by a factor of 10 or more is practical. With a large amount of feedback the source voltage follows—i.e., almost equals—the input voltage, and the voltage gain by Eq. (4.21) is determined by the resistance ratio. If the voltage gain G'_v of Eq. (4.21) can be represented accurately enough for practical purposes by the resistance ratio:

$$G'_v = \frac{R_L}{R_S} \tag{4.23}$$

then the feedback is called **adequate**. The term *adequate* means only that the loop gain $B_v G_v$ is so large that the term *one* in Eq. (4.21) is negligible. However, with adequate feedback we can find the expected amplifier gain by examining the resistors in the feedback circuit, and we do not need to know the open-loop gain.

4.10 SOURCE FEEDBACK: AN EXAMPLE

As a practical example of source feedback, consider the amplifier in Fig. 4.15 with the measured signal voltages normalized to a 1 mV gate-to-source voltage. The 1 mV gate-to-source signal may be easily set up by using a differential oscilloscope connected to show this voltage differentially. A comparison of the input circuit voltages shows that the source voltage is only 0.6 of the input voltage. Hence, the amplifier fails to have what we have called **significant** feedback. The voltage gain without feedback is 10 and the resistance ratio R_L/R_S is 6.7. To provide significant feedback, we must make the resistance ratio 3.3 by Eq. (4.17). Changing R_L does not change the amount of feedback because the voltage gain changes also. Increasing the source resistor does increase the feedback, but the increased dc bias adversely affects the Q-point, and a positive gate bias is required to restore the Q-point conditions. Evidently the 3.3 kΩ resistor is used only for bias.

In a laboratory study of amplifiers the amount of local feedback may be easily estimated by comparing the input circuit voltages and using Eq. (4.16) as the test for significant feedback. When there are several stages with overall feedback, the amount of feedback has to be determined by comparing the

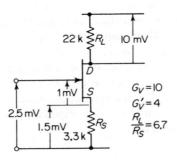

Figure 4.15. *Feedback amplifier lacking minimum significant feedback, i.e., $e_S = 0.60 e_I$ and $G'_v = 0.6 (R_L/R_S)$.*

input signal with the gate-to-source signal or its equivalent. The amount of feedback, which is the gain reduction produced by feedback, is then the ratio:

$$F_B = \frac{e_I}{e_I - e_S} \tag{4.24}$$

With a large amount of feedback the input and the source voltages tend to become alike, except for the distortion and noise introduced by the amplifier. For this reason the distortion and noise in the gate-to-source voltage, the difference $(e_I - e_S)$, may make it difficult to obtain a precise value of the denominator in Eq. (4.24). Likewise, the signal observed in the early stages of a feedback amplifier may have a poor waveform. However, the difference (or error) signal often gives a clue to feedback or servo system difficulties.

4.11 HIGH-GAIN FET AMPLIFIERS

The amplifiers shown in Fig. 4.16 illustrate the advantage of using high figure of merit FETs, i.e., low V_P devices. A high g_m to drain current ratio makes it possible to use a high valued drain resistor and to obtain a high voltage gain. In addition, because the dc drain voltage must be at least as high as V_P, a high figure of merit implies that the drain may be operated at low voltages or from a low supply voltage. The stage using a 2N3086 FET requires a low drain resistor with 6 V on the drain. This stage has a voltage gain of only 2, but with a negative gate bias the drain resistor can be increased until the voltage gain is about 10. The 2N3687 FET stage uses a ten times higher drain resistor and with zero bias has a voltage gain of 25. A disadvantage of both amplifiers in Fig. 4.16 is that they become inoperative if the drain supply voltage is lowered without lowering the drain resistor. This difficulty can be minimized by using a lower value drain resistor and accepting a lower gain.

Figure 4.16. *J-FET RC amplifier with resistor load.*

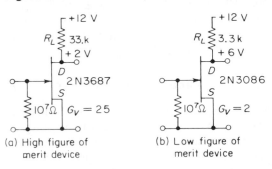

(a) High figure of merit device

(b) Low figure of merit device

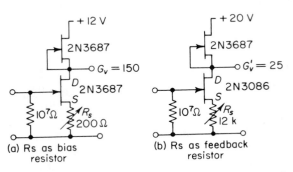

Figure 4.17. *J-FET amplifier with FET constant current load.*

The amplifiers in Fig. 4.17 show that higher voltage gains may be obtained by using a second FET as a constant current load in place of the drain resistor. When a pair is used as in Fig. 4.17, the FETs should be selected so that they are reasonably alike. The small source resistor R_S is used to balance the drain currents and to adjust for a maximum voltage gain. If maximum voltage gain is obtained with the source resistor shorted out, the FETs should be interchanged in order to use the higher I_{DSS} device as the amplifier and to permit drain current balancing. If the source resistor has to be made so high that the gain is reduced by feedback, a pair of FETs even more alike must be selected.

The amplifier in Fig. 4.17a is remarkable for its high voltage gain, high temperature stability, and low noise characteristics. With similar FETs the collector Q-point will show little temperature drift. The voltage gain is one-half the amplification factor of the device. This amplifier operates well with supply voltages of only 4 V, but the voltage gain varies approximately in proportion to the supply voltage. By replacing the lower FET with a high drain current device (Fig. 4.17b), the source feedback resistor may be increased to provide significant feedback. Using a 2N3687 FET in combination with a 2N3086 FET and a 12 kΩ source resistor, we can reduce the voltage gain to 25 from a no feedback gain of 125. With a 15 V to 30 V drain supply—i.e., well above the device pinchoff voltage—the gain changes are reduced significantly by the feedback.

4.12 FREQUENCY RESPONSE OF FET AMPLIFIERS

Field-effect devices have high voltage gain bandwidth products, 100 MHz or more. This means that an amplifier with a voltage gain of 10 can be made to have this gain up to about 10 MHz. If the voltage gain is reduced to 2 by reducing the circuit impedances, the frequency response should be flat up to about 50 MHz. This limiting of the bandwidth is caused by the very small interelectrode capacities in the FET and in the externally connected circuits. One should note, however, that the gain bandwidth of a device usually cannot be attained unless that amplifier is suitably designed. For example, we cannot have both a high-impedance level and a high gain-bandwidth product.

4.13   THE MILLER EFFECT (Refs. 9, 10)

The frequency response of an amplifier having a high voltage gain is especially sensitive to degradation by small capacitors between the gate and drain of the FET. This phenomenon, known as the *Miller effect*, is simply an increase of input capacitance caused by a multiplication of the gate-to-drain capacitance C_{gd} due to the voltage gain of the stage. The Miller effect is easily explained using Fig. 4.18, which shows a FET stage having a gate to drain capacitance C_{gd}. The stage is assumed to have a voltage gain A and the usual input-to-output 180° phase inversion. Because the input and output voltages are in series aiding, the input current will be $(1 + A)$ times larger than the current that would flow if the input generator were driving C_{gd} alone. It follows that the input source is loaded as if the resulting input capacitance were $(1 + A) C_{gd}$.

When allowance is made for the gate-to-source capacitance C_{gs}, the total input capacitance becomes:

$$C_I = C_{gs} + (1 + A)C_{gd} \qquad (4.25)$$

When the voltage gain is large, the relatively small gate-to-drain capacitance causes the principal input loading of a stage. The 2N3687 FET used in Fig. 4.17 will have a gate-to-source capacitance of 1.5 pF. The gate-to-drain capacitance is 1.4 pF, but the external wiring capacitance will add at least another 1.1 pF. With the voltage gain of 150 shown in Fig. 4.17, the equivalent input capacitance becomes:

$$C_I = 1.5 + (151)2.5 = 380 \text{ pF} \qquad (4.26)$$

A capacitance of 380 pF has a reactance of 1 MΩ at 400 Hz. Therefore, with a 1 MΩ generator impedance the amplifier in Fig. 4.17 might have a high-frequency cutoff as low as 400 Hz. As this example shows, the equivalent input capacitance of a high-gain amplifier may be many times larger than the input capacitance of the FET, and the wiring capacities must be carefully reduced in high-gain amplifiers intended for high-frequency applications.

Figure 4.18. The Miller effect.

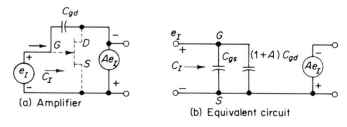

(a) Amplifier   (b) Equivalent circuit

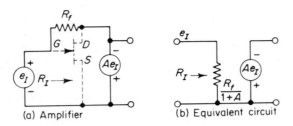

Figure 4.19. *Miller feedback.*

The cutoff frequency of the amplifier in Fig. 4.17 may be raised to 2.4 MHz by reducing the voltage gain to 25 and lowering the generator impedance to 1000 Ω.

For RF applications above 100 MHz, dual gate MOS FETs offer drain-to-gate input capacitances as low as 0.01 pF, with a transconductance exceeding 5000 μMho. With appropriately reduced electrode voltages these FETs may be used to replace most high-frequency pentodes.

Amplifiers are sometimes biased by connecting a resistor between the gate and the drain, as shown in Fig. 4.19(a). This connection introduces signal feedback that is called *Miller feedback* because it reduces the input resistance in the same way the input impedance is reduced by capacitance feedback. With a drain signal A times larger than the gate signal, the input current is increased by the feedback resistor R_f just as if the input were shunted by a resistor $R_f/(1 + A)$. As shown in Fig. 4.19(b), the effects of Miller feedback on the amplifier gain and input impedance are the same as the effects produced by the equivalent resistor $R_f/(1 + A)$. The quantitative explanation of this effect is obtained by substituting resistors for the capacitors in Fig. 4.18 and following an analysis similar to that used in developing Eq. (4.25). We must understand that the input impedance of a high-gain amplifier is greatly changed whenever an impedance is connected between the input and a point having an amplified signal. Stray capacitances and stray leakage resistances must be expected to cause disturbing changes in the characteristics of high-gain amplifiers. Similar disturbances are caused by common ground resistances in high current gain amplifiers.

SUMMARY

Field-effect semiconductors require simple bias techniques and have high-input impedance and high-output impedance. By neglecting the internal output impedance, the voltage gain may be calculated by assuming the FET is a current source $g_m e_g$. Because the mutual conductance g_m varies with the drain current I_D, the g_m must be corrected for calculation by using Eq. (4.10).

The nonuniformity of FET devices forces the use of a significant degree of feedback. *Significant feedback is defined as that feedback which reduces the stage gain by a factor of 4. Significant feedback exists in an operating amplifier if the series feedback voltage is at least $\frac{3}{4}$ of the input voltage.* When the feedback reduces the gain by a factor of 10 or more, the feedback is called *adequate*, meaning that the gain of the amplifier may be determined by considering only the resistors of the feedback circuit.

The *Miller effect*, caused by shunt voltage feedback, makes the input capacitance of an amplifier increase with the voltage gain of the input stage. The Miller effect tends to reduce the gain bandwidth product of any amplifier, but is particularly troublesome in high input impedance amplifiers. Miller feedback through an external gate-to-drain impedance reduces the gain and the input impedance of an amplifier in a manner analogous to the Miller capacitance effect.

PROBLEMS

4-1. The FET in the amplifier of Fig. P-4.1 has a mutual conductance of 800 μmho. (a) If the internal impedance is high enough to be neglected, what is the voltage gain of the amplifier? (b) If the internal impedance of the FET is so low that the voltage gain is only two-thirds the gain calculated in (a), what is the magnitude of the internal impedance?

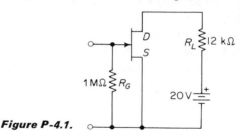

Figure P-4.1.

4-2. The FET in the amplifier shown in Fig. P-4.2 has a transconductance of 600 μmho when operated as shown. (a) If the measured drain voltage is 22 V, what is the expected voltage gain of the amplifier? (b) What is the voltage gain calculated from Eq. (4.13)? (c) What must be done to realize the gain predicted by Eq. (4.13)?

Figure P-4.2.

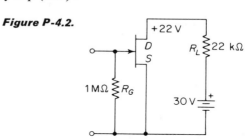

4-3. A given FET has a figure of merit of 1670 μmho per mA. (a) What is the indicated pinchoff voltage if the zero bias drain current is 1 mA? (b) What is the voltage gain if this FET is operated in the amplifier in Fig. P-4.2 by biasing the FET to make the drain voltage 22 V, as shown? (c) What is the voltage gain if the FET is used in the same circuit but with zero bias?

4-4. Refer to the amplifier in Fig. 4.8 and assume the FET has a very high internal impedance. (a) If the source resistor is doubled and a 2 V battery is connected in series with the resistor so that the Q-point conditions do not change, what is the gain of the amplifier with feedback? (b) Does the stage have significant feedback?

4-5. Suppose that the FET in the amplifier in Fig. 4.8 is replaced by a device having twice the transconductance but the same drain current and electrode voltages shown in the figure. (a) What is the open loop voltage gain? (b) What is the gain with feedback? (c) Does the stage have significant feedback?

4-6. Suppose that replacing the FET in the Fig. 4.15 amplifier causes the amplifier to have significant feedback. (a) What is the open loop voltage gain? (b) What is the gain with feedback?

4-7. (a) If both FETs in Fig. 4.16 have a gate-to-drain capacitance of 2 pF and a gate-to-source capacitance of 3 pF, what is the input capacitance of each stage? (b) If the driving generator has an internal resistance of 0.33 MΩ, what is the high-frequency cutoff in each amplifier?

4-8. Show that the FET characteristics shown in Figs. 4.2, 4.3, 4.5, and 4.7 conform to Eq. (4.14).

4-9. Show that the MOS amplifier shown in Fig. 4.9 has significant feedback when the source impedance is 4.5 MΩ.

CHAPTER FIVE

Multistage Amplifiers

High-gain amplifiers are constructed by connecting several stages in a series or cascade. Amplifiers are generally characterized by the coupling network because the network determines the frequency characteristic and controls the Q-point stability. The early types of transistors could not be direct-coupled easily. Therefore, most multistage circuits have followed the method of coupling vacuum tubes. The preferred methods have used resistance-capacitance (RC) and transformer coupling between stages.

Since 1961 the field-effect and planar transistors have made direct-coupling a practical way of simplifying circuits and of eliminating components. Present-day FETs and silicon transistors are used most effectively as direct-coupled pairs because the pair has a high ratio of active to inactive elements.

This chapter describes a few of the many possible ways of coupling FETs and transistors to form multistage amplifiers. The multistage dc amplifier is not considered here, partly because high-gain, direct-coupled amplifiers are becoming readily available as integrated circuits which can be used in applications nearly as easily as a single-stage amplifier.

5.1 METHODS OF COUPLING AMPLIFIER STAGES

Amplifier stages may be coupled in a series, providing the coupled impedance levels are alike and the interstage network supplies correct bias for the adjacent stages. If the coupled-stage impedance levels are approximately the same, the output of each stage can be connected to the next by using a capacitor to provide dc isolation and prevent disturbing the biasing. Higher gains per stage may be obtained by using a high collector load impedance and by using a transformer to match the unequal impedance levels and provide the required dc isolation. Capacitor coupling is usually the simplest and least expensive, whereas transformers are usually heavy, expensive, and more prone to contribute reliability problems.

Multistage dc amplifiers can be constructed by direct-coupling the collector of one stage to the following input if dc voltage and current sources (bias) are provided to establish the Q-point conditions required in successive stages. With each additional stage a dc amplifier becomes increasingly difficult to design and adjust. However, there are distinct advantages in direct-coupling transistors in pairs with dc isolation provided only between every other stage. The direct-coupled pairs (sometimes called *compounds*) offer the advantage of requiring fewer components, particularly the bulky interstage capacitors, and offer higher per stage gains. The direct-coupling also simplifies the application of feedback. Hence, two-stage feedback is often employed in lieu of single-stage local feedback.

Multistage amplifiers may have three or more direct-coupled stages, but the problems with Q-point instability make such amplifiers too complicated except for special purposes—e.g., high-gain dc and operational amplifiers. Field-effect transistors have a low temperature sensitivity and can be direct-coupled to transistor pairs to form simple, high-input-impedance, three-stage dc-amplifiers.

5.2 MULTISTAGE GAIN CALCULATIONS

The overall gain of a series of iterated (alike) stages of transistor amplifiers cannot exceed S for each stage added to the chain. For example, the transistor gain-impedance relation shows that the gain of a single stage will be:

$$G_v = S\frac{R_L}{R_I} \tag{5.1}$$

and for two stages will be:

$$G_v = S_1 S_2 \frac{R_L}{R_I} \cong S^2 \frac{R_L}{R_I} \tag{5.2}$$

and so on. The R_L/R_I factor represents gain obtained from the ratio of the output to the input impedance, and this ratio is not changed by adding an intermediate stage. Hence, the gain obtained from each successive stage cannot exceed one additional β factor. In a practical amplifier we cannot obtain more than a factor of 20 for each added stage because it is difficult to operate each stage so that it has a high S-factor. Furthermore, the collector resistor and the following base resistor bypass part of the signal current to ground. Therefore, a series of stages cannot have a current gain much in excess of 20 or 30 per stage. This is not a serious practical limitation but means that two transistor stages may be needed in order to give as much voltage gain as a single vacuum tube stage. This requirement is more than offset by the relative simplicity of a transistor stage. Very often we must accept output stage load values R_L that are several orders of magnitude smaller than the required input impedance. These impedances make R_L/R_I in Eq. (5.2) very small, and the required number of stages may seem unduly large until one realizes that current gain is easily obtained in a series of direct-coupled CC stages known as *Darlington compounds*.

Multistage amplifiers use various combinations of local and overall dc and ac feedback. Sometimes it is better to use local feedback to stabilize the Q-points and to use overall ac feedback over several stages to stabilize the ac gains. In other amplifiers high ac gain may be needed more than a lower but stable ac gain, in which case emitter bypass capacitors are used to restore the ac gain to the no-feedback value. Whenever the emitter capacitor is used, the input impedance of a stage falls to βh (in parallel with R_B), and the input impedance becomes an order of magnitude smaller than R_B, the value with feedback. With the relatively low input impedance, the entire ac input current enters the base, and the current gain of the stage becomes β, the no-feedback current gain. However, in a series of iterated (repeated) stages the emitter bypass capacitors are not likely to increase the overall ac gain as much as might be expected from the TG-IR.

Bypassing the emitter requires the use of large emitter and coupling capacitors; hence, it usually proves more economical to leave the emitters unbypassed and to add an extra stage to make up the lost gain. For such RC-coupled amplifiers, the coupling capacitor need only have a reactance smaller than the series resistance of the prior stage collector resistor and the following stage base resistor. (The output impedance of a stage is that of the collector resistor.)

5.3   RC-COUPLED AMPLIFIERS (Ref. 1)

The amplifier shown in Fig. 5.1 has two stages that are identical except for the use of an *npn* transistor in the first stage and a *pnp* transistor in the second stage. The stages are coupled by 10 μF interstage capacitors. As long as coupling capacitors are used, both stages could be like the first stage or both like the second. Each such stage is like the RC stage in Fig. 3.4 except that the collector and the emitter resistors have been reduced to 510 Ω and 51 Ω, respectively. Each stage by itself has a voltage gain of 10 because R_L/R_E is 10. However, with the stages coupled, the ac load on the first stage is 510 Ω in parallel with 510 Ω and 5.6 kΩ. Hence, we expect a voltage gain of about 4.8(10) = 48. By the S-factor calculation, Eq. (5.2), the input impedance is 510 Ω in parallel with 10 kΩ, or 485 Ω. The ac S-factor of the second stage is 4.8, and the TG-IR gives the voltage gain as 10(4.8)510/485 = 50. The measured voltage gain is 40, which implies an error of only 10 percent in each stage.

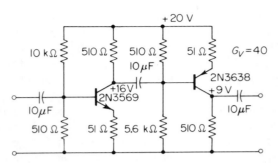

Figure 5.1. *Two-stage resistance-capacitance coupled npn-pnp amplifier.*

The errors have come from a number of approximations that have greatly simplified the gain calculation. We have recognized the principal elements that determine the gain, and we understand the circuit better than by using a complicated formula. We accept a 10 to 20 percent error as the price paid for simplicity and understanding.

The two-stage amplifier can be simplified by eliminating the coupling capacitor and making the amplifier direct coupled. Surprisingly, the coupling capacitor can be shorted out with almost no change in the overall performance, provided the first stage bias resistor is increased about 30 percent (13 kΩ). The first stage bias current must be reduced because the first stage collector current is supplying part of the second stage bias current. Finally, as shown in Fig. 5.2, the second stage bias resistor and one of the interstage resistors can be removed entirely if the first stage bias current is decreased again. These changes increase the amplifier's overall voltage gain to 70 or 80.

For the two-stage direct-coupled amplifier, the calculated gain is 100 and the measured gain is an additional 15 percent lower than expected. The first stage emitter current is 3 mA, so h is more than 8 Ω. Thus, the calculated gain with $h = 0$ is 15 percent high. A clue to this error is found in the relatively low value of the dc emitter voltage. If the emitter voltage is less than 0.25 V, h is more than 10 percent of R_E, and the voltage gain is corre-

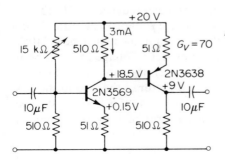

Figure 5.2. *Two-stage direct-coupled amplifier.*

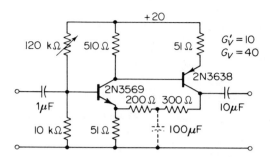

Figure 5.3. *Direct-coupled amplifier with feedback.*

spondingly reduced. Similarly, the low emitter voltage may indicate that the first stage lacks significant feedback.

Because the direct-coupled amplifier uses opposite type transistors, *npn—pnp*, the output stage collector resistor may easily be returned to the input stage emitter so that the amplifier has feedback and a voltage gain of 10. The dc feedback increases the dc voltage of the first emitter; hence, with feedback we must lower the bias resistor to 6.2 kΩ. However, the feedback increases the input impedance, and the first stage bias and base resistors can be increased by 20 times. This amplifier, as shown in Fig. 5.3, has good Q-point and gain stability and an excellent wide-band frequency response. The calculated gain with feedback is $R_L/R_E = 551/51 = 11$. The open-loop gain is 70, so there is more than significant feedback.

As indicated in Fig. 5.3, the negative feedback can be removed from the amplifier by adding a capacitor to bypass the ac feedback. At 40 Hz the 100 μF capacitor has a reactance of 40 Ω. Therefore, the loss in the feedback network is increased about 7 times, and at frequencies well above 40 Hz the feedback loss is proportionately higher. The overall gain with the feedback removed is 40, which is nearly a factor of 2 lower than the open-loop gain, since the capacitor has reduced R_L.

It should be observed that the direct-coupled pair in Fig. 5.2 not only uses 3 fewer components than the amplifier in Fig. 5.1, but it has a higher voltage gain. The amplifier in Fig. 5.3 illustrates the use of feedback to exchange voltage gain for improved performance. Feedback makes the amplifier useful over a wide range of the supply voltage, and the gain is independent of the transistor βs. For many applications the feedback amplifier in Fig. 5.3 makes a very satisfactory single-sided amplifier.

5.4   THE DARLINGTON COMPOUND (Ref. 1)

Unquestionably the most widely used transistor compound—an emitter follower driving a second emitter follower—is well known as the ***Darlington***

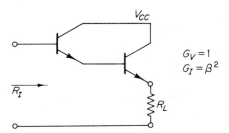

$$G_V = 1$$
$$G_I = \beta^2$$

Figure 5.4. *Darlington compound.*

connection. This compound, shown in Fig. 5.4, provides one of the simplest ways of direct-coupling two transistors and is useful as an impedance transformer that has a very wide-band frequency response. The Darlington connection has slightly less than unity voltage gain but gives a very high current gain (β^2) so that the input impedance approaches:

$$R_I = \beta^2 R_L \tag{5.3}$$

The input transistor should be selected for its ability to provide a high current gain at $\frac{1}{10}$ to $\frac{1}{100}$ the emitter current of the second stage. Often Darlington pairs will use identical transistors with the second stage operated at an emitter current that gives a maximum current gain. A common form of the Darlington amplifier is shown in Fig. 5.5.

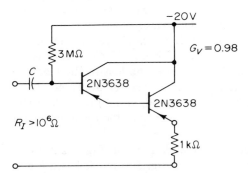

Figure 5.5. *Darlington CC amplifier.*

The Darlington connection operates very well as a high-gain CE amplifier stage, and often a transistor amplifier can be improved quickly and easily by replacing a single transistor by a Darlington pair. The Darlington pair is sometimes made a triplet by adding a third stage.

For wide temperature range applications and for uniformity or stability of overall current gain we should provide resistors that limit the S-factor of both stages. With silicon transistors the Q-point then becomes relatively insensitive to temperature changes. The practical form of the Darlington amplifier shown in Fig. 5.6 has an overall S-factor of 200 and an input imped-

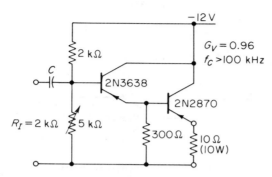

Figure 5.6. *Practical CC-CC amplifier. All resistors can be scaled 100 times larger for high impedance applications.*

ance of about $S^2R_L = 200(10) = 2000\ \Omega$. By increasing all the resistors by a factor of 100, the load impedance becomes $1000\ \Omega$ and the input impedance is about $200\ k\Omega$. For either form of the amplifier the voltage gain is just a little less than 1, which means that the compound is providing a power gain of nearly 200.

Darlington pairs of silicon transistors can be used as if the pair were a single high-gain transistor. This device is often used to increase the current gain of power transistors. The pair can be used with a collector load resistor, so that the device becomes a high current gain CE amplifier biased by the familiar emitter feedback biasing circuits. However, as we shall show in the following section, there are advantages in using collector feedback biasing whenever the signal level is small enough to permit operating the collector at low dc levels.

5.5 CE COLLECTOR FEEDBACK PAIR

The direct-coupled collector feedback amplifier in Fig. 5.7 offers rather useful and interesting performance characteristics. The amplifier uses a Darlington pair as a direct-coupled CE amplifier having a $5\ k\Omega$ collector load resistor. Because there are two emitter diodes in series, the input base operates at about 1.2 V above ground. The collector feedback tends to hold the collector Q-point voltage only about 0.2 V above the base voltage. If the collector voltage falls below 1.4 V, the current gain of the first stage falls and the collector voltage rises. If the collector voltage rises above 1.4 V, the bias current and the current gain increase, causing the collector voltage to fall. The collector voltage regulation is so effective that the collector is maintained at about 1.4 V above ground for 10 to 1 collector supply voltage changes and for a considerable change in the ambient temperature.

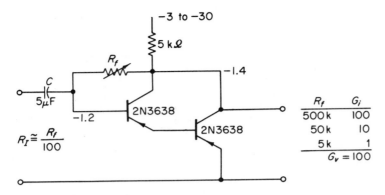

Figure 5.7. *Collector feedback CC-CE pair amplifier.*

The regulation also makes it possible to change the Darlington current gain as much as 1000 to 1 by varying the feedback resistor without adversely changing the Q-point. This feedback resistor provides one of the simplest ways of changing the current gain of a transistor amplifier. The voltage gain varies with the supply voltage and is approximately 100. Because of internal feedback this gain is not easily calculated, and the low input impedance makes the voltage gain of little meaning. With a 5 kΩ source the iterated voltage gain is equal to the S-factor R_f/R_L, as given by the TG-IR.

The amplifier in Fig. 5.8 is a three-stage version of the collector feedback pair. In this amplifier advantage is taken of the relatively fixed collector voltage of the first stage in direct coupling a third stage. The third stage increases the voltage gain by 10, so the overall voltage gain is 1000. The input impedance of the amplifier is 800 Ω, but we may easily add series resistance as an exchange of voltage gain for a higher input impedance. If the series input resistor is made 0.1 MΩ, the voltage gain drops to about 5 (14 db). The calculated voltage gain is $S_1 S_3 R_L / R_I = 10(170)600/100,000$ We note that S_3 exceeds β of the last stage, so we estimate a corrected current gain $S_3 = 70$. Since the calculated gain varies with β_3, we may expect an overall voltage gain of 4 to 5.

If high temperature stability is not required, the feedback resistor can be increased to 500 kΩ and R_I increased to 1 MΩ. The amplifier then has a voltage gain of about 10. The amplifier allows great flexibility in the choice of gains and impedances as long as output signal levels below 1 V rms can be tolerated.

The collector feedback amplifier is easy to use, requires few components, and for room temperature applications can be direct-coupled to a fourth stage. If the Darlington pair shown in Fig. 5.8 is replaced by a single CE stage, the direct-coupled CE-CE amplifier provides a simple replacement for increasing the gain of an existing CE stage without adversely degrading the

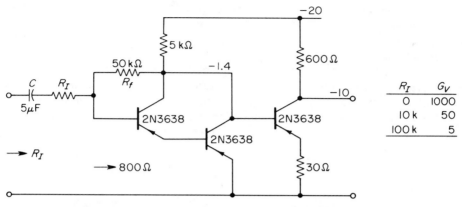

Figure 5.8. *Three-stage CC-CE-CE collector feedback amplifier.*

R_I	G_V
0	1000
10 k	50
100 k	5

Q-point stability. The complete three-stage integrated module is available at low cost in a TO-5 package and is temperature compensated for wide temperature range applications [GENERAL ELECTRIC Co. 4JPA113].

5.6 POWER PAIRS

The amplifier in Fig. 5.9 is an interesting 5 W Class A power amplifier. The output stage uses a *pnp* germanium power transistor for a maximum power gain at low cost. The driver stage is a CE-CE pair direct-coupled to the output transistor. The input transistor is a high current gain silicon transistor used because it offers a very low I_{co} (leakage) current. The second stage is an *npn* 1 W silicon transistor having a high current gain at 1 A emitter current levels. Each transistor offers the characteristics required for its place in the amplifier, and all are low-cost devices.

The two input transistors operate as an emitter follower with TR-2

Figure 5.9. *A 5 watt Class A amplifier, G_V = 10.*

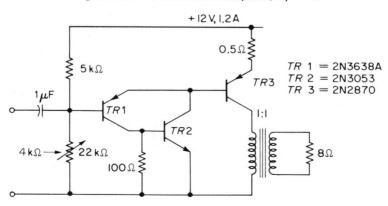

increasing the current gain of TR-1. All three transistors operate as a single stage, and the emitter voltage of TR-3 supplies overall feedback. The voltage gain of the amplifier is $R_L/R_E = 8/0.5 = 16$. The measured gain is lower, mainly because of the transformer loss.

The amplifier has 55 db power gain, unusually good waveform, and is thermally stable as long as the output transistor is mounted on a small heat sink. The amplifier operates equally well for low power applications with TR-2 removed—provided the bias current is reduced and the output transformer is changed to present a higher impedance to the collector.

5.7 FET-TRANSISTOR COMPOUNDS (Ref. 1)

Multistage amplifiers often use FETs and transistors in alternate stages or direct-coupled as FET-Transistor compounds. The FETs make it possible to use high interstage impedance levels—i.e., smaller coupling and filter capacitors—and allow the use of high-impedance potentiometers as controls. Transistors supply high transconductances at low dc collector currents. Thus, amplifiers combining FETs with transistors can be expected to exhibit the advantages of both devices. One of the most useful combinations uses a FET direct-coupled to a CE transistor amplifier.

The direct-coupled three-stage amplifier in Fig. 5.10 offers the high input impedance of the FET and a voltage gain of 100 with the Q-point and the overall voltage gain stabilized by significant feedback. These direct-coupled stages are easy to construct and offer high input impedances to above 100 kHz. High figure of merit FETs allow the use of high resistance interstage bias resistors and offer high voltage gains. FETs requiring a high drain current have to be loaded by a low-valued bias resistor, and the voltage gain is reduced. Thus, the amplifier may not have significant stabilizing feedback unless, by using a larger valued source resistor, the feedback is increased and the gain is reduced further.

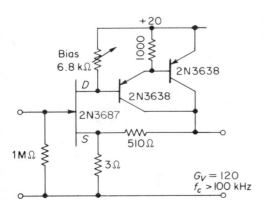

Figure 5.10. *High input impedance FET-transistor pair amplifier.*

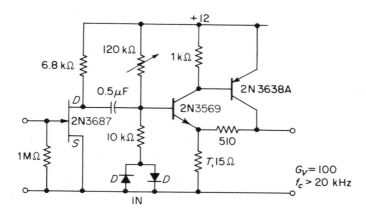

Figure 5.11. *Temperature compensated FET-transistor amplifier.*

The amplifier in Fig. 5.10 is satisfactory for applications at room temperatures, but the high dc gain makes it unsatisfactory for wide temperature range applications. With adequate feedback the expected voltage gain is $510/3 = 170$, and with significant feedback the gain should be $0.75 (170) = 127$. Since the measured gain implies significant feedback, we can expect the gain to increase about 4 times when the emitter resistor is bypassed.

The amplifier in Fig. 5.11 is designed to be used in ambient temperatures from about $-25°C$ to $+60°C$. The wide temperature range is achieved by eliminating the direct-coupling between stages and by introducing temperature compensation in the transistor amplifier to control the Q-point drift. The diodes in series with the transistor base resistor reduce the base bias voltage as the temperature increases and in this way control most of the Q-point drift.

The temperature drift can be further reduced by using a positive temperature coefficient emitter feedback resistor and by increasing the feedback to reduce the voltage gain. Silicon resistors are available that have a 0.7 percent/°C positive temperature coefficient. When the voltage across a silicon emitter resistor is 0.15 V, the net voltage change is 1.0 mV per °C. This voltage-temperature change can be adjusted easily by constructing the emitter resistor as a series or parallel combination of a carbon and a silicon resistor.

The equivalent resistance of the interstage network may be estimated as 5 kΩ. By the TG-IR the voltage gain of the transistor stages with the emitter bypassed is at least $50^2(510)/5000 = 255$. The gain with adequate feedback is $510/15 = 34$, which shows there is enough open-loop gain for adequate feedback. The gain of the input stage is the product of the interstage resistance and g_m of the FET. If g_m is 600 μmho, the first stage gain is 3 and the overall gain is 100. Since the overall gain varies with g_m, we should expect the overall gain to depend on the FET selected for the input.

5.8 INSULATED GATE FET AMPLIFIERS

The enhancement mode, insulated gate FETs make it possible to construct simple direct-coupled multistage amplifiers. Two such three-stage
amplifiers are shown in Figs. 5.12 and 5.13. The amplifier in Fig. 5.12 uses
three identical enhancement mode FETs with a feedback resistor connecting
the output stage drain back to the input gate. With like drain resistors the dc
feedback holds the Q-points alike and independent of the supply voltage.
Because of the feedback the input impedance of the amplifier is equal to the
feedback resistor divided by the overall voltage gain, i.e.:

$$R_I = \frac{R_f}{G_v} \tag{5.4}$$

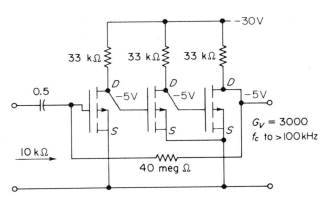

Figure 5.12. *High-gain wide-band insulated gate amplifier; all transistors 2N4352.*

If the amplifier is driven by a source having an impedance that is less than the
input impedance R_I, then the feedback is effectively bypassed and the gain is
3000. With an input capacitor having a reactance small compared with the
input impedance, the amplifier has a high ac gain and a frequency response
extending to about 100 kHz.

The amplifier in Fig. 5.12 becomes a dc amplifier having a voltage gain
of 1000 if the input capacitor is replaced by a 40 kΩ resistor R_g. DC operation
requires that the input terminal be connected to a point 5 V negative with
respect to ground, and requires a well-regulated dc voltage supply. As a dc
amplifier, the voltage gain with feedback equals the ratio of the feedback
resistance R_f divided by the series input resistance R_g. The open-loop gain is
about 3000 when R_g is zero.

A similar three-stage amplifier (Fig. 5.13) uses a unipolar or a depletion
mode FET in the input stage in order to permit dc amplification from a
grounded source. Feedback is introduced by connecting the first and third

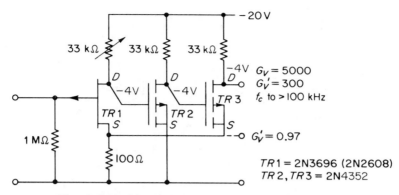

Figure 5.13. *High-gain low-noise low-drift dc amplifier.*

FETs to a common source resistor. In this amplifier the input impedance is determined by the gate resistor and the gate-to-drain capacitance. If the signal generator has an impedance of 1 MΩ, the upper cutoff frequency is 20 kHz. The amplifier requires a well-regulated voltage supply unless the dc gain is reduced by increasing the feedback resistor. The first stage drain resistor serves as a bias adjustment for the second and third stages and will vary with the figure of merit of the FET in the first stage. The better FETs will tolerate a high valued drain resistor and give a voltage gain of at least 15 in all 3 stages.

5.9   NONBLOCKING AMPLIFIER

Many amplifiers are required to have capacitor coupling at the input or between stages. The capacitors are needed either to eliminate low frequencies and noise or to shape the signal. However, if a capacitor is connected to the base of a transistor or to the gate of a J-FET, the transistor or FET rectifies the incoming signal when large signals are applied.

The rectification of large signals charges the capacitor and biases the amplifier at cutoff or beyond. Although the capacitor is charged rapidly through the rectifying diode, the capacitor usually discharges much more slowly when the rectifier is reverse biased. An amplifier that is cut off by a large signal is said to be **blocked**, and the characteristically long time during which the amplifier is cut off is called the **blocking time**.

The amplifier shown in Fig. 5.14 is nonblocking because the insulated gate FET is unable to rectify large input signals. The amplifier has a voltage gain of 1000 without feedback, and with the capacitor *C* removed feedback reduces the gain to 10. The amount of ac feedback is easily changed by inserting a resistor in series with the capacitor *C*. Several FETs may have to be tried to find one that supplies the correct bias for the transistor amplifier.

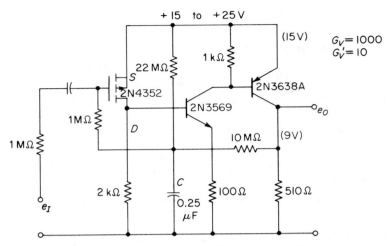

Figure 5.14. Nonblocking high-gain amplifier.

The open-loop voltage gain of the amplifier is calculated by assuming the capacitor C is very large. The voltage gain of the FET stage is $g_m R_D$. The current gain of the transistor section is $S_2\beta_3$, and the voltage gain is $S_2\beta_3 R_L/R_D$. Thus, the open-loop voltage gain is

$$G_v = g_m S_2 \beta_3 R_L \qquad (5.5)$$

By substitution we obtain $G_v = 0.0007(20)(100)510 = 700$.

Equation (5.5) shows that the gain depends on the semiconductors in the first and last stages and there is local feedback in the second stage. The measured gain of 1000 suggests that g_m may be about 1000 μmho. The calculated gain e_o/e_I with the capacitor removed is 11, so there is adequate dc feedback for Q-point stability.

5.10   INTEGRATED CIRCUITS

Integrated circuits are complete semiconductor circuits fabricated as monolithic (single block) bars of silicon by the use of diffusion techniques to form the transistors, diodes, resistors, and capacitors. The elements of a circuit are interconnected by depositing a metallic lead pattern over the oxide surface of the bar. The leads make contact only where windows are left in the oxide. The finished circuit is packaged for protection. The entire block of many components is supplied with input and output terminals and with connections for the power supply. The user need only connect the power supply and the terminal circuits.

The design of integrated circuits is an art and a specialty of its own. The user is free of all concern for the circuit design, except to understand the capabilities and limitations of the completed circuit. For these reasons the application of integrated circuits tends to approach the simplicity of using a single transistor.

The advantages of integrated circuits (ICs) are that excellent performance and high reliability are furnished in a small package. The disadvantages stem from the limited power capabilities of any device in a small package, the high cost, and the lack of control over the circuit parameters that the user may require for some applications. Because large capacitances and inductances cannot be made in a small volume, such components must be outside the integrated circuit package with connections provided. In general, integrated circuits offer little advantage in circuits that require a series of capacitor- or transformer-coupled stages, as in a multi-stage tuned RF amplifier. An IF amplifier constructed with an IC uses a single high-gain amplifier preceded by a filter that has the response characteristics normally provided by the tuned interstage transformers.

There are problems in the manufacture of different kinds of semiconductors, even *pnp* and *npn* devices, on the same substrate. Furthermore, even the resistors of an integrated circuit cannot be made to differ more than two or three orders of magnitude. Yet the circuit design must allow for component tolerances as high as ± 30 percent. These problems of construction limit the yield of integrated circuits which meet the customer's requirements, and low yields make the unit cost high.

Integrated circuits are widely used in digital computers and are steadily becoming more readily available at reasonable cost for low power linear circuit applications. Some integrated circuits comprise only one or two transistors (matched pairs, Darlington pairs, or a differential stage). Others may have three or four transistors with resistors that make up logic packages or low-gain amplifiers. Some integrated circuits are complete differential amplifiers that have ten or more transistors and twice as many resistors. In some units heat is developed and used with a thermal feedback path to hold the amplifier at a constant temperature, thereby making the circuit insensitive to external temperature changes.

5.11   APPLICATIONS OF INTEGRATED CIRCUITS (Ref. 2)

The applications of integrated circuits are infinitely varied but are limited at the present time either to systems that need many identical circuit packages or to systems in which the high cost is justified by the need for reliability and the need to conserve space. Digital systems can be designed with a few basic building-block circuits. Thus, a selection of perhaps a half dozen basic elements may be used in quantities of 100,000 by a single computer manufacturer. Except for mass-produced circuits used in radio and TV receivers, the design of analog systems generally lacks the simplicity and uniformity needed for profitable manufacture of an integrated circuit. A color TV may have five or more separate IC packages. These may comprise the IF and video amplifiers, the FM and audio amplifiers, the chroma demodulator, and user-operated control circuits.

Linear integrated circuits are generally used either as high-gain, low-frequency amplifiers or as operational amplifiers with a feedback network to control the response. These integrated circuits have a high power gain and are differential amplifiers so that the output can be direct-coupled to the input.

The method of connecting a typical IC amplifier is illustrated by the circuit shown in Fig. 5.15. Usually, equal positive and negative power supplies are required with the center of the power supply grounded, and the amplifier may or may not have a terminal for connection to ground. The power supply need not be regulated and is customarily bypassed to ground by 0.05 or 0.1 μF capacitors connected close to the amplifier. If the supply is not shared with other equipment, the bypass capacitors may not be needed.

Both inputs of the amplifier must be given a dc connection to ground, either through a resistor or the signal source. Since the amplifier bias current

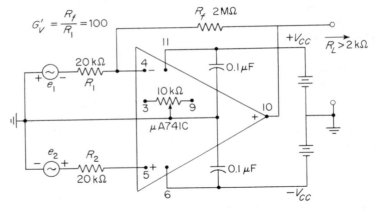

Pin numbers are for TO-116 flat case.

Figure 5.15. *Typical linear IC amplifier hookup.*

flows through these resistors, the voltage drop of the current is an offset signal that is amplified by the gain with feedback. For this reason the resistors should be approximately equal and not more than 100 kΩ. If the resulting shift of the dc output voltage is not acceptable, the gain or the resistors must be lowered. Some amplifiers have terminals for connecting an external offset adjustment. For the amplifier shown in Fig. 5.15 an offset adjustment is obtained by connecting a 10 kΩ potentiometer to unbalance the input-stage emitter resistors.

The gain of the amplifier with feedback is equal to the ratio of the

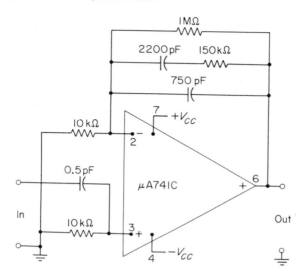

Figure 5.16. *Phono IC preamplifier with RIAA response.*

feedback resistor R_f to the input resistor R_1 and is usually between 1 and 1000. Because the gain without feedback is several orders of magnitude greater than the input-to-output gain with feedback, the amplifier input signal e_i is small compared with e_1 and e_o. Thus, we may think of the amplifier as holding the minus input at ground potential and making the signal e_s see R_1 as the input impedance. Because the plus input does not have feedback, the signal e_2 sees the relatively high input impedance of the amplifier. The voltage gain of the amplifier on the non-inverting side (+) is $1 + R_f/R_1$, one more than the gain on the inverting side (−).

The μA741 amplifier is shown with a feedback network that gives the amplifier 40 dB voltage gain. Because the gain-bandwidth is 1 MHz, the response with feedback should be flat up to 10 kHz.

The control of amplifier frequency response by a complex feedback network is nicely illustrated by the phono-preamplifier shown in Fig. 5.16. The input is connected to the plus side of the amplifier, so the pickup is loaded by 10 kΩ. The low-frequency gain of the preamplifier is $10^6/10^4 =$ 100. The two feedback capacitors and the 150 kΩ resistor reduce the gain with increasing frequency except for a constant gain between 100 and 2000 Hz. Above 2000 Hz both the closed- and open-loop gains decrease with frequency. Since the open-loop gain of the amplifier is high compared with the gain with feedback, the response is determined only by the feedback network. And the feedback elements are precisely specified by the RIAA frequency standard.

5.12 HYBRID INTEGRATED CIRCUITS

Hybrid ICs are microcircuits in which discrete components are connected to a monolithic integrated circuit, and the entire assemby is usually sealed in a small plastic package. Hybrid circuits can incorporate FET input stages and better devices than are available in integrated circuits, and they allow greater control over the characteristics of components. Hybrid circuits are more versatile, as they can be made in smaller quantities for a specific application. They are capable of operation at frequencies and power levels not yet achieved with monolithic ICs. High-performance operational amplifiers (OP amps) are commonly manufactured as hybrid circuits that are supplied in an epoxy-filled package. Low-power, Class-B audio amplifiers are supplied as hybrid ICs with metal ears for connecting a heat sink or contacts for connecting external power transistors.

Integrated circuit arrays of transistors and diodes are available for the design and manufacture of hybrid circuits. These semiconductors match each other and have similar temperature characteristics because they are made in close proximity on a single chip. Circuits using these arrays sometimes have characteristics that cannot be obtained with discrete components.

SUMMARY

A pair of direct-coupled transistors is shown to provide more voltage gain and a simpler circuit than when capacitor-coupled. Transistor compounds such as the Darlington CC-CC pair, the CC-CE pair, and the CE-CE pair may be operated essentially as a single high β transistor. A practical form of a Darlington compound is shown to operate as a 200-to-1 impedance reducing transformer with unity voltage gain.

A three-stage collector feedback pair offers a wide choice of input impedances or an easily controlled voltage gain. A direct-coupled three-stage amplifier is described that operates as a 5 W power amplifier.

The TG-IR shows that each additional stage of a multistage *transistor* amplifier increases the overall current or voltage gain by the current gain of the stage. With local feedback the current gain of an n-stage amplifier is approximately S^n where S is the typical S-factor. With either overall series-current feedback or shunt-voltage feedback the gain of a multistage amplifier may be calculated from values of the feedback resistors as if for a single-stage amplifier.

Transistors and field-effect devices are easily operated as pairs with feedback applied over both—as with a single transistor. Field-effect transistors in combination with bipolar transistors provide high-impedance circuits, high gain, and low Q-point drift with change in temperature. Insulated gate

FETs make useful high-impedance amplifiers that have only one FET and one resistor per stage.

Integrated circuits are generally direct-coupled amplifiers (or switching circuits) that require many transistors and few auxiliary components. The application of an integrated circuit requires that the user understand the overall performance characteristics and the manner of connecting external equipment, but the internal circuit details are rarely, if ever, the concern of applications personnel. Hybrid ICs may incorporate high-performance devices for small quantity applications, but they are more expensive and less reliable than a monolithic circuit.

PROBLEMS

5-1. Assume that the amplifier in Fig. 2.7 is to be capacitor-coupled to an identical stage. (a) Find the voltage gain of each stage if the transistors have a current gain of 48. (b) Repeat the calculation, assuming both transistors have a current gain of 20.

5-2. (a) For the two-stage amplifier in Fig. P-5.2 write the TG-IR for each stage separately and combine to find the overall voltage and current gains. Neglect R_A and assume $h = 20\ \Omega$. (b) Repeat with the 10 kΩ resistor removed.

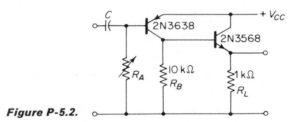

Figure P-5.2.

5-3. (a) Show how the first stage in Fig. P-5.2 could be changed to a CC stage, and (b) find the gain-impedance relations, as in Prob. 5-2.

5-4. (a) For the amplifier in Fig. P-5.4 find the dc collector and emitter voltages of the first stage. (b) Find the overall dc current gain. (c) Find

Figure P-5.4.

the overall ac voltage gain. (d) What is the input impedance? (*Hint:* Start with the *Q*-point given for the second stage.)

5-5. What is the overall voltage gain of the two-stage amplifier shown in Fig. P-5.5?

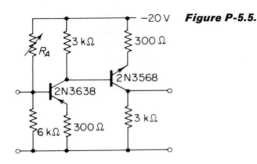

Figure P-5.5.

5-6. Repeat Prob. 5-5, assuming that the emitter resistors are bypassed using a large capacitor in series with a 15 Ω resistor.

5-7. Explain how the overall current and voltage gain of the amplifier in Fig. P-5.4 is changed if the output terminal is moved to the second-stage emitter.

5-8. Find the overall current and voltage gains of the amplifier in Fig. 5.9. Give several reasons why the indicated voltage gain is not as high as the calculated value.

5-9. If the overall voltage gain of the amplifier in Fig. 5.11 is 100, what voltage gain would you expect to measure in the FET stage?

5-10. An amplifier is required with $R_L = 500 \ \Omega$, $R_I = 10^6 \ \Omega$, and a voltage gain of 100. (a) What *S*-factor is required if the amplifier has 3 stages? (b) Would it be practical to use only two stages? (c) Does the *S*-factor indicate that 4 stages may be too many? Ans: (a) $S = 60$. (b) No. (c) $S = 15$, maybe.

5-11. A 2-stage amplifier has $S = 60$ in both stages, $R_L = 500 \ \Omega$, and $R_I = 10 \ k\Omega$. What is the expected voltage gain with and without significant feedback? Ans: 45 and 180, (33 dB and 45 dB).

CHAPTER SIX

Power Amplifiers and Transformers

Power amplifiers are required to deliver a large amount of signal power with low distortion and a high power efficiency. The output transistors of power stages are usually high-current devices designed to dissipate the collector circuit power that is converted to heat. Since the dynamic characteristics of transistors are especially well suited for Class B operation, an audio power amplifier usually has a Class B power stage driven by a direct-coupled low-power Class A amplifier.

The power output of an amplifier is limited by the power capability of the transistor (and costs), and to develop maximum power a transistor must be loaded by a particular value of load impedance. Power amplifiers are usually coupled to the load by an output transformer. This arrangement keeps the dc collector current out of the load and by using a particular turns ratio, the transistor load is made the value which uses the transistor characteristics for the best advantage.

In this chapter we outline the characteristics of transformer-coupled Class A and Class B power amplifiers and present a practical description of audio-frequency transformers.

6.1 TRANSFORMERS FOR MAXIMUM POWER OUTPUT

When the signal input to an amplifier is indefinitely increased, a point is finally reached where the power output cannot exceed the current or voltage capability of the last stage. The current and voltage limits of an amplifier are fixed by the output transistors and, indirectly, by cost or space limitations. When either a voltage or current limit is reached, some advantage can be obtained by using a transformer between the output transistor and the load. For example, suppose the output transistor is driven to its upper current limit but is still capable of withstanding a higher output voltage. If the load impedance is increased, then for the same current the new load receives a higher peak voltage. As a result, the new load receives higher power from the amplifier—a higher power which can be transferred at high efficiency to the original load by using a transformer that is capable of receiving power at one impedance level and delivering it at another impedance level.

By properly selecting the impedance ratio of the transformer, the load can be made to utilize the transistor at both its peak current and peak voltage ratings. In this way a given output transistor is made to deliver a maximum power to the load. If, by a slight change in the impedance, a transistor can be made to deliver a much better waveform at a slightly reduced power output, this load impedance might be indicated in a specification as providing a maximum undistorted power output. Output transformers almost never match impedances.

A low-level amplifier is usually thought of as either a voltage amplifier or a current amplifier that receives signals from a source resistance R_G and delivers an amplified signal to a load resistance R_L. The transistor gain-impedance relation (TG-IR, Eq. 2.4) shows that with fixed source and load resistances a given voltage gain corresponds to a particular current gain, so that a transistor amplifier is characterized by the value of either gain. When we work with an amplifier, the simplest way to measure the gain is to measure ac voltages with an oscilloscope, and we tend to work with the various voltage ratios which we call *voltage gains*. With the TG-IR the voltage ratios can be converted to current gains, which are more meaningful in power amplifiers where the load impedance may be several orders of magnitude smaller than the input impedance.

The reader should remember that changing the input impedance or the load resistance will change the observed voltage and current ratios, and, therefore, these impedances should be specified. (See dB table in Appendix.) However, for a power output stage the best indicator of the design effectiveness is the *power gain* $G_v G_i$, which is independent of the impedances.

6.2 SINGLE-SIDED CLASS A AMPLIFIER

A transformer-coupled class A power amplifier using a 2N2869 medium power transistor is shown in Fig. 6.1. The amplifier is called *Class A* to indicate that the dc collector current is a substantially fixed value that does not vary with the ac signal.

The design of a reliable power amplifier is difficult and time consuming, even for an amplifier as simple as the one in Fig. 6.1. However, optimum component values needed to construct practical amplifiers can be obtained by consulting the manufacturer's transistor handbooks. For example, the 2N2869 data suggest that a 5 W power amplifier can be constructed to

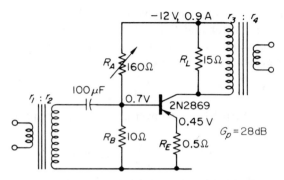

Figure 6.1. *Single-sided Class A 4 watt amplifier.*

operate on a 14.4 V dc supply, provided the collector load impedance is 15 Ω and the base input impedance is 10 Ω. These values designate the tentative component values shown in Fig. 6.1 for R_L and R_B.

The load resistor R_L is shown on the primary side of the output transformer, and the secondary is open. The resistor is placed on the primary because it is easier in power amplifier analysis to work with the equivalent primary load resistance and to introduce the transformer turns ratio after the optimum load is selected. In practice filament transformers are usually available that will carry the collector current, thereby expediting the experimental analysis of a power amplifier. A transformer can be sought later when the required impedance ratio is firmly established.

The emitter resistor in Fig. 6.1 provides enough feedback to reduce distortion and prevent thermal runaway (Ref. 1). Power transistors usually require a low S-factor emitter feedback design, but the 2N2869 can be safely operated with a 0.5 Ω emiter resistor (i.e., $S = 20$).

The performance characteristics that can be expected of the amplifier may be calculated in the following way. The handbook data suggest a 900 mA emitter Q-point current; hence, the base voltage is about 0.7 V. Assuming the amplifier is to be operated on a 12 V dc supply, the bias adjusting resistor R_A will be something less than:

$$R_A < \frac{12 - 0.7}{0.7} R_B = 160 \ \Omega \tag{6.1}$$

Calculating the gain-impedance characteristics of this amplifier using the methods of Chap. 3, we find that with the resistance values given in the circuit $G_v = 15/0.5 = 30$ and $G_i = 10/0.5 = 20$. The indicated power gain is $\dot{G}_p = G_v G_i = 600$, or 28 dB. The power output is calculated by finding the peak sinusoidal output voltage across the load. With a good power transistor the peak load voltage occurs when the instantaneous collector-to-emitter voltage is approximately zero. Thus, we find that the peak load voltage is almost equal to the supply voltage, 12 V. The emitter resistor and resistance in the transistor and the power supply may be expected to limit the peak signal to about 11 V. Converting the peak signal to rms volts, we find that the expected power output is:

$$R_O = \frac{e^2}{R_L} = \frac{[0.7(11)]^2}{15} = 4 \text{ W} \tag{6.2}$$

Dividing the power output by the power gain gives the required power input, as follows:

$$P_I = \frac{P_O}{G_p} = \frac{4}{600} \cong 10 \text{ mW} \tag{6.3}$$

The power output divided by dc power input, or efficiency, gives:

$$\frac{P_O}{P_{DC}} = \frac{4}{11} \quad \text{or} \quad 37\% \tag{6.4}$$

For a practical application we have enough information to assemble a trial amplifier. The input transformer has to be selected to couple the given signal source to the 10 Ω base resistor. For test purposes either a 500 Ω to voice coil output transformer or a 120 V to 12 V filament transformer is a perfectly satisfactory input transformer. Suitable output transformers are available for coupling to voice-coil impedances, but a reasonably large iron core is required to prevent core saturation by the 1 A Q-point current. The bias resistor should be adjusted to give a compromise between low dc collector power input and good waveform at the required peak output signal.

The amplifier should be tested carefully to evaluate its performance and to find its limitations. The performance test should include the frequency response and a power output measurement. The peak collector voltage and current should be measured to ensure that the transistor is not operated too close to the peak ratings. Because the peak-to-peak output signal across the load is about 22 V, the collector voltage is about −1 V on the positive-going peak and about −23 V on the negative-going peak. The peak collector current is about 0.75 A above the 0.9 A Q-point current. The 2N2869 data sheet shows a collector breakdown voltage rating of 50 V and a peak current rating of 10 A. Therefore, the transistor is being operated well within the peak voltage and current ratings. Because the power output increases as the square of the supply voltage, a 14 V supply should permit 5 W output as given in the manufacturer's data.

6.3 LOAD LINES

The load line method of evaluating transistor power amplifier performance uses the static collector current-voltage characteristics on which a line is placed to represent the collector load resistance. The collector characteristics for the 2N2869 transistor are shown in Fig. 6.2. The nearly horizontal

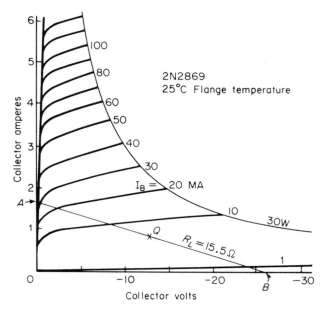

Figure 6.2. *Collector current-voltage characteristics for a typical power transistor.*

curves in the figure exhibit the collector current as a function of the collector voltage for fixed values of the base current. Observe that the curves are widely spaced at low collector currents and that they tend to crowd together as the collector current exceeds about 2 A. This crowding means that the current gain is lower at high collector currents than at low currents and suggests the possibility of waveform distortion.

The load line for the amplifier in Fig. 6.1 is shown as the line AB in Fig. 6.2. Point A represents the 1.65 A peak collector current at the instant the collector is shorted to the emitter. The point B, at the extension of the line AQ through the Q-point, usually falls at about twice the dc supply voltage. The slope of the load line is the total ac resistance in the collector circuit and should include the external emitter resistance and the power supply resistance.

The load line method of analysis is based on the incorrect assumption that the amplifier performance can be evaluated by examining the relation between the collector curves and the points followed by the instantaneous changes of the collector current and voltage along the load line locus. The error in the method is that a Class A transistor amplifier must have considerable emitter feedback. Hence, the crowding together of the base current lines is not a useful measure of the distortion. Furthermore, the transistor cannot be biased by using the base current value on the curve passing through the

Q-point. The collector curves are different from one transistor to another because the current gain is so variable, and the curves shift extensively with the junction temperature. With feedback biasing the stage current gain is fixed by the base and emitter resistors and the base current values are not useful, except possibly to estimate the current gain at high collector currents. In practice the bias resistor is selected by the methods of Chap. 3 or by trial.

Usually it is simpler and better to observe the performance characteristics of an amplifier by bench measurements than to struggle with characteristic curves, load lines, and feedback calculations. And for performance tests we should observe that the power transistors meet the manufacturer's specifications by measuring the collector current as the base current is varied, while using a dry cell to hold the collector voltage at 1.5 V. The low collector voltage makes it possible to obtain this transfer curve without a heat sink.

A single-sided Class A amplifier has the disadvantage of requiring a large output transformer and of producing considerable second harmonic. These disadvantages are somewhat reduced by using two transistors to drive the load push-pull so that the alternate signal peaks are more alike than they are with a single transistor.

6.4 PUSH-PULL AMPLIFIERS

A push-pull amplifier is formed by combining two single-sided amplifiers, as shown in Fig. 6.3. The push-pull Class A amplifier offers about three times the power output of a single-sided amplifier with less distortion. It possesses the advantage of having dc collector current flowing in opposite directions relative to the transformer core, reducing the likelihood of saturating the iron.

Figure 6.3. *Push-pull power amplifier, 2 watts Class A or Class B.*

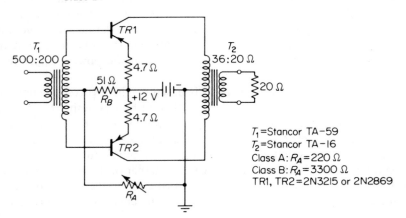

T_1=Stancor TA-59
T_2=Stancor TA-16
Class A: R_A=220 Ω
Class B: R_A=3300 Ω
TR1, TR2=2N3215 or 2N2869

The push-pull amplifier shown in Fig. 6.3 is designed for low-power speech or music applications, using a 12 V automobile battery as the power supply. The input transformer is selected to present a 200 Ω source from base-to-base when the primary is driven from a 500 Ω generator. The output transformer provides a voltage step-down of about 1.3 to 1, and the maximum output is 18 V peak-to-peak across the 20 Ω load, or 2 W. The dc power input is 7 W, making the overall efficiency 28 percent.

The output transformer used in this amplifier is built on a small core, and the loss in the dc resistance of the windings and the emitter resistor is about 30 percent of the total output power. Thus, the collector circuit is developing about 3 W at over 40 percent efficiency. The maximum theoretical efficiency is 50 percent. With the transformers shown in Fig. 6.3, the amplifier has a power gain of 250 times, or 24 db. Transformers built on a larger core would increase the power gain to about 30 dB. The collector winding of the output transformer should have a collector-to-collector impedance given by:

$$R_{CC} = \frac{2V_{CC}^2}{P_O} \tag{6.5}$$

Equation (6.5) indicates correctly that high power output is obtained by reducing the load resistance. However, lowering the load resistance requires that the emitter resistance be lowered to keep the ac power loss in the resistor about 10 to 20 percent of the total output power. When operated Class A, the power transistors require a heat sink that will dissipate nearly four times the maximum output power.

A disadvantage of the Class A amplifier is that the collector current flows at all times and the transistor dissipation is highest when no ac signal is present. Push-pull Class A transistor amplifiers are seldom used, even for low-distortion amplifiers, because transistors operate almost ideally in Class B service.

6.5 CLASS B AMPLIFIERS (Ref. 2)

A class B amplifier is biased in such a way that the collector current is nearly zero when no signal is applied. With a signal applied, one transistor amplifies the positive side of the input signal and the other amplifies the negative side. These half signals are then combined by the transformer to restore the original waveform. The reader should observe that a Class B amplifier uses only one side at a time of both transformers, so the winding impedances must be calculated by using their value between one side and the center tap. The nominal full winding impedance is four times the required side impedance.

Ideally, a transistor in Class B service should be biased to cut off the collector current under no-signal conditions. The transistor remains cut off at very low signal inputs because transistors have low current gain at cutoff and turn on abruptly with a larger signal. This nonlinear response gives the output signal a waveform like that shown in Fig. 6.4 and is called **crossover distortion**. However, a small forward bias causes a small collector current to flow at low signal levels, suppressing most of the distortion. The residual distortion is further reduced by the negative feedback usually provided by the high-gain amplifier used to drive a Class B output stage.

A typical Class B amplifier has the same circuit configuration as shown in Fig. 6.3. The resistor R_B should be no more than one-fourth the transformer impedance on one side, and R_A is adjusted to minimize the crossover distortion while being careful to avoid too much dc power input under no-signal

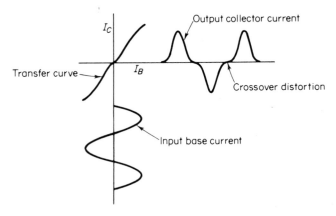

Figure 6.4. *Class B amplifier: crossover distortion, waveforms, and transistor transfer curve.*

conditions. Because the base-emitter voltage of a transistor varies with temperature, the bias required to maintain the zero bias condition is sometimes made to vary with temperature by using a forward-biased diode in place of R_B, or a thermistor in place of R_A.

Class B amplifiers are commonly used in portable radios and public address amplifiers because music and speech signals are highly intermittent, and small transistors will deliver the required power without needing heat sinks. The emitter resistors provide feedback and prevent thermal runaway at sustained high signal levels. While Class B operating efficiencies are somewhat higher than those of Class A, the real advantage is that very little standby power is required under no-signal conditions.

6.6 EXAMPLE OF A CLASS B AMPLIFIER

The amplifier shown in Fig. 6.3 becomes a typical Class B amplifier merely by decreasing the forward bias. Increasing the bias resistor to 3300 Ω

reduces the collector dc stand-by power to 0.8 W but does not change the maximum output power of 2 W. Full output is obtained with a dc power input of 5 W, which indicates a significantly higher efficiency—56 percent—than that of the Class A amplifier. The maximum theoretical efficiency is 78 percent without crossover biasing. The transformer loss reduces the overall efficiency to 40 percent.

The Class B amplifier requires about twice the drive voltage of a Class A amplifier, so the power gain is only about 20 db. With speech and music signals the Class B amplifier does not require a heat sink, and the 2N3215 transistors can be replaced by a low-power type, e.g., 2N3638.

The power output of the Class B amplifier is determined by the collector load resistance in much the same way the output is determined by Eq. (6.5) for the Class A amplifier. Since the Class B transistor drives the collector load on only alternate half cycles, the nominal winding impedance on one side is given by:

$$R'_L = \frac{V_{CC}^2}{2P_O} \tag{6.6}$$

Because the full-winding collector-to-collector winding impedance is four times the value given by Eq. (6.6), the nominal transformer impedance for the Class B amplifier is the same as that given in Eq. (6.5) for the Class A amplifier. In other words, with a given transformer, a given collector supply voltage, and a fixed peak collector current, the same output power is obtained by Class A or Class B operation.

An amplifier operated on a fixed supply voltage is changed from Class A to Class B by reducing the bias and the dc Q-point power. The peak voltages impressed on the windings do not change, but the transistors are forced to share the load differently. Class B amplifiers are generally preferred to Class A push-pull transistor amplifiers, especially with overall feedback to control the crossover distortion.

6.7 THE SINGLE-ENDED CLASS B AMPLIFIER

The single-ended Class B amplifier shown in Fig. 6.5 has the advantage of Class B operation without the need for an output transformer. The interstage transformer windings are polarized so that the transistors conduct on alternate half cycles. The series connection of the power transistors causes the signal current to flow through the load in opposite directions on alternate half cycles. The load is driven as if from a push-pull amplifier.

The power transistors are biased to reduce the crossover distortion, and the driver transistor is biased to operate Class A. For high-fidelity applications, as in a stereo amplifier, shunt feedback is provided by connecting a resistor from the load to the base of the driver transistor.

If a Class B power amplifier is operated with the load short-circuited,

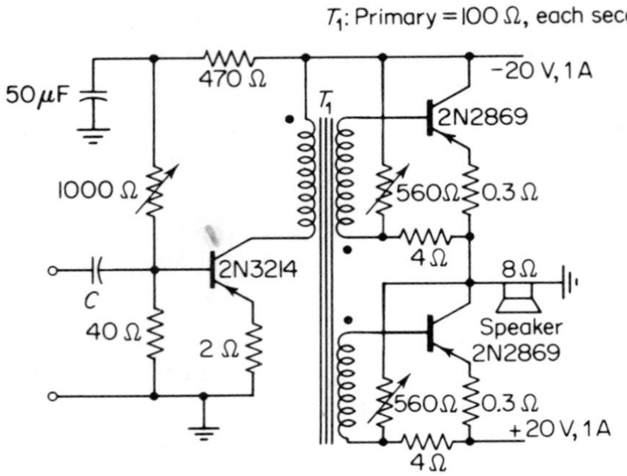

T_1: Primary = 100 Ω, each secondary = 8 Ω

Figure 6.5. *Single-ended 15 watt Class B amplifier.*

the transistors are sometimes destroyed by second breakdown.* Otherwise, the Class B transistor power amplifiers give excellent service and a high quality of sound reproduction.

6.8 COMPLEMENTARY-SYMMETRY AMPLIFIERS

A low-distortion (high-fidelity) amplifier is usually some form of a direct-coupled complementary-symmetry power amplifier. In these amplifiers the power transistors are effectively connected in parallel and are made to conduct on alternate half cycles, as in a Class B amplifier. The advantage of the complementary-symmetry circuits is that coupling transformers are not required and with direct-coupling it is relatively easy to use 20 to 40 dB of overall feedback.

A low-gain, 3-W complementary-symmetry Class B audio amplifier is illustrated in Fig. 6.6. The power transistors are the series-connected complementary (pnp and npn) 2N5194 and 2N5191 silicon transistors. Both the output stage collectors are connected to the load through a large capacitor. The power stage is driven by the parallel-connected 2N1304 and 2N1305 germanium transistors. Because the emitters of the driver stage are coupled to the load, the two-stage output section has a voltage gain of 1 and the load is driven as if by a low-impedance emitter follower.

The 2N3569 CE input stage provides the amplifier voltage gain which

---

*Second breakdown is a sudden and destructive channeling of the collector current into a localized area of the transistor; it occurs when high currents and high voltages exist simultaneously in the collector circuit. The speed and the destructiveness of the phenomenon complicate its control. Second breakdown is usually prevented by limiting the collector current-voltage product. Curves showing the safe current-voltage product of power transistors are usually given in the manufacturer's data sheets.

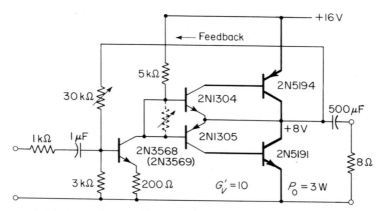

Figure 6.6. *Complementary-symmetry Class B amplifier and input stage.*

is used partly for overall feedback. The bias resistor couples the output back to the input base and is adjusted to make the dc voltage at the output just half the supply voltage. Because the bias resistor supplies both dc and ac feedback, the Q-point is under feedback control, and additional ac feedback must be obtained by a separate capacitor-coupled resistor.

The amplifier shown in Fig. 6.6 has a power gain of 40 dB, and the input impedance is 1 kΩ. Since the current required of the power supply varies with the signal, the supply should be regulated or have a large filter capacitor.

The germanium transistors in the driver stage have a low turn-on voltage, and with emitter feedback there is negligible cross-over distortion. If the germanium transistors are replaced by silicon devices, a forward-biasing diode or resistor must be connected between the two bases, and a satisfactory reduction of the cross-over distortion is not usually obtainable.

A quasi-complementary-symmetry amplifier uses alike power transistors (both npn or pnp). With this change one side of the driver and amplifier connects to the load as a Darlington pair and anti-cross-over bias is provided by a series of two or three diodes connected between the driver stage bases.

The circuit of a complementary-symmetry amplifier is more complicated when higher gain and more feedback are required. The circuits of amplifiers with better performance characteristics are described in Refs. 2 and 12.

6.9 CLASS C AMPLIFIERS

Radio-frequency (RF) power amplifiers are generally Class C "amplifiers" because a linear relation is not required between the output and input, and the transistor can be used as a high-speed switch.

A Class C amplifier is biased beyond cutoff so that the collector current in the transistor is zero when no signal is applied, and the current flows for appreciably less than one-half of each cycle when an ac signal is applied. The collector load is in parallel with a tuned circuit that sustains a sinusoidal wave across the load, even though the energy is supplied in short pulses. Class C amplifiers are characterized by high collector circuit efficiencies (70 percent) because the collector current flows only when the instantaneous voltage across the transistor is low. The transistor operates as a switch, and the collector supply delivers energy to the amplifier only when the largest portion of the energy will be transferred to the load. A Class C amplifier is a good harmonic generator when a sharply tuned circuit is adjusted to resonate at a harmonic of the exciting frequency. Examples of Class C amplifiers are described in Chapter 12.

A linear Class C amplifier is one so adjusted that the output is essentially proportional to the exciting voltage applied to the input. The linear amplifier is used as a power amplifier of amplitude-modulated waves because the amplitude changes vary so slowly with time that the output amplitude follows the input signal, even though the tuned circuit is resonant at the exciting frequency. The amplitude modulation is controlled by the audio frequency information signal and varies so slowly that the amplitude changes are not significantly modified by circuits that are tuned at the radio frequency. The linear amplifier is the tuned equivalent of a Class B audio amplifier, but differs in that the linear amplifier need not be push-pull to produce a sine wave signal. The sinusoidal waveform is maintained by using a tuned collector load known as the **tank circuit**. Push-pull Class C amplifiers are used when a higher output power is required than can be obtained from a single transistor.

6.10 TRANSFORMERS (Ref. 10)

A transformer is simply two or more coils of wire (**windings**) that are closely coupled by a magnetic core. If both windings have the same number of turns and a signal is applied to one, most transformers are so efficient and well-coupled that it is difficult to find by a simple measurement any difference between the signal on the input winding and the signal on the output winding. If the windings have unequal numbers of turns, then the voltages are proportional to the winding turns ratio. A transformer is primarily a device that changes the input signal voltage e_I to an output voltage e_O. When the secondary is loaded, as shown in Fig. 6.7, the currents i_I and i_O in the windings are inversely proportional to the windings turns ratio. The primary voltage and

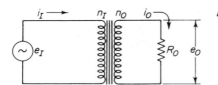

Figure 6.7. *Transformer.*

current are simply related to the secondary voltage and current because the power input to the primary winding is transferred to the secondary without a significant power loss.

In summary, the main characteristics of a transformer can be represented by two windings that transfer power from the input to the output at high efficiency. Hence:

$$e_o i_o = e_I i_I \tag{6.7}$$

where

$$e_o = \frac{n_o}{n_I} e_I \tag{6.8}$$

and

$$i_o = \frac{n_I}{n_o} i_I \tag{6.9}$$

In addition, the input resistance (impedance) R_I is given by:

$$R_I = \frac{e_I}{i_I} = \left(\frac{n_I}{n_o}\right)^2 R_o \tag{6.10}$$

Equation 6.10 shows that the **impedance ratio** of a transformer is the **square of the turns ratio**.

Consider an ac voltage source having an internal series impedance R_G coupled by a transformer to a load R_L, as shown in Fig. 6.8. The power delivered to the transformer input is:

$$P_I = i_I^2 R_I = \frac{e_G^2}{(R_G + R_I)^2} R_I \tag{6.11}$$

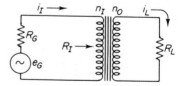

Figure 6.8. *Transformer with generator and load.*

Observe that the input power is zero when either $R_I = 0$ or $R_I = \infty$. When $R_I = R_G$, the transformer input is said to match the generator impedance, and the power delivered to R_I is a maximum. With matched impedances, the voltage across the transformer primary is one-half the open circuit generator voltage e_G. The matched impedance is obtained by adjusting the load R_L, or the transformer turns ratio, in order to make $R_I = R_G$. Equation (6.10) shows that the matched input impedance exists when:

$$R_G = \left(\frac{n_I}{n_o}\right)^2 R_o \tag{6.12}$$

An exact impedance match is not required if we are willing to accept a small power loss. If the impedances R_I and R_G are unequal by as much as a factor of 3, the power loss is only about 25 percent or 1 db. A 6 to 1 mismatch reduces the power output to only one-half maximum. Impedance matching is used when the signal level is low, either to prevent reflected waves on long lines or to obtain as much signal power as possible from an essentially weak signal source. However, the highest amount of power output from an amplifier, without regard to the required input signal, is obtained only by *mismatching* the impedances in order to operate the transistor at the best combination of high current and high voltage output. A 5 W amplifier driven to deliver 1 W will usually deliver *more* than 1 W if the load impedance is changed to match the transistor output impedance more closely. The same amplifier driven at a 5 W level will deliver *less* than 5 W if the impedances are matched.

A transformer can be constructed to operate remarkably well over a wide frequency range if the practical limitations that make it impossible to require simultaneously high efficiency, large frequency range, versatility, and small physical size are recognized. The most important limitation in a communication transformer, such as the one used in amplifiers, comes from the failure of the transformer to transfer power at low frequencies. The coil of a transformer has a finite value of self-inductance, and at low frequencies the inductive reactance of a winding is small compared with the impedance across the winding. The low winding reactance loads the generator and drops the input voltage.

As in Fig. 6.9, a transformer may be represented as an ideal transformer T in parallel with a finite inductance L_P. The inductance L_P represents the inductance of the primary winding. The ideal transformer represents the turns ratio and is assumed to have an infinite inductance. The secondary

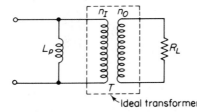

Figure 6.9. *Transformer with equivalent shunt inductance.*

Ideal transformer

winding will have a different value of inductance, L_S. Because the inductance of an iron core winding is proportional to the square of the turns, the inductances are related by:

$$L_S = \left(\frac{n_O}{n_I}\right)^2 L_P \qquad (6.13)$$

The reactance looking into the primary winding is ωL_P. The reactance ωL_S, as given by Eq. (6.13), is the reactance seen looking back into the secondary winding. In other words, the inductance of the transformer can be represented either as L_P on the primary side or as L_S on the secondary side.

The effect of a finite inductance is to cause a signal loss at low frequencies. If the generator has an impedance R_G and R_L is open, i.e., $R_I = \infty$, then the amplitude-frequency response of the transformer is down to 70 percent, or 3 db, when:

$$X_L = R_G \tag{6.14}$$

i.e.,

$$f_c = \frac{R_G}{2\pi L_P} \tag{6.15}$$

If the load matches the source, then $R_I = R_G$, and the cutoff frequency is at one-half the frequency given by Eq. (6.15). This result indicates correctly that the low-frequency cutoff of a transformer depends on both the generator and the load impedance. In fact, if R_P is the resistance equivalent of R_G and R_I in parallel, then the cutoff frequency is given by:

$$f_l = \frac{R_P}{2\pi L_P} \tag{6.16}$$

A manufacturer usually specifies a cutoff frequency f_l for a particular impedance termination. We can estimate easily how the cutoff frequency will change whenever we find a change in the load impedance desirable. Often a transformer can be used satisfactorily between impedances different from those specified by the manufacturer, provided a corresponding change in the low-frequency cutoff can be tolerated.

We can determine the high-frequency cutoff of a transformer in a similar manner from the equivalent series inductance and the shunt capacitance of the windings. The equivalent transformer circuit in Fig. 6.10 has a T-network in the primary that contains primary and secondary equivalent series leakage inductances, the equivalent shunt capacitance, and the equivalent primary inductance. The high-frequency cutoff f_h is approximately the frequency at which the series leakage reactance equals the total series resistance seen by the primary signal. Thus, when the equivalent circuit and the load are referred to the primary, and the copper and core losses are neglected, the high-frequency cutoff is

$$f_h \approx \frac{R_G + R_I}{2\pi(L_{1p} + L'_{1s})} \tag{6.17}$$

Figure 6.10. *Transformer equivalent circuit.*

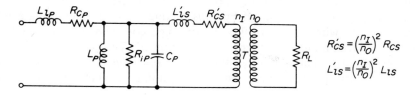

If a transformer is terminated by lower impedances than specified, the high cutoff frequency is lowered. If the terminating impedances are higher than specified, the high cutoff frequency is raised. Because transistor impedance levels are low, we can, as a rule, easily meet audio-frequency high cutoff requirements without undue care in using the rated impedances. The winding capacitance usually causes a resonance near or above the high-frequency cutoff.

In the mid-frequency range a transformer consumes some power. At low signal levels the principal loss is in the dc resistance of the windings. The primary resistance is represented in Fig. 6.10 by the resistance R_{CP}, and the secondary resistance R_{CS}, referred to the primary, is:

$$R'_{CS} = \left(\frac{n_I}{n_O}\right)^2 R_{CS} \tag{6.18}$$

At high signal levels the core loss can be represented by an equivalent shunt resistance, R_{iP}.

At audio frequencies the winding resistance of a medium-size (1 W) transformer will be about 1/20th of the nominal impedance. Hence, the impedance can be estimated as 20 times the dc resistance. Small transformers (0.1 W) have nominal impedances about 5 times the dc resistance. The small transformers do not have adequate window space for the windings and are usually rated to operate with high flux densities in the core. Because small transformers have high copper and core losses, care must be taken that the transformers are terminated in the specified impedances, and a loss of as much as 6 db may be expected. Small transformers are generally acceptable only for speech frequencies (200 Hz to 3000 Hz) and upward.

6.11 AC POWER TRANSFORMERS

Power transformers are designed to operate at line frequency (60 Hz) and at power levels of 10 W or more. In other respects, power transformers are not very different from inexpensive types of audio-frequency communication transformers. The power rating of a transformer is fixed by the amount of heat that can be dissipated over a period of time. At the rated power levels the paper in the transformer is heated to above 100°C, and the useful life of a transformer is determined by the rate at which the insulation deteriorates.

A well-designed power transformer has a power loss that is about one-half in the winding resistance (copper loss) and one-half in the core (core loss). Copper loss is minimized by constructing a transformer using a few turns of large wire. The core flux density, i.e., core loss, is minimized by winding a large number of turns per volt applied. The compromise in the number of turns that minimizes the power loss is usually expressed as a number of turns per volt impressed on the winding. A 60 Hz filament transformer having a 35 VA (volt-ampere) load rating has 1 square inch area of core through the winding and is wound with about 6 turns per volt. Larger transformers and higher operating frequencies permit a smaller number of turns per volt.

A power transformer is so closely designed that it cannot be operated above rated voltage. A transformer can be operated below rated voltage provided the load power is reduced so as to avoid exceeding the rated winding currents, i.e., copper loss. The power rating of a power transformer can be estimated by using the volt-ampere rating of a filament transformer having the same core dimensions and intended for the same operating frequency. For interpolation, the VA rating increases approximately as the cube of the core dimensions.

The power rating of an audio transformer is determined by distortion and frequency response requirements, and the winding temperature rise is not often a consideration. An audio output transformer is usually rated for about one-fifth the power output of a power transformer having the same size core.

SUMMARY

The characteristics of a power amplifier are determined by the power capability of the output transistors, provided the transistors are loaded by the correct impedance. The power capability of the transistors and typical circuits are usually given in the manufacturer's transistor handbook.

A given power transistor is caused to deliver a maximum power by using that particular load impedance which operates the transistor at its peak current and voltage capabilities. The correct load impedance may be determined by a load line on the collector curves, but the correct bias is best determined by experiment. Power transistors are operated with low dc S-factors; otherwise, the high operating temperatures cause excessive Q-point drift.

Impedance matching using transformers permits the transfer of maximum power from one impedance to a different impedance. Transformer coupling is used where necessary to prevent power loss caused by a high ratio of the circuit impedances or where the available power is limited, as in power amplifiers. Transformers change impedances by the square of the turns ratio and change voltages by the turns ratio. The power is transferred with a 5 to a 25 percent loss, although the loss in very small transformers may exceed half the input power.

The low-frequency cutoff of a transformer is the frequency at which the inductive reactance of a winding equals the equivalent shunt resistance that exists when the source and the load are connected to their respective windings.

The equivalent circuit of a transformer is easily determined for guidance in using the transformer at impedances that are different from those specified by the manufacturer.

Linear Class A and Class B amplifiers are usually assembled from data supplied by the transistor manufacturer. An original design requires experience and a thorough understanding of many interrelated factors. Class A amplifiers are used either for power outputs of less than a watt or as Class B driver stages.

Class B power amplifiers are capable of linear amplification when biased to eliminate the crossover distortion. These amplifiers have a low standby current, and for audio applications may be used either without heat sinks or with much smaller heat sinks than are required for Class A amplifiers.

A complementary-symmetry amplifier is a parallel-connected form of Class B amplifier that does not require coupling transformers and permits the use of considerable feedback.

Class C amplifiers have high power efficiency but require a tuned (resonant) load of the kind used in radio-frequency circuits. The transistor in a Class C amplifier converts dc power to ac power by switching a tuned load circuit to the power supply once each cycle of the RF operating frequency.

PROBLEMS

6-1. Assume that the amplifier in Fig. 6.1 is driven by connecting the input to a collector circuit current source. Calculate the expected voltage gain and current gain of the stage.

6-2. Derive Eqs. (6.5) and (6.6) of the text.

6-3. Explain why more input power is required to drive an amplifier that is operated Class B than when operated Class A.

6-4. Plot a curve showing how the power input to a transformer, as given by Eq. (6.11), varies with the input impedance R_l. Mark the points on the curve that are referred to in the text.

6-5. A transformer with a primary inductance of 2 H (henries) and a secondary inductance of 50 H is to be used at audio frequencies. If the low-frequency 3 db cutoff is at 100 Hz, what are the nominal primary and secondary impedances?

6-6. Assume the transformer in Prob. 6-5 has a primary dc resistance of 50 Ω and a secondary dc resistance of 625 Ω. Draw an equivalent circuit representing the transformer.

6-7. If the output transformer in Fig. 3.11 is required to have a 3 db cutoff at 33 Hz, what inductance should the primary have? What dc resistance would you expect this winding to have?

6-8. A transformer with a 3 to 1 turns ratio is intended to be operated between a 450 Ω source and a 50 Ω load. The nominal frequency range is given as 100 Hz to 25 kHz. What would the frequency range become if the source had 300 Ω and the load had 100 Ω?

6-9. A transformer has three identical windings. What is the impedance looking into one winding when: (a) The other windings are in series and are connected to 100 Ω? (b) Each of the other windings is connected to its own 100 Ω resistor? (c) Only one of the windings is connected to 100 Ω? (d) Two windings are connected in parallel to a single 100 Ω resistor?

6-10. Explain the difference between coupling a 500 Ω signal generator to a 50,000 Ω load using: (a) A transformer having a 1 to 10 turns ratio or (b) A CE transistor amplifier having an input impedance of 500 Ω.

6-11. If the complementary-symmetry amplifier shown in Fig. 6.6 has significant feedback what is the voltage gain from the input base to the output load? Ans: 6, 15 dB. By what factor is the dc gain reduced by feedback and is this significant feedback? Ans: The factor is 3.1, 10 dB, which is almost significant.

CHAPTER SEVEN

DC Power Supplies and Regulators

This chapter describes power supplies and regulators that are suitable for transistor equipment. Practical design rules are given for Zener diode regulators, and the characteristics of a feedback regulator are examined.

The useful power output of most electronic circuits actually comes from the dc source that powers the transistors. Occasionally electronic equipment can be made to operate on ac power, but linear and small-signal amplifiers must have a source of dc power. The amount of power required is usually from 5 to 20 times the amplifier output power. The dc should be a constant voltage, free of ripple, and should present a low impedance to the amplifier. Packaged ac-operated power supplies are available for almost any need, but usually each piece of electronic gear has its own power supply. With a little care, most electronic equipment can be designed to operate on a simple power supply. More often than not, equipment that requires a high-performance power supply is poorly designed. In many cases, a simple change or the addition of a Zener diode will relax the demands on the dc supply.

7.1 BATTERY POWER SUPPLIES

For portability and convenience, equipment is sometimes required to operate on batteries. At other times, especially for temporary service, a battery may be a simpler energy source than a power line operated rectifier and filter. The useful life of a battery depends on its load current, the minimum useful voltage, and the time schedule of its use. Since the battery life is nonlinearly related to each of these factors, predicting battery life accurately is practically impossible. The most unpredictable factor in the use of batteries comes from an operator's neglect to turn off a battery at the end of a day's use. For this reason a battery is often not practical unless it can provide several days of continuous service. On the other hand, transistors require such low dc power levels that it is sometimes possible to install a battery without a switch and still obtain one or two years of continuous service.

It is difficult to compare batteries of different kinds and constructions. In a general way, however, the ampere rating of a battery depends on the volume of the unit cell, on its chemical constituents (i.e., type), and on the number of cells in parallel. The voltage rating of a battery depends on the cell type and on the number of cells in series. A rough comparison of the most commonly used types of cells is shown in Table 1.

Table 1. *Approximate Characteristics of No. 2, D-Size Cells*
(3.2 cm. diam. × 6.1 cm. long)

Type	Initial Voltage	End Voltage	Milliampere Hours
Zinc-carbon	1.55	0.9	4,000
Alkaline	1.55	0.9	7,000
Mercury	1.35	1.25	12,000
Nickel-cadmium	1.35	1.15	3,000
Lithium	2.6	2.2	15,000

From the table it can be seen that mercury cells, during their useful life, have about 3 times longer life and about 6 times better voltage regulation than ordinary dry cells. The nickel-cadmium cells have about the same ampere-hour rating as dry cells but offer 3 times better regulation during discharge and can be recharged repeatedly; they withstand abuse and can be stored either charged or discharged. Storage batteries usually have a very low impedance, especially when being charged. Therefore, they are often used to substitute for very high capacitance electrolytic capacitors. Two such batteries connected in parallel can be charged as one but used as if separated. The high effective capacity of the battery and the resistance of the paralleling leads will isolate equipment that otherwise could not share a single battery in common.

7.2 AC POWER SUPPLIES (Ref. 5)

Line-operated power supplies are easily made which convert available ac power to whatever form of low voltage dc power is required by amplifiers and similar equipment. A typical circuit used to convert ac power to dc power is shown in Fig. 7.1. The power transformer is required both to isolate the ac line from the dc equipment and to change the line voltage to a low voltage ac that can be rectified and reduced by filtering to steady dc. The four diodes

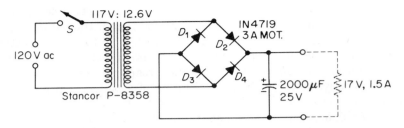

Figure 7.1. *Bridge rectifier power supply.*

are connected in such a way that the alternate half cycles of current supplied by the transformer flow always in the same direction into the filter capacitor and the load. This configuration of rectifiers is called a *full-wave bridge rectifier*. Because of the simplicity of semiconductor rectifiers, it is better to use a full-wave bridge rectifier in preference to one of the simple forms often found in vacuum tube equipment.

The full-wave rectifier produces a series of half sine waves at double the input frequency, so that the lowest frequency that must be removed by the hum filter has twice the supply frequency. The ac transformer has to supply an amount of ac power that is equal to the dc power delivered to the load plus a sizable component of reactive power that circulates in the filter capacitor and the transformer inductance. With a full-wave rectifier the volt-ampere (VA) input to the transformer can be conservatively estimated to be one-fourth to one-third larger than the dc load power. Half-wave rectifiers are seldom used because they require a transformer having about twice the VA rating needed for a full-wave rectifier. The full-wave rectified power, having double the ac input frequency, is more easily filtered than half-wave power at the input frequency. As a result, both the filter and the transformer for a full-wave rectifier are smaller, lighter, and less expensive than for an equivalent half-wave rectifier.

The VA rating of a transformer is usually specified by the manufacturer for either a 60 Hz or 400 Hz input. The ampere rating of a transformer is fixed by the wire size—i.e., the copper loss—and is therefore independent of frequency. The voltage rating is fixed by the reactance of the winding and is therefore approximately proportional to the operating frequency. In other words, a 400 Hz power transformer can be used at a lower input frequency if the input voltage is reduced in proportion to the frequency reduction. A 60 Hz transformer can be used at 50 Hz if the load current is reduced a little to offset the increased reactive input power. A low-frequency power transformer can be used at a higher frequency if the winding voltages are not increased quite in proportion to the frequency increase. Therefore, the power capability of a given transformer increases approximately as the square root of the operating frequency.

A rectifier is followed by a filter in order to reflect the ac components of the rectifier output back to the power line and transfer the dc component unimpeded to the load. The filter in a power supply is called a ***smoothing filter***. As shown in Fig. 7.1, the filter is often only a large electrolytic capacitor, in which case the dc voltage is about 1.3 times the rms ac input voltage.*

The size of the capacitor and the dc load resistor determine the extent to which the rectifier ripple is reduced. The rms ripple voltage is measured as a percentage of the dc load voltage. For a 60 Hz full-wave rectifier the percent ripple r is approximately:

$$r = \frac{200,000}{RC} \tag{7.1}$$

where R is the equivalent dc load resistance in Ω, and C is the capacitance in μF (microfarads). [A half-wave rectifier has twice the percent ripple given by Eq. 7.1.] If the load has 100 Ω and the capacitor is 1000 μF, the ripple is about 2 percent. Because the large capacitor required for low ripple is expensive and causes high peak rectifier currents, we usually find that the input capacitor is chosen to allow about 5 percent ripple. A further reduction of ripple is obtained more economically by adding either a second stage RC filter or a voltage regulator.

A large capacitor across the load also presents the low ac output impedance needed to prevent undesirable crossover and feedback when high-gain amplifiers share a common power supply. The capacitor is continually discharged by the connected load but is recharged in brief current impulses whenever the peak rectifier output voltage exceeds the capacitor voltage. These current pulses can be 10 to 100 times the average load current and may easily exceed the peak current rating of the rectifiers. The current peaks can be limited by inserting a resistor anywhere in series with the transformer. The transformer used in Fig. 7.1 has 2 Ω dc resistance, so the addition of a resistor was not considered necessary. The current peaks were found to be about 4 times the dc load current, and semiconductor rectifiers usually have peak current ratings of about 10 times their average current rating. A resistance in series with the secondary winding that is equivalent to 10 percent of the load resistance reduces the dc load voltage about 20 percent, but reduces the peak rectifier currents by a much larger factor. Unfortunately, with a capacitor input the peak rectifier currents are not easily predicted, so it is difficult to generalize the prevention of peak current difficulties. Furthermore, the input capacitor itself should be conservatively rated when it is subjected to current peaks. If the capacitor has a low capacitance, the ac ripple voltage may be quite high. Hence, the capacitor should have a dc voltage rating at least one-third higher than the dc load voltage.

---

*The reader may find it helpful to compare the 24-V power supply in Fig. 7.14 with the 12-V supply in Fig. 7.1.

7.3 THE ZENER DIODE

A Zener diode is a semiconductor rectifier having a useful reverse breakdown characteristic. With forward conduction the Zener diode is like an ordinary diode. With a reverse voltage below the junction breakdown voltage, the Zener diode is practically an open circuit and still behaves like an ordinary diode. However, if the reverse voltage approaches the breakdown rating, the current increases abruptly. At voltages greater than the breakdown value, the voltage drop across the diode is essentially constant for a wide range of currents. This is the Zener control region where the diode acts as a constant voltage reference or as a control element useful for many different purposes. Zener diodes are obtainable with breakdown ratings upward from about 3 V.

The volt-ampere characteristics for a typical 1 W Zener diode are represented in Fig. 7.2. The characteristics show the familiar forward conduction curves in the first quadrant. At the Zener knee in the third quadrant the reverse current abruptly increases when the current exceeds a minimum of about 1 mA. In the essentially constant voltage region the Zener voltage increases about 0.01 V per mA, thus indicating an equivalent internal resistance of 10 Ω. When the Zener voltage is 10 V and the current is 100 mA, the diode must dissipate 1 W—its maximum room temperature power rating. Low voltage Zener diodes have a more rounded knee than the Zener shown in Fig. 7.2, and high voltage Zeners of a given type are usually much like a 10 V diode.

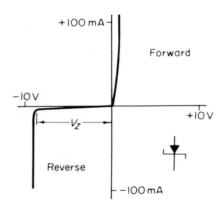

Figure 7.2. *Zener diode forward and reverse current curves.*

Figure 7.3 shows expanded scale curves for 10 V and 8.2 V, 1 W, Zener diodes. The resistance of the diode in the breakdown region is obtained from the upward slope of the curve. Notice that the voltage rises as the input current increases and that the 10 V diode, for example, is like a battery being charged which has 6 Ω internal resistance. The equivalence between the diode

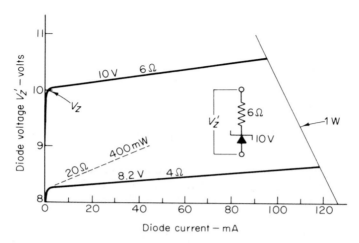

Figure 7.3. *Zener diode terminal voltage vs. current*
(*1 watt diodes*).

and a battery, as suggested in Fig. 7.4, exists only as long as the charging
current is maintained. Because the diode tries to maintain a constant voltage,
even for high input currents, it must be protected from damage by preventing
the power source from raising the terminal voltage more than a few tenths of

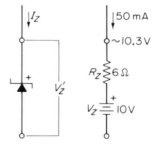

Figure 7.4. *A Zener diode and equivalent*   (a) Zener diode   (b) Equivalent diode
circuit.   with input current

a volt above the Zener rating. To prevent damage, a Zener diode is usually
connected to a power supply, as shown in Fig. 7.5, with a series resistor R_S
which limits the input current. As will be seen later, the series resistor is usually several times as large as the internal Zener resistance. Hence, to explain
how the series resistor limits the current, we can represent the diode as a
constant voltage (battery) and assume the Zener resistance R_Z to be negligible.

  Consider a Zener diode connected to a power supply and a series resis-

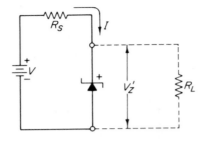

Figure 7.5. *Zener regulator circuit.*

tor, as represented in Fig. 7.5. The Zener diode is reverse biased, which means that V is greater than V'_z and is of the same polarity. The current in R_s is determined by the voltage **difference** $(V - V'_z)$ and by the resistance R_s. Taking the circuit in Fig. 7.6 as a concrete example, suppose that V is supplied by a 12 V automobile battery. Call the highest expected terminal voltage V_H and the lowest voltage V_L. If the automobile battery is being discharged, V_L may be only 11 V; if charged, V_H may be 15 V. Suppose now that a 10 V Zener diode is selected to maintain 10 V across a 300 Ω load. The voltage difference $(V - V_z)$ varies from a high of 5 V to a low of 1 V. If the 10 V, 1 W Zener diode represented in Fig. 7.3 is used as the regulator, then R_s must be selected to ensure that the Zener current at the highest input voltage, 15 V, is less than the maximum current rating of the diode, i.e., 100 mA. With 15

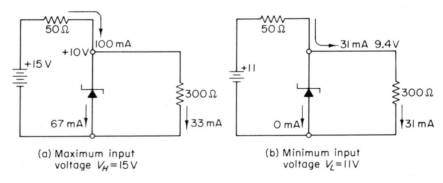

(a) Maximum input
voltage $V_H = 15$ V

(b) Minimum input
voltage $V_L = 11$ V

Figure 7.6. *Zener circuit calculations trial with $V_Z = 10$ V*

V input, a 50 Ω series resistor limits the current input to the diode and the load to a value:

$$I = \frac{V - V'_z}{R_S} = \frac{15 - 10}{50} \cong 100 \text{ mA} \qquad (7.2)$$

With the load in place the diode current will be 67 mA, as shown in Fig. 7.6. However, if the load is removed, the diode current increases to 100 mA and the diode operates at its 1 W rating. At the low input voltage, 11 V, the total input current is only:

$$I = \frac{11 - 10}{50} \cong 20 \text{ mA} \qquad (7.3)$$

This input current, 20 mA, is not enough to supply the 33 mA load current necessary to maintain 10 V across the load, so the load voltage drops below the Zener knee. Because the diode is nonconducting, the load current drops to 31 mA, as shown in Fig. 7.6(b). This example has shown that a 10 V, 1 W, Zener diode cannot hold the load voltage constant and stay within its power rating with such a large change of the input voltage.

Suppose that the load is changed so that 8.2 V are required at 40 mA—i.e., the same 0.33 W load power—and suppose also that an 8.2 V, 1 W, Zener diode is used, as shown in Fig. 7.7. The 8.2 V diode will tolerate 120 mA. Suppose that the series resistor is still 50 Ω. With the maximum input voltage, the input current is:

$$I = \frac{15 - 8.2}{50} \cong 136 \text{ mA} \qquad (7.4)$$

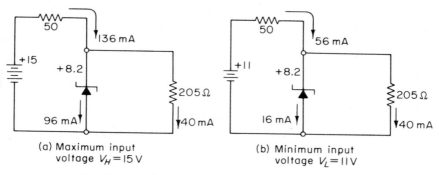

(a) Maximum input
voltage $V_H = 15$ V

(b) Minimum input
voltage $V_L = 11$ V

Figure 7.7. Zener circuit calculation trial with $V_Z = 8.2$ V.

and we may assume that the diode could withstand the 13 percent overload when the load happens to be removed. With the load in place the diode current is a reasonable 96 mA. The maximum power that must be dissipated in the resistor is:

$$P = I^2R = (0.136)^2(50) = 0.9 \text{ W} \qquad (7.5)$$

Consider the conditions existing with the low input voltage. The minimum input current will be:

$$I = \frac{11 - 8.2}{50} \cong 56 \text{ mA} \qquad (7.6)$$

Subtracting the 40 mA demand of the load, we are left with 16 mA to supply the minimum Zener current. Hence, from the minimum 11 V input to the maximum 15 V input, the Zener diode voltage will increase about 0.4 V, as shown by Fig. 7.3. This increase shows that the voltage regulation, as seen by the load, is improved about 10 times.

The foregoing example shows two important characteristics of a simple Zener regulator circuit. When the input voltage is variable, we must use a Zener voltage that is considerably lower than the minimum supply voltage; and the Zener current at the maximum supply voltage may be several times larger than the load current. Both characteristics mean that a Zener regulator may be quite inefficient. Because the Zener regulators operate at low efficiencies, they are not often used at power levels exceeding a few watts.

The 8 V, 1 W, Zener diode of the example has an equivalent internal resistance of 4 Ω. The regulator in Fig. 7.7 has a 50 Ω series resistor; hence, voltage changes and power supply ripple are reduced by a factor of 12. The outstanding advantage of a Zener regulator is that, unlike an electrolytic capacitor, the Zener diode is effective at very low frequencies and even for direct current! This advantage has contributed much to the simplification and improved performance of dc and low-frequency amplifiers.

Zener regulators are used so frequently that a procedure for regulator circuit design is desirable. As is often the case in circuit design when we are required to make a compromise between several conflicting desires, a simple procedure cannot be outlined that meets a wide variety of conditions. A design is particularly difficult and perhaps impractical if the input voltage varies more than 30 percent. If the input voltage is essentially fixed and the load is never removed, we need only to select a Zener voltage one or two volts below the input voltage and select a series resistor that will carry both the Zener current and the load current. For a moderately difficult set of design requirements we can often obtain a satisfactory trial regulator design by the following procedure:

1. Select a Zener voltage V_Z about $\frac{2}{3}$ of the lowest expected input voltage V_L:

$$V_Z \cong \tfrac{2}{3} V_L \tag{7.7}$$

2. The power P_Z dissipated by the Zener diode is approximately the product of the Zener voltage and maximum Zener current. Assuming that P_Z may be as high as 3 times the power input to the load alone, we have:

$$P_Z \cong 3 V_Z I_L \tag{7.8}$$

3. The Zener current rating I_Z will be about $3I_L$; therefore, the load current can be neglected in finding a trial value for the series resistor R_S. The series resistor must limit the Zener current to its maximum rated current when the input voltage is highest. Hence:

$$R_S \cong \frac{V_H - V_Z}{I_Z} = \frac{V_H - V_Z}{P_Z} V_Z \tag{7.9}$$

4. The design is checked by finding the input current at the maximum input voltage. From the input current we find the power rating of the series resistor, the maximum diode current, and the maximum power input to the Zener diode. From the minimum input voltage we determine the minimum input current and the minimum diode current.

If the design is unsatisfactory, the easiest adjustment is to double or halve the size of the series resistor and recheck the performance. The reader will find that the description given of the 8 V regulator in Fig. 7.7 is essentially an outline of the calculations required for a design check.

7.4 EXTENDING REGULATOR POWER RANGE

Higher powers can be dissipated in a Zener regulator by using several diodes in series or in parallel (Fig. 7.8). A parallel connection cannot be satisfactory unless the Zener diodes are forced to share the current by adding resistors in series with each diode. The resistors degrade the regulation. A series connection of diodes is simpler because all diodes carry the same current.

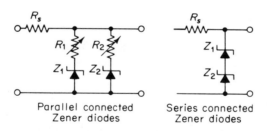

Parallel connected
Zener diodes

Series connected
Zener diodes

Figure 7.8. *Zener diodes, parallel and series connected.*

The temperature change of the Zener voltage is about $+0.05$ percent per °C when the Zener voltage is above 10 V. That is, as the device heats up, the Zener voltage becomes larger. Devices having Zener voltages below 5 V have negative temperature coefficients; above 5 V the diodes have positive temperature coefficients. By selecting the diodes for a series string, the overall temperature coefficient can be made somewhat smaller than that of a single Zener diode. Alternatively, the positive temperature coefficient of a Zener diode can be offset by connecting forward biased diodes in series with the Zener diode. The forward biased diodes have a negative temperature coefficient of about $-2\,\text{mV}$ per °C. The emitter diode of a 2N3638 transistor makes a good 6 V Zener diode when we connect the collector to positive, the emitter to negative, and leave the base open. The forward-biased collector diode provides temperature compensation.

7.5 TRANSISTOR REGULATORS

A better way of extending the power rating of a regulator is to use a power transistor as a variable resistor (on a heat sink) and to control the transistor by using the voltage difference between the load voltage and that of a low-power reference diode. The circuit in Fig. 7.9 is designed as a replacement for the 1 W Zener regulator in Fig. 7.7 except that the resistors are lowered by a factor of 10 to increase the output power by a factor of 10. We see that the transistor is caused to conduct when the Zener diode conducts; hence, the transistor tends to maintain the load voltage about 0.3 V above the Zener voltage. This regulator has a no-load to full-load regulation of 1 percent, or an equivalent internal output resistance of 0.2 Ω, and the load voltage changes are reduced to 7 percent of the input voltage changes. This type of regulator can be built to control a load power of about 10 times the power rating of the Zener diode. The disadvantage of the shunt regulator is that it continuously loads the power supply at full load, and power is lost in both the series resistor and in the transistor.

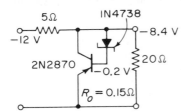

Figure 7.9. *Shunt transistor regulator. Regulation: 5 percent.*

A regulator that uses the transistor as a variable series element is more efficient at all loads than a shunt regulator. For this reason electronic regulators are usually a form of series regulator, as shown in Fig. 7.10. The Zener diode holds the base at a fixed voltage with reference to ground, and, in following the base, the emitter holds the load voltage about 0.3 V below the Zener voltage. The series resistor R_S must be adjusted to supply the base with about $\frac{1}{20}$ of the maximum load current when the input voltage is lowest. The transistor must have at least 1 V or 2 V from collector to base at the minimum input. The maximum input is determined either by the voltage rating of the transistor or by the amount of power that can be dissipated.

The operating characteristics of the series regulator are essentially the same as those for the shunt regulator, except that the load voltage changes are reduced to about 1 percent of the input voltage changes.

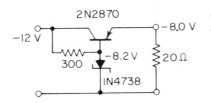

Figure 7.10. *Series transistor regulator. Regulation: 1 percent.*

7.6 FEEDBACK REGULATORS

High-performance electronic power regulators are usually a form of direct-coupled feedback amplifier in which the output voltage variations are returned as feedback to oppose the input changes. By constructing the regulator as a high-gain feedback amplifier, the voltage change seen by the load can be made negligible compared with the voltage changes at the input. To demonstrate this statement, the feedback regulator in Fig. 7.11 is arranged to show its form as a two-stage feedback amplifier. Observe that the output stage drives the load as an emitter follower and the output is fed back to the first stage base by a dc connection. The Zener diode in the first stage emitter substitutes for a low resistance emitter resistor, so that the amplifier has high effective gain. The series transistor TR-1 is turned ON by current in the base-to-collector resistor, and the load voltage rises until TR-2 reduces the bias current available to TR-1. The amplifier seeks a quiescent condition that makes the base of TR-2 about 0.3 V higher than the emitter voltage determined by the Zener diode. Because the load voltage is fixed by the Zener diode, we call Z the *reference diode*.

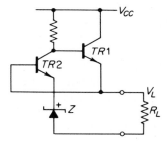

Figure 7.11. *Regulator shown as a feedback amplifier.*

Without feedback TR-1 acts as a series resistor element that reduces the supply voltage to the desired load voltage without regulating the output. With feedback the regulator is improved approximately in proportion to the amount of feedback, subject, of course, to practical limitations and stability problems.

A practical form of the two-stage feedback regulator is shown in Fig. 7.12. Observe that this circuit requires a silicon power transistor and that the output voltage is about 0.5 V higher than the Zener voltage because the input stage has a silicon transistor. This circuit provides about 1 percent regulation as long as the supply voltage is at least 2 V or 3 V above the output voltage. The power dissipation in TR-1 is the product of the load current and the difference between the supply voltage and the load voltage. Thus, large voltage differences require that the power transistor have a low thermal resistance and a large heat sink. The circuit in Fig. 7.12 is hampered by the difficulty that the collector voltage of the transistor TR-2 cannot exceed the

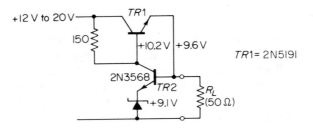

Figure 7.12. *Practical feedback regulator using silicon transistor.*

base-emitter voltage drop of the power transistor. A more practical form of this regulator usually has the base of *TR*-2 connected by a voltage divider to a point 1 V or 2 V below the output terminal. This regulator also has the disadvantage of requiring a power *npn* (silicon) transistor.

The regulator shown in Fig. 7.13 uses a power *pnp* (germanium) transistor driven by a CB amplifier stage. If the base of *TR*-2 is connected to a Zener diode, the output voltage is maintained about 0.5 V below the Zener voltage, and the load sees a low output impedance. If the base is connected to a potentiometer, as at *B*, the output voltage can be adjusted, although the regulator output impedance is somewhat degraded. By connecting an electrolytic capacitor at *B* from the base to ground, the output impedance and the ac rectifier ripple of the voltage supply can be reduced as effectively as if a capacitor 50 times larger were connected across the load. In this way *TR*-2 operates as a capacitance multiplier up to the transistor β cutoff frequency (Ref. 1). For a low-impedance output at high frequencies, a mylar capacitor should be connected directly across the load.

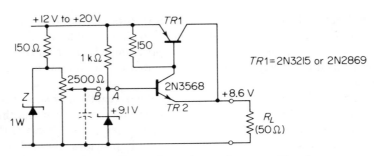

Figure 7.13. *Feedback regulator using germanium transistor.*

A power supply and hum reducing regulator suitable for small audio amplifiers is shown in Fig. 7.14. The power supply uses a full-wave bridge rectifier and delivers 33 V at 0.5 A. With a 1000 μF capacitor, as shown, the

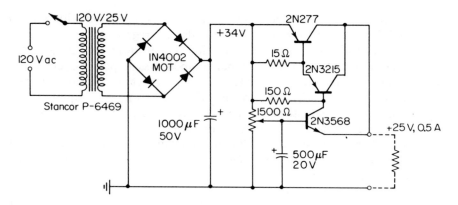

Notes:
   Rectifier ripple = 2 %
   Output ripple = 0.01 %

Figure 7.14. Regulated dc power supply (adjustable voltage).

rms ripple voltage (0.7 V) is about 2 percent of the dc output voltage. It should be noted that almost any regulator has to be supplied with a reasonably smooth input voltage, so this power supply must have at least 500 μF across the full-wave rectifier. The regulator in Fig. 7.14 is similar to the regulator in Fig. 7.13, except that the feedback gain has been increased by making the power transistor a Darlington pair. This regulator (Fig. 7.14) is inexpensive, easy to construct, and delivers up to 15 W with less than 0.01 percent ripple. When the potentiometer is set for a low output voltage, the ripple is as low as 0.001 percent. For low output impedance or for a constant voltage output independent of the input voltage, the capacitor should be replaced by a Zener diode regulator.

The manufactured regulated power supplies are usually high-performance series-type regulators. Their circuits are complicated by the need for a wide range of adjustable output voltage and for a means of limiting the short-circuit output or load current. The feedback amplifier may have 4 or 5 direct-coupled stages that require internally regulated and temperature compensated supplies. In many applications it is much better to use several 2-stage voltage regulators and Zener regulators where needed in the circuit in preference to a central high-performance regulated supply.

7.7   INTEGRATED CIRCUIT REGULATORS

Monolithic ICs are available which simplify the construction and assembly of high-performance regulators. With an inexpensive IC and a few external components one may obtain 0.1 per cent regulation with a 1 per cent stability. Except for low-power loads, IC regulators are packaged with a power transistor to dissipate the inevitable power loss. Because the IC regulators need only a few external components, dc power circuits may be regulated on the individual cards of a complex system. This arrangement symplifies the design, wiring and repair of the central power supply.

For even higher stability, precision regulators are built using OP amps with a discrete Zener diode as the reference and a separate power transistor. These high-stability regulators are capable of providing 0.01 per cent regulation for line, load, and temperature changes, and 120 dB rejection of the line ripple.

7.8 ZENER DIODES IN AMPLIFIERS

Zener diodes are sometimes used as voltage-dropping coupling elements in direct-coupled amplifiers. The amplifier in Fig. 7.15 is an interesting example of this use of a Zener diode. The amplifier is direct-coupled throughout and has a single stage voltage gain of 12. An interesting feature of this

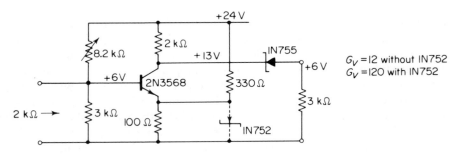

Figure 7.15. Zener-coupled amplifier.

amplifier is that the input and the output terminals have the same voltage of 6 V. Therefore, several such stages can be cascaded by direct-coupling. Furthermore, both the input and the output can be brought to ground potential by connecting the common bus to −6 V. One should notice that the emitter is returned to ground through a low resistance in order to obtain a reasonably high voltage gain. The 330 Ω resistor supplies current to raise the emitter to about +5.5 V. A stage gain of 120 can be obtained by replacing the emitter resistor by a 5.5 V Zener diode (1N752). With the Zener diode the 330 Ω resistor can be raised to 1 kΩ. This resistor cannot be eliminated because the Zener diode has too high an incremental resistance if the diode current is as low as the emitter current. Replacing the emitter resistor by a Zener diode is not usually satisfactory unless the stage is operated from a well regulated power supply.

7.9 ZENER DIODE NOISE

Zener diodes can be an undesirable source of random noise in high-gain amplifiers. In general, the amount of noise is variable and unpredictable. Low noise Zener diodes have a noise level of about 2 μV per square root Hz. An amplifier having a bandwidth extending from 100 Hz to 10,000 Hz has a bandwidth of 10,000 Hz. The square root of the bandwidth is 100 square root Hz. As seen through this amplifier, the Zener diode will produce 2 $\times$ 100 μV or 0.2 mV rms noise. Ordinary Zener diodes may produce 100 times as much noise. As the frequency range of the amplifier is extended below 100 Hz, the noise is inversely proportional to frequency (1/f noise). Therefore, a diode may cause high noise levels in low-frequency and dc amplifiers. The noise in the audio range can be reduced by shunting the Zener diode with a capacitor having a reactance, at the lowest frequency of interest, that is small compared with the incremental resistance of the diode.

SUMMARY

A full-wave capacitor-input power supply has a dc output voltage that is approximately one-third higher than the nominal ac voltage of the power transformer. The value of the input capacitor is usually selected by Eq. 7.1 to make the ripple about 5 per cent of the dc voltage. When better characteristics are required, the capacitor is followed by an additional filter or a feed-back regulator.

Zener diodes are useful as auxiliary dc regulators and as replacements for electrolytic capacitors. The Zener regulator reduces power supply ripple by a factor of 10 and provides the advantage of effective filtering at very low frequencies. The efficiency of a Zener regulator is only about 25 percent.

Zener diodes are useful dc coupling elements for use in amplifiers, but produce noise that is particularly strong at frequencies below 100 Hz.

A transistor regulator is a form of dc feedback amplifier that extends the current capacity of the Zener diode used as the reference element. A single stage transistor regulator provides about 1 percent regulation—a factor of 100 improvement. Because transistor regulators operate as linear amplifiers, the dc power input must be filtered by capacitors that reduce the ripple to about 10 percent.

PROBLEMS

7-1. A full-wave power supply furnishes 1 A at 24 Vdc. The power supply is operated with a 6000 μF input capacitor, as in Fig. 7.1. What is the approximate VA rating required for the transformer? What is the expected percent ripple? What would be a reasonable value of series winding resistance?

7-2. A 10 W Zener diode is used to regulate a 0.4 A load at 30 V. A 10 Ω series resistor protects the diode. (a) With the load in place what is the maximum permissible input voltage? (b) At what input voltage is the regulator ineffective?

7-3. The 2N2870 transistor in Fig. 7.9 has a heat sink that dissipates 8 W. Over what range of input voltage can the regulator be used?

7-4. The 2N2870 transistor in Fig. 7.10 has a heat sink that dissipates 12 W. Over what range of input voltage can the regulator be used?

7-5. Why is the voltage gain of the amplifier in Fig. 7.15 only 12 and not 30?

7-6. The Zener regulator in Fig. P-7.6 holds the load at 10 V with 15 V across the capacitor and at 11 V with 20 V input. With a full-wave rectifier, what percent ripple should we expect at the load?

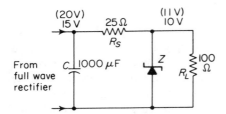

Figure P-7.6.

7-7. Discuss the characteristics of the regulator shown in Fig. P-7.7.

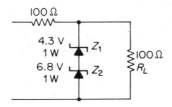

Figure P-7.7.

7-8. Discuss the characteristics of the regulator shown in Fig. P-7.8. Specify power ratings for each component, assuming the maximum input voltage is 40 V.

7-9. Specify approximate current and voltage ratings for all components in Fig. 7.14 except the capacitors. Also give power or VA ratings.

Figure P-7.8.

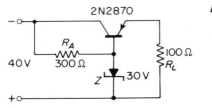

CHAPTER EIGHT

Transistor and FET Switches (Switching Circuits)

Transistors have important applications as switches. At the command of an electrical signal, a transistor is capable of opening or closing circuits at kilowatt power levels or at high operating speeds. Transistor switches offer small size, high reliability, low power losses, nanosecond switching speeds, and silent operation. They are used in computers and control devices, as relay replacements, and in switching amplifiers.

Field-effect transistors provide excellent switching characteristics and are generally preferred for low-current applications. The channel resistance may be changed from tens of megohms to tens of ohms by a gate-control signal that is measured in tens of microwatts. The high gate impedance of FETs provides better isolation between the gate-control signal and the channel. Therefore, FETs may be used as series (or parallel) switching elements where transistors cannot be easily employed. With field-effect devices the low capacitance between the gate and the channel permits fast switching while coupling only a small transient disturbance into the circuit being switched. FETs have greatly simplified and improved logic and digital circuits such as are found in computers, instruments, and communication circuits.

This chapter presents examples of semiconductor switches and includes multivibrators and related trigger circuits which are essential elements of pulse-operated systems.

8.1 TRANSISTORS AND RELAYS

For switching applications the collector and emitter terminals of the transistor are the terminals of the switch. If the base is connected to the emitter or reverse biased, the collector circuit is like an open switch, except for a leakage current that is usually negligible. With a forward biasing base

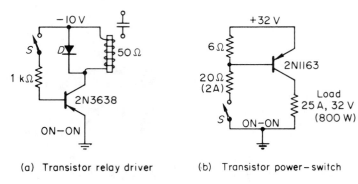

(a) Transistor relay driver   (b) Transistor power-switch

Figure 8.1.

current the collector circuit is like a closed switch for collector currents that are less than β times the base current. For moderate currents, the voltage drop in the transistor is often less than 0.15 V, the voltage drop in a good quality mechanical switch.

One of the common applications of transistor switches is for the control of power relays, as shown in Fig. 8.1(a). A transistor driver is used to operate the relay where the amount of control power is limited or where the electronic circuit must be isolated from the contact circuit. The relay in the figure is operated by supplying about $\frac{1}{2}$ W to the coil, and the contacts turn on another device (such as a small motor). The transistor operates as a simple switch— not as an amplifier—because the base current is much higher than that required for an amplifier. The advantage of using a transistor to operate the relay is that the required base control power is less than $\frac{1}{10}$ the relay power and the switch contacts are not subjected to high-current inductive arcs. Observe that the transistor must be protected from high-voltage inductive spikes by shunting the relay coil with a diode.

The circuit in Fig. 8.1(b) illustrates the use of a high-current power transistor for switching nearly a kW (kilowatt) of power. This load power is 20 times the power dissipation rating of the transistor at 60°C, and the load current is 10 times the current in the operating switch. By adding a driver stage, the operating current can be lowered by a factor of 10. Higher load currents can be switched by paralleling transistors, if transistors with higher current ratings are not available.

8.2 TRANSISTOR SWITCHING CHARACTERISTICS

A switching transistor is usually operated in the CE configuration. Therefore, to explain switching we refer to the CE collector characteristics shown in Fig. 8.2. The load line AB in the figure represents the 50 Ω dc resistance load of the relay. When the transistor is turned OFF by reducing

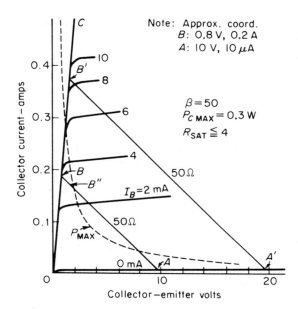

Figure 8.2. *Collector characteristics of a switching transistor.*

the base current to 0, the relay current is negligible, the collector voltage equals the 10 V supply voltage, and the *Q*-point is at *A*. When the relay is turned ON, by applying a base current of 4 mA or more, the *Q*-point is at *B*, and the collector voltage is only a few tenths of a volt. At both ends of the load line, *A* and *B*, the power that must be dissipated by the transistor is small. At *A* the voltage is high but the current is low, while at *B* the current is high but the voltage drop is low. In switching from OFF to ON, the *Q*-point moves along the load line from *A* to *B*. At the center of the load line the power input to the transistor is usually higher than the transistor can withstand continuously. This power loss is minimized by switching the transistor rapidly and infrequently between *A* and *B*.

The load line *A'B'* in Fig. 8.2 represents the conditions existing when the same 50 Ω load is switched, using a 20 V supply. By doubling the supply voltage, the power input to the transistor is increased nearly 4 times. A calculation of the power dissipated when the transistor is left ON at the *Q*-point *B'* shows that the transistor would be forced to dissipate more than its rated power. Operation on a 20 V supply is therefore impractical.

At the point *A* on the 10 V load line the transistor is turned OFF either by reducing the base current to a negligible value or by connecting a voltage that reverse biases the base emitter diode. When turned OFF, the collector leakage current of a small switching transistor is only a few microamperes.

At the point B the transistor is turned ON by applying a base current of at least 4 mA, and the Q-point stays at B for any base current that exceeds the turn-ON current. When the operating point is at B, we say that the transistor is *saturated*.

A close examination of the CE collector curves shows that the line OC is the locus of the collector current when the base current is high enough to hold the transistor in saturation. The reciprocal of the slope of the line OC is called the *saturation resistance* $R_{(SAT)}$. The transistor represented by the curves in Fig. 8.2 has a saturation resistance of about 4 Ω. As long as the base current is high enough to hold the transistor Q-point at B, the voltage drop is typically less than 0.4 V. If the base current is decreased or the load requires a higher current, the point B may move into the active region. Moving the ON point into the active region increases the ON voltage drop, and the switching is called *nonsaturated*. Nonsaturated operation produces a several times faster switching, but it is difficult to ensure that the point B'' remains closed to the $R_{(SAT)}$ line where the ON dissipation is low.

For a fixed load current I_L saturated operation requires only that the operating base current is given by:

$$I_B \geqq \frac{I_L}{\beta} \tag{8.1}$$

Observe that a high-current gain is desirable for switching because the turn-ON base current and the control power are reduced by a high β.

Transistor parameters vary an order of magnitude with temperature, with the Q-point, and because manufacturing tolerances cannot be closely controlled; therefore, a circuit should be designed to operate even when these parameters are the most unfavorable. This is called *worst case design*. The design of switching circuits is a specialty beyond the present discussion, but simple rules are sometimes effective for the analysis of switching difficulties. It is customary to assume in design that the current gain of a transistor may be one-half the typical β and to use a base current that is about 3 times the value calculated in Eq. (8.1). Three times the calculated current ensures that the ON point B remains in the saturated region, even under adverse operating conditions. In many applications a transistor can be switched ON by applying a base current of $\frac{1}{10}$ the maximum expected collector current and can be switched OFF either by removing the base current or by connecting the base to the emitter.

8.3 SWITCHING AMPLIFIERS

Figures 8.3 and 8.4 show examples of 2-stage transistor switching amplifiers. In these circuits both transistors operate as switches, either full ON or full OFF, and the input switching currents are about $\frac{1}{10}$ the current required for a single-stage transistor switch. Fig. 8.3 requires a pair of *npn*

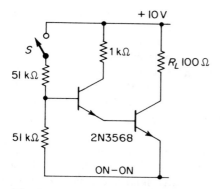

Figure 8.3. CC-CE switching amplifier (alike transistors).

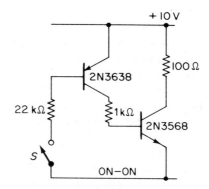

Figure 8.4. CE-CE switching amplifier (opposite transistors).

transistors and Fig. 8.4 uses alternate *pnp* and *npn*. Otherwise, there is little difference between the two circuits. Both circuits have the emitters essentially connected to one side or the other of the power supply, and both have a 1 kΩ resistor in the first-stage collector circuit in order to limit the second-stage base current. By observing these practices, we may easily add one or more power stages and switch much higher power levels.

The two-stage switching amplifier in Fig. 8.5 is sometimes used as a way of reversing the switching action, so that closing the switch *S* turns OFF the load current. In this circuit the second-stage transistor is OFF when the first stage is ON.

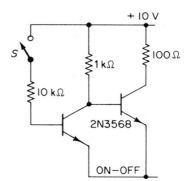

Figure 8.5. CE-CE switching amplifier (alike transistors).

Circuits similar to those in Figs. 8.3, 8.4, and 8.5 are often used as switching amplifiers or as squaring amplifiers; i.e., converting irregular waveforms to square waves. The purpose of such amplifiers is to produce a square wave or a series of variable-width square waves that are under the control of a sine wave or a series of pulses. The amplifiers shown produce a square wave signal across the load when driven by a 5 to 10 V rms signal.

8.4 SWITCHING TIME

Whenever a transistor is turned ON, electric charges are stored in the base-emitter capacitance, and the charge must be removed before the transistor will turn OFF. In a similar way charge must be stored in the capacitance before a transistor turns ON. The sum of the turn-ON and the turn-OFF times is the total switching time of the transistor, a figure of merit. A high-speed computer may require a total switching time of less than 0.1 μs (microsecond).

The switching time may be reduced by using small high-frequency transistors that have a low base-emitter capacitance. The turn-ON switching time is reduced by shunting the series resistor with a small speed-up capacitor, shown as C in Fig. 8.6. The turn-OFF time is reduced by applying an excess of reverse biasing voltage to accelerate the removal of stored charge. The reverse bias is also needed to prevent turn-ON at high operating temperatures. For both reasons switching circuits usually have positive and negative supply voltages, as shown in Fig. 8.6.

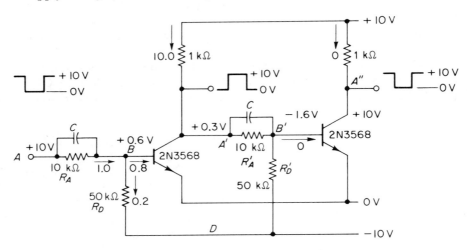

Figure 8.6. *Two-stage inverter-inverter. Volts and milli-amperes shown for +10 V at input. Arrows designate mA.*

8.5 SWITCHING INVERTERS (Ref. 1)

Each switching stage in Fig. 8.6 is usually called an *inverter* because a square step applied at the input appears inverted at the output. Switching inverters are a basic component of most pulse and digital circuits, and one must recognize that an inverter is quite different from an amplifier. The analysis of the operating conditions is much simpler than it might appear.

Consider first the input conditions of the first stage when the transistor is turned ON by applying a $+$ 10 V step at A. Because the base is forward biased, the base-emitter voltage is negligible. The current supplied from A is simply the current in R_A when B is grounded, and the current in R_D is the current existing when B is grounded. The net base current is the difference between the two input currents. When the input and the bias voltages are obtained from equal magnitude supplies, as in Fig. 8.6, the bias resistor is usually about 5 times the input resistor and the base turn-ON current is 80 percent of the input current.

Consider now the conditions when the transistor is turned OFF by reducing the voltage at A to 0. The second stage illustrates the proposed condition because the collector voltage at A' is essentially 0. The negative bias voltage acting alone reverse biases the base B'. Because the base is an open circuit when reverse biased, the base voltage is the bias voltage at D reduced by $R_{D'}$ and $R_{A'}$. For the resistor values given we find that the base voltage is about $\frac{1}{6}$ of the bias voltage. The resistor $R_{A'}$ is made small to prevent the transistor leakage current from offsetting the base voltage at high operating temperatures.

An inverter converts an ON signal to an OFF and vice versa. The two-stage inverter is useful as a source of both equal and out-of-phase switching steps. Perhaps the most common use of a two-stage inverter is in flipflops that are constructed by connecting the output terminal A'' back to the input A. The resulting feedback is a positive feedback that enhances the input signal. A flipflop locks itself in one state and can be switched to the other by coupling a pulse of the correct polarity into either collector. By coupling the pulse into both collectors via steering diodes, a flipflop switches alternately for each input pulse; in this way two positive input pulses produce one positive output pulse, and we have a scale-of-two divider or counter.

Flipflops and inverters form an important group of switching circuits in which transistors are triggered by a pulse to switch from ON to OFF and vice versa. Some of these circuits, known as **monostable triggers,** can be used to form a pulse of predetermined amplitude and time duration when triggered by a lower amplitude pulse. Other circuits use inverters as switching amplifiers in logic circuits that respond to a required combination of inputs. These, for example, may turn ON when any one of a group of inputs is turned OFF. Flipflops, triggers, and logic circuits are the basic building blocks of computers, scale circuits, and switching devices.

8.6 FLIPFLOPS (Ref. 12)

A flipflop is a positive feedback amplifier in which one transistor is OFF when the other is ON until an overriding pulse forces the transistors to exchange roles. The exchange of states produces an output pulse having a fixed magnitude and a steep wave front. Because successive pulses have alternate polarities, a succeeding flipflop can be made to respond only to every other pulse as a binary counter.

With positive feedback a part of the output signal is returned to the input and phased to increase the amplifier signal. If a circuit *loop* from the input to output and back to the input is capacitor coupled, positive feedback makes an amplifier oscillate. If the feedback loop is direct coupled throughout, positive feedback moves the collector Q-points to a full ON or OFF state, as in a flipflop. Similarly, an amplifier that cannot be biased in the active region may have positive direct-coupled feedback.

A flipflop illustrates a most useful application of positive feedback. Feedback makes a circuit independent of the transistors within the feedback loop. In positive feedback circuits the transistors are driven so hard ON or OFF that the output signals have precise levels and are largely independent of the transistors and of the input signal distortion. This characteristic of a positive feedback amplifier stabilizes the pulse-forming and counting circuits used in computers and pulse-operated instruments and makes these devices feasible. Positive feedback is usually introduced by a combination of two methods. A two-stage CE-CE amplifier in which the output is coupled back to the input has positive feedback, and a two-stage CE-CE amplifier in which the emitters share an unbypassed emitter has positive feedback.

The flipflop in Fig. 8.7 is clearly very similar to the two-stage inverter (Fig. 8.6) except that an emitter resistor is shared by the emitters and the base resistors are lowered so that they can be returned to ground. Triggering is initiated by applying a steep-sided pulse or square wave that moves the Q-points of both stages into the active region. Because the positive feedback makes the system unstable, each Q-point moves the other until one is driven

Figure 8.7. *Saturated flipflop with emitter trigger.*

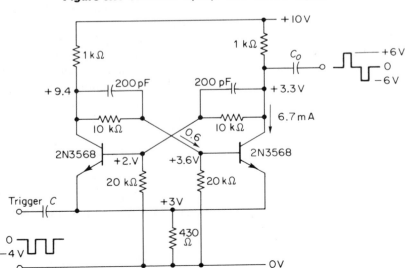

full ON and the other full OFF. The states exchange because the capacitor
on the OFF-side initially has several times the charge of the other capacitor,
and the ON transistor is turned OFF by the predominating charge. Although
both emitters are subjected to the same pulse, the initial charge on the feedback capacitors forces the system to return to the opposite state, provided
that the trigger pulse has sufficient energy to move the Q-points into the
active (amplifier) region and is short compared with the decay time of the
coupling capacitors.

The Q-point conditions of a saturated flipflop can be approximated by
assuming that one transistor is removed from the circuit and that the other
has the base, emitter, and collector shorted together. The base current may be
neglected. The student should confirm this statement by comparing calculated
voltages with the measured values given in Fig. 8.7. The calculation shows
that the base drive current of the ON stage is about 0.6 mA and the collector
current is less than 7 mA. Therefore, the circuit requires a minimum dc current gain of at least 12. Each time the circuit is pulsed the voltages shown on
one transistor interchange with the voltages on the other. The calculated
collector voltages show that the output pulse will have a peak amplitude of
6 V. Observe also that the Q-point at the midpoint of the switching cycle
may be checked by connecting the two collectors to each other.

The design of a good flipflop is complicated by the need to ensure that
the transistors can be turned OFF at high temperatures when the collector
leakage current is high, and by the need to ensure that the ON transistor is in
saturation at low temperatures when the current gain is the lowest. The size
of the capacitors is a compromise between being large to ensure reliable
switching and small to permit fast switching.

Flipflops may be triggered by applying a pulse to both emitters, both
bases, or both collectors. The trigger pulse may be coupled through a resistor,
capacitor, or diodes. Depending on its polarity, the trigger may turn both
transistors ON or OFF, and the collector-to-base capacitors in discharging
make the transistors exchange states.

8.7   NONSATURATING FLIPFLOPS (Ref. 12)

High switching speeds are attained by using high-frequency types of
transistors, low impedance levels, and nonsaturating Q-points. The need for
high switching speeds, particularly in computers, has produced many nonsaturating switching circuits that are usually complicated. The flipflop in
Fig. 8.8 is an example illustrating some of the problems of a nonsaturating
design. The circuit is made nonsaturating by lowering the base resistors so
that the turn-ON base voltage cannot exceed 2 V. The emitter resistor limits
the emitter current to the value at which the emitter just follows the base. By
limiting the emitter current, we also limit the voltage drop in the collector

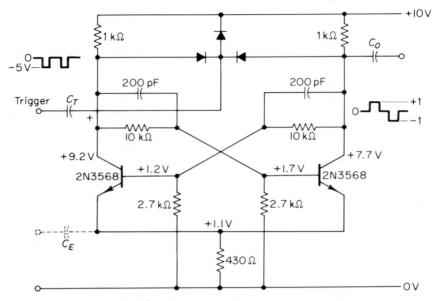

Figure 8.8. *Nonsaturated flipflop with collector trigger.*

resistor. For the circuit in Fig. 8.8 the maximum possible voltage across the emitter resistor is about 0.6 V less than the maximum base voltage, i.e., 2 V — 0.6 V, or 1.4 V. Therefore, the maximum voltage drop across the collector resistor cannot be more than $\frac{1000}{430} \times 1.4 = 3.3$ V. In this way saturation is avoided because the collector-to-emitter voltage cannot drop below 10 — 3.3 — 1.4 = 5.3 V, and the Q-point must be in the active region.

The difficulty with this circuit is that the output signal is less than 2 V, and both transistors remain in the active region which makes the circuit sensitive to electrical noise. Collector triggering by the use of steering diodes is illustrated in Fig. 8.8. The polarity of the diodes is such that the negative input pulse turns OFF both transistors, and a positive pulse is rejected.

The nonsaturating flipflop in Fig. 8.9 uses diodes to clamp the collector voltage to the base drive, so that the collector voltage cannot fall below the base voltage. The diode, operating as a switch, clamps the collector to the bias resistor when the collector drops to 2 V. By diverting the base current, the diode prevents the collector from falling below 2 V and keeps the transistor from being driven into a saturated Q-point. The advantage of the diode clamp is that the collector voltage change is reasonably large and is a fixed value.

Base triggering, illustrated in Fig. 8.9, is more sensitive than the other trigger methods but requires an accurately controlled trigger pulse.

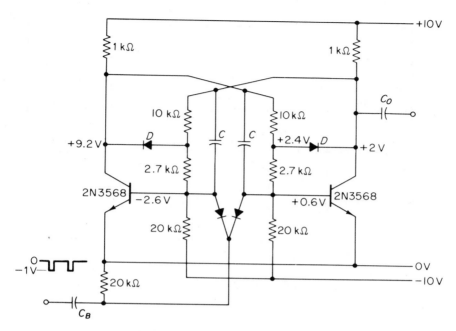

Figure 8.9. *Nonsaturated flipflop with base trigger and diode clamp.*

8.8 MULTIVIBRATORS (Ref. 12)

A multivibrator is a 2-stage amplifier that is connected back to itself so that there is an excess of positive ac feedback. A multivibrator, as shown in Fig. 8.10, appears on the surface to be the same as a flipflop. However, the dc feedback in the multivibrator is reduced by separating the emitters so that the static Q-points are in the active region and are stable. The capacitors that cross-couple the collectors to the bases increase the ac loop gain and force the transistors to switch alternately ON and OFF between temporarily stable states. The astable, or free-running multivibrator, generates a square wave that is useful as a clock timing frequency, and the square wave is easily converted to a sawtooth by driving an RC integrator or converted into a pulse by an RC differentiator (see Chap. 10).

The switching period of a multivibrator is determined by the coupling capacitors and the resistors of the circuit. In turn, each capacitor is charged through the forward biased base-emitter diode in less time than is required to discharge the other, so that the half period of a cycle is determined only by the discharge times. Since the transistor is reverse biased, the discharge time

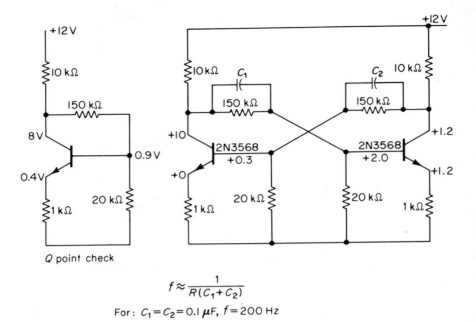

$$f \approx \frac{1}{R(C_1 + C_2)}$$

For: $C_1 = C_2 = 0.1\,\mu F$, $f = 200$ Hz

hence: $R = 25$ kΩ; $f = \dfrac{40 \times 10^{-6}}{C_1 + C_2}$; $f \propto V_{CC}$

Figure 8.10. *Multivibrator.*

is determined mainly by the base resistor. Because switching is a nonlinear process and a calculation of the multivibrator period is only an approximation, the period shown in Fig. 8.10 is an experimentally measured time. It is pointless and misleading to suggest that the frequency relation can be accurately calculated by practical approximations. It is always better to construct a circuit, adjust the frequency, and then study the performance carefully with the application problems and conditions in mind. A simple formula does not take into consideration the fact that the frequency of a multivibrator is nearly independent of the collector voltage when the emitter resistors are not bypassed, nor that the frequency is almost proportional to the supply voltage when the emitter resistors are bypassed or omitted.

8.9 MONOSTABLE MULTIVIBRATORS

A monostable multivibrator, or one-shot, is a flipflop that has one base ac cross-coupled and the other dc-coupled, as shown in Fig. 8.11. This coupling arrangement produces a stable and an unstable state. When triggered by a pulse, the mono switches to the unstable state and is held there until the coupling capacitor discharges. The circuit returns to the initial, or relaxed,

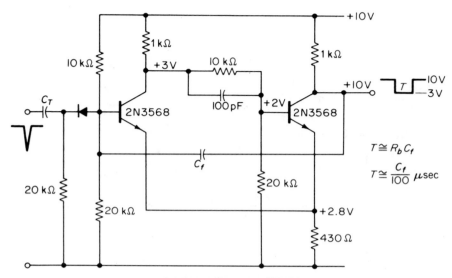

Figure 8.11. *Monostable flipflop or frequency divider.*

conditions that are determined by the resistors of the circuit. The duration of the pulse output is determined by the capacitor C and the resistors through which it discharges, so that the duration of the output pulse is always the same. In the circuit shown in Fig. 8.11 the capacitor discharges through both bias resistors as if they were in parallel (by the Thevenin equivalent). The voltages indicated in the figure are those existing under relaxed conditions with the input transistor normally ON.

A mono is used to produce a pulse of predetermined amplitude and shape each time the mono is triggered. Monos are called ***pulse shapers*** and ***pulse restorers***. They are used to produce a measured time delay after a triggering pulse and sometimes are used as frequency dividers.

8.10 SCHMITT TRIGGERS

A Schmitt trigger is a flipflop that has one RC feedback connection removed, as shown in Fig. 8.12. Because the emitters share a common resistor, the amplifier has a large positive dc feedback that drives the transistors into a stable ON or OFF state. When the input terminal is grounded, the input stage is OFF and the output stage is ON. As the input voltage is increased, there is no change until the voltage reaches a threshold level—about 6 V— when both Q-points move into the active region. Because the emitters are regeneratively coupled, the gain is high: a small increase of the input voltage drives the output transistor rapidly from ON to OFF, and the input transistor

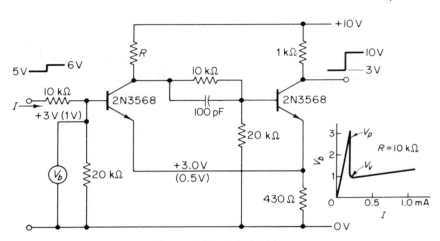

Figure 8.12. *Schmitt trigger.*

is turned ON by the positive feedback. If the input voltage is lowered, nothing happens until the voltage drops to a reset level—about 5 V—where the transistors switch back to their initial states. The difference in voltage between which the switching is initiated by increasing and decreasing signals is called the *hysteresis voltage*, or *backlash*. By making the collector resistor 1 kΩ in the first stage and 10 kΩ in the second stage, we reduce the hysteresis voltage to 0.1 V. The small coupling capacitor steepens the wave front and speeds the switching.

A Schmitt trigger is useful for detecting a particular voltage level, for restoring waveforms, or for converting sinusoidal and nonrectangular signals to square waves—a process called *squaring*. Schmitt triggers are easily constructed to operate at frequencies above 1 MHz.

If the collector resistor in the first stage of a Schmitt trigger is increased to about 10 times the second stage collector resistor, switching produces a change in the common emitter voltage and a corresponding change in the input base voltage. The effect of this voltage change, shown in Fig. 8.12, is to reduce the input base voltage about 2 V at the trigger point as the input current is increased. A drop in voltage with an increase in current means that there is a negative resistance, which in this case is produced by positive feedback. Between the peak voltage V_p and the valley voltage V_v, the input resistance is a high negative resistance. Negative resistances are sometimes used to offset a positive resistance, as in an active filter, or to reduce the loss in a long transmission line.

8.11 FIELD-EFFECT TRANSISTOR SWITCHES

A field-effect transistor operated in the ohmic region, i.e., $|V_D| < |V_P|$ is an ohmic resistor that can be varied by changing the gate voltage. When the gate voltage is zero, the drain-to-source resistance is nearly $1/g_m$ of the FET and an FET having a g_m of 4000 μmho will have a resistance of $10^6/4000$, or 250 Ω. Increasing the gate voltage toward drain current cutoff increases the drain resistance to very high values, 10 to 100 MΩ, and makes these devices useful as variable resistors for gain and AGC control or as switches.

The need for FETs with a low ON resistance has made high g_m devices readily available with ON resistances as low as 4 Ω. Because operation in the ohmic region implies that the magnitude of the drain voltage is always less than the pinchoff voltage, the peak signal level across the resistance must not exceed the pinchoff voltage. The gate switching power is small, even with high pinchoff voltages; hence switching FETs may have high pinchoff ratings.

A common use of FET switches is for chopping low-frequency or dc signals in order that these signals may be amplified in an ac amplifier instead of in a more complicated dc amplifier. FETs for chopper applications must have low gate-to-drain capacitance and may not have high enough g_m for applications as a switch.

Two uses of complementary FETs as switches are illustrated in Figs. 8.13 and 8.14. In both figures the terminals X and Y are available for connecting an external circuit in the manner suggested by the relay in the figure. The FETs are shown as being controlled by a switch A that connects either to -12 or $+12$ V. In a practical application it is only necessary that A connect to a point where the required voltage change is available. The FETs replace contacts that are presumed unavailable in a switch. In both figures when the point A is at -12 V, the point X is connected to the arm of the switch Y. When A is at $+12$ V, Y is connected to Z, and the left side FET is open in both figures. Similarly, a pair of alike FETS can be used as in Fig. 8.15 if one gate is connected to one side of a flipflop with the other gate connected to an opposite side, and the flipflop connects alternately to $+12$ V and to -12 V.

Figure 8.13. *FET switch.*

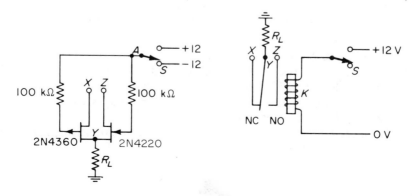

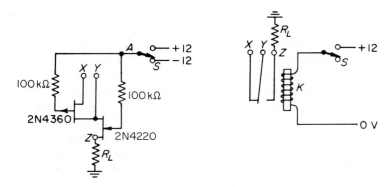

Figure 8.14. FET switch.

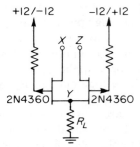

Figure 8.15. FET switch (alike FETs).

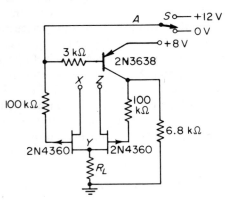

Figure 8.16. FET switch with transistor inverter.

The FETs shown in Figs. 8.13, 8.14, and 8.15 have turn-ON resistances of about 200 Ω. These resistances can be lowered by connecting several FETs in parallel, or low resistance devices can be substituted. Operation in the pinchoff region requires that the signal and the dc voltages applied at X and Y be between +4 and −4 V.

The circuit shown in Fig. 8.16 shows how a *pnp* transistor inverter is used to operate a pair of alike FETs from a 0 to $+12$ V switching signal.

FET switches are used in flipflops and multivibrators especially when a high input impedance is desirable. FETs can be used in most vacuum tube switching circuits by lowering the supply voltage and the plate resistor. The choice between the use of transistors or FETs in integrated circuits depends mainly on the relative cost, reliability, and the power requirements.

8.12   MOS-FET SWITCHES

Because of manufacturing improvements, the MOS FETS are rapidly replacing bipolar transistors in many applications, particularly as memories and switches in computers. The MOS FETs are almost ideal switches because they permit bilateral current flow in a channel that can be rapidly changed from a few tens of ohms to more than 1000 megohms. The MOS control gate has an input resistance of approximately 10^{14} ohms, which is high enough to permit temporary storage of charge in the gate capacitance of the device.

The MOS FETs are particularly attractive for use in large-scale integrated circuits (LSI) because their small physical size and low-power dissipation permits the use of many more devices on a silicon chip than is possible with bipolar transistors. Moreover, MOS FET switches do not require well regulated supply voltages and can be made relatively immune to noise signals. The earliest applications of MOS FETs in computer logic and memories used p-channel devices because of the difficulty of manufacturing n-channel devices, particularly on the same chip. Circuits using only p-channel devices are known as PMOS circuits.

Any of the JFETs shown in the switching circuits, Figs. 8.13 through 8.16, may be replaced by MOS FETs without affecting the circuit behavior. The circuits shown earlier in this chapter may be used equally well with MOS FETs, except that the timing and waveforms may be changed by the characteristic that a MOS gate does not conduct the turn-on signal as with a transistor or a JFET. The substitution of MOS FETs for the JFETs offers an interesting study for the reader.

The advantages of using complementary MOS FETs in IC switching and memory applications has been limited to military and aerospace applications where relatively high costs can be accepted. Manufacturing improvements have increased yields, and a growing number of manufacturers have reduced costs, so that complementary MOS devices have replaced transistors and PMOS in many IC applications.

8.13   COMPLEMENTARY MOS SWITCHING CIRCUITS

Complementary MOS (CMOS or COS/MOS) switching and memory circuits are rapidly becoming accepted as superior means of performing many memory and switching operations. CMOS circuits have the advantage of

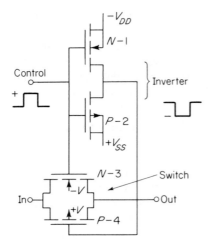

Figure 8.17. *CMOS bilateral switch (switch ON with control positive).*

requiring low operating power, and, because pull-up resistors are not required, as many as 1200 devices can be placed on a 150 × 150 mil chip that has space for only 150 transistors with pull-up resistors. The MOS devices are almost perfect switches, and a single MOS may fan out to several hundred other MOS devices, depending on the capacitance load that may be tolerated at the switching speeds desired. Many new applications are being found for CMOS switching because of the low-power requirements and because CMOS circuits can operate on a wide range of supply voltage (3 to 15 V) and the supply voltage does not need to be regulated.

The bilateral analog switch shown in Fig. 8.17 illustrates the low-power advantages of a CMOS switch. The series connected CMOS devices N-1 and P-2 are used as an inverter to supply gate control voltages for the parallel-connected switches N-3 and P-4. With 5 V control and supply voltages the NMOS, N-1, is ON and the PMOS, P-2, is OFF. Thus, the gate of N-3 has +5 V, and the gate of P-4 has −5 V. The +5 V and −5 V gate voltages make the switches operate as a bilateral switch for any signal that is between plus and minus 5 peak volts. Because the ON resistance of the switch is only a few hundred ohms and the OFF resistance is about 10^{11} ohms, there are only a few microwatts of power dissipation for the signal levels used in digital equipment. Similarly, there is no significant power loss in the inverter because the series circuit is opened by either N-1 or P-2, except for a few microseconds during the switching cycle when both are ON.

CMOS devices have similar power and space-saving advantages when used for information storage in computer memories. As with the CMOS switch, a memory cell does not need pull-up resistors, so most of the chip space can be used for FETs, but the maximum chip size is somewhat limited by the many terminals required to select a particular memory cell.

SUMMARY

Transistors act as closed switches when the base current is sufficient to hold the collector Q-point on the R_{SAT} line. They are as open switches when the emitter diode is reverse biased. FETs built for switching applications have low ON resistances and are switched OFF by applying gate voltages that exceed the pinchoff value.

Inverters, flipflops, multivibrators, and trigger circuits are the basic structures of digital computers. Most of these circuits use positive feedback that makes the transistors operate as fast switches. Nonsaturating switching keeps the transistors in the active region in order to prevent an excess of stored charge and delayed switching.

Switching circuits are analyzed by determining that the circuit has a high loop gain in the active region and by examining the Q-points when the transistors are alternately ON and OFF.

Transistors and FETs may be used as high-speed switches or relays. Transistors are used in high-current power applications and as high-speed computer switches. JFETs are used as switches in discrete circuit applications where costs may be a primary consideration. MOS FETs are used in LSI computer switching and memory circuits where a lower operating speed is accepted to gain more compact circuits and reduced operating power.

PROBLEMS

8-1. Refer to the switching amplifiers in Figs. 8.3, 8.4, and 8.5. Assume that the transistors have current gains of 100 and that the saturation voltages are negligible. For each figure show what currents and voltages you would expect to find in the circuit when the switches are closed.

8-2. Assume that the transistor in Fig. 8.1(b) has a current gain of 30 and a saturation voltage of 0.5 V at 15 A collector current. Construct a load diagram representing the load conditions shown in the figure. Calculate the collector circuit power loss, the required base control power, and the maximum instantaneous collector input power during switching.

8-3. Assume that the emitters in the saturated flipflop of Fig. 8.7 are momentarily grounded. Show that the initial charge on the capacitors will cause the transistors to exchange their ON and OFF states that existed just before the emitters were grounded.

8-4. Explain how a multivibrator operates and show that the switching is regeneratively sustained.

8-5. Show what voltages you would expect to find in the multivibrator circuit in Fig. 8.10 if one collector is connected to the $+12$ V supply.

Show what voltages you would find if the collectors are connected to each other. (*Hint:* Try $V_c = 6, 7, 8, 9, 10$ V.)

8-6. Explain why a Schmitt trigger behaves differently when the collector resistors are alike than when they are quite different. Consider the case when one resistor is 10 times larger than the other and vice versa.

8-7. Show that the stage current gain in the multivibrator of Fig. 8.10 exceeds the stage current loss. What are the stage current gains and current losses in the inverter of Fig. 8.6?

8-8. Show that an emitter resistor shared by adjacent CE stages produces regenerative feedback. Calculate the loop gain for the Schmitt trigger in Fig. 8.12 (difficult).

8-9. When operated as a switch, what is the zero bias ON resistance of each of the FETs for which a drain current-drain voltage curve is given in Chap. 4?

CHAPTER NINE

Diodes As Switches; Logic Circuits

There are many interesting applications of diodes as switching elements in which the diodes are more than just rectifiers. In some applications diodes steer a signal one way or another, depending on the signal polarity. In other applications, where a signal exceeds a given magnitude, diodes close to protect sensitive meters or to damp out inductive transients that cause interference and damage switch contacts. In still other circuits diodes operate as voltage-controlled gates that open and close a channel under the control of a clock signal (an oscillator), or switch the channel so that the signal is periodically inverted, as in a synchronous rectifier. Electronic computers operate by responding to signal combinations that can be described by using conjunctives such as AND, OR, NOT, and NOR. For example, a circuit that responds only when signals *A and B* are present is called a *logic circuit*. The circuits of computers that respond to logical operations are some form of a voltage-operated switch. Both diodes and transistors are frequently used in logic circuits. This chapter reviews some of the uses of diodes as signal-sensing and voltage-actuated switches.

9.1 DIODE SWITCHES

A silicon diode is practically an open circuit when reverse biased; when forward biased, its resistance is in the tens or hundreds of ohms. The dynamic (ac) resistance of a diode without a dc bias varies with the signal level (unless the signal voltage is less than 10 mV) and the temperature. For these reasons diodes are not often considered useful except when biased or used as a switch. One-half of a volt is sufficient either to reverse bias or forward bias a small diode, and the switching power required is only a few milliwatts. Many applications use diodes as voltage or polarity sensitive switches. Surprisingly,

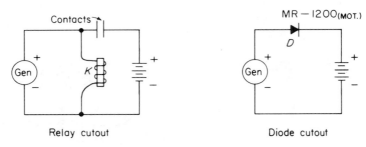

Figure 9.1. *Generator cutout circuits.*

the power loss in a diode at rated current is not more than twice the loss in a typical power switch after half the normal operating life. The power loss in good silver switch contacts is about $\frac{1}{5}$ the loss in a diode, but a diode is free from wear, contamination, noise, and vibration problems.

An automobile type of battery charger must have a voltage sensitive cutout to prevent the battery from discharging though the generator whenever the generator voltage is less than the battery voltage. The relay cutouts commonly used in automobile charging systems often cause a burned-out generator if the contacts stick or otherwise accidentally close. An inexpensive MR 1200 diode, connected as shown in Fig. 9.1, protects the generator from reverse current and is more reliable than a relay cutout.

A diode acts as a simple device for lowering the power supplied to ac-operated soldering irons, incandescent lamps, and similar devices. As illustrated in Fig. 9.2, a diode opens the power line every other half cycle and reduces the power input to a resistance load to one-half the full-wave input. The fact that the half-wave signal has a dc component is generally unimportant. Lamps, motors, and loads that vary with the voltage input ordinarily take about $\frac{3}{4}$ the full-wave power. Because the diode conducts only half time, and the forward voltage drop of the diode is about 0.6 V at rated current, the power rating of the diode need by only $\frac{1}{2}(0.6/120)$ times the power rating of

Figure 9.2. *Power reducer, low loss.*

Figure 9.3. *Power amplifier with reversed dc supply. Compare with Fig. 6.1.*

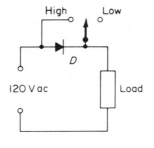

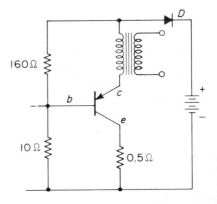

the 120 V load. A 0.5 W diode is suitable if the load does not exceed 400 W and has the advantage that the power loss and the heat that must be dissipated are negligible. Observe that for the half-power input to resistance loads the rms voltage is 70 percent of the line voltage.

Diodes are useful as automatic switches for protecting power amplifiers and inverters from damage when a power supply is accidentally connected with the polarity reversed. The circuit shown in Fig. 9.3 represents a power amplifier using a *pnp* transistor that has the battery connected with the incorrect polarity, and the collector is forward biased, as indicated by the arrow. The amplifier is protected from the reverse current because the diode *D* is reverse biased. Observe that if the diode is removed—i.e., replaced by a short—the transistor base current flows from the base to the negative side of the battery. In normal operation the base current is limited by the 160 Ω resistor, but with the battery reversed, the base current is limited only by the 10 Ω resistor, and the collector current may be 10 times normal. Unless protected, power transistors are usually destroyed by accidental reversal of the power supply. If the power is switched ON by a relay, protection may be assured by connecting a low-power diode in series with the relay coil.

9.2   DIODE TRANSIENT SUPPRESSORS

A common use of semiconductor diodes is the suppression of transients caused by the opening of relay control circuits. When the operating current of a relay is interrupted, the energy stored in the magnetic structure produces a voltage across the coil that is about 100 times the original operating voltage. A diode connected as shown in Fig. 9.4 turns ON when the switch is opened and provides a conduction path for the coil current so that the magnetic energy is dissipated more slowly in the dc resistance of the coil. The diode holds the terminal voltage to about one-half volt and must be rated to carry the relay operating current.

A good rule is to use diode suppressors on all electromagnetically operated devices, either to prevent transients in sensitive amplifiers or to prevent high-voltage damage to the magnetic devices themselves. Power

Figure 9.4. *Inductive transient suppressor.*

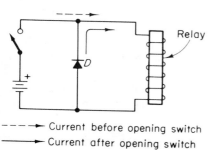

- - - → Current before opening switch
——→ Current after opening switch

Figure 9.5. *Transient suppressor diode.*

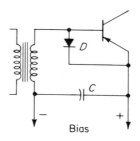

transistors are sometimes protected by connecting a diode across the base-emitter junction, as shown in Fig. 9.5. When the transistor is transformer-coupled to a driver stage, large signals may reverse bias—i.e., open—the base-emitter diode and produce destructive voltages or annoying transients. Small transformers have enough resistance in the winding to protect the transistor, but a suppressor diode should be used whenever the transformer has a large core and a low winding resistance. Sometimes the diode D is replaced by a Zener diode to limit the forward drive. By limiting the forward drive, the Zener diode protects the transistor from most collector circuit overloads.

9.3 DIODE METER PROTECTORS

A sensitive d'Arsonval type of dc meter usually reads full scale when the terminal voltage is about 0.1 V. A pair of silicon diodes connected back-to-back across the meter, as shown in Fig. 9.6, presents a high resistance up to full scale readings of the meter and introduces a negligible error. If a high voltage is accidentally applied to the metter, the forward biased diode limits the terminal voltage to about 0.6 V—a low multiple of the full scale voltage, even for input currents of several amperes. A sensitive meter cannot be protected by a fuse, but the shunt diodes increase the overload current exponentially and will open the fuse in time to protect the meter. In effect, the diodes act as fast and reliable switches.

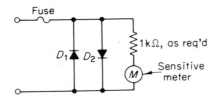

Figure 9.6. *Diode meter protector.*

9.4 DIODE STEERING CIRCUITS

Diode steering circuits are found in communication and indicator applications where a signal is to be steered to one or another load, depending on the signal polarity. On the other hand, the full-wave diode bridge circuit shown in Fig. 9.7 uses diodes as polarity-sensitive switches to ensure that the input voltage is always connected to the load with the same polarity. The diode bridge may be used where a circuit is to be triggered regardless of the incoming signal polarity, or where an instrument is to be operated without specifying the polarity of the power supply.

The rectifier-type voltmeter shown in Fig. 9.8 illustrates the use of a diode bridge to steer an ac signal to a dc load so that the signal is inverted

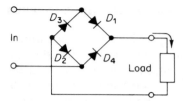

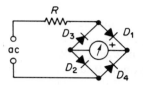

Figure 9.7. *Diode steering bridge.*   **Figure 9.8.** *Rectifier-type voltmeter.*

every other half cycle. The voltage to be measured is applied through the range resistor R to opposite corners of the bridge. The dc meter reads the average of the rectified current, but the meter is usually calibrated to read the rms value of a sine wave. As a result, such instruments do not read correctly when used with nonsinusoidal waveforms. Because each diode requires about one-half of one volt for switching the diode, steering and rectifier circuits are not effective at low signal levels. The lowest range on a rectifier-type voltmeter is usually about 2 V, and this range requires a specially calibrated scale. A filter capacitor is not required across the meter because the meter responds too slowly to follow the ac components of the rectified signal.

9.5 VOLTAGE DOUBLER

An ac rectifier can be made to deliver a no-load voltage that is nearly twice the peak ac input voltage by the circuit shown in Fig. 9.9(a). Depending on the ac polarity, the diodes switch the transformer alternately to each of the series-connected capacitors, and each is charged to the peak input voltage. The circuit is characterized by poor regulation unless the capacitors are very large, and the diodes must withstand an inverse voltage that is twice the peak ac input voltage. Similar diode switching circuits, called **voltage multipliers**, produce high multiples of the input voltage. A voltage quadrupler is illustrated in Fig. 9.9(b).

Figure 9.9. (*a*) *Diode voltage doubler.* (*b*) *Voltage quadrupler (doubler shown by solid lines).*

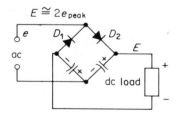
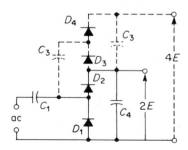

9.6 DIODE MODULATORS AND CHOPPERS

Circuits carrying low-level signals are often switched by using diodes driven by a separate power source. The circuit shown in Fig. 9.10 shows a diode bridge that connects the input to the output, either through D_1 and D_2, or through D_3 and D_4, depending on the polarity of the square wave switching signal. If the dc is applied as an input e_0 on the right, the switching signal converts the dc to an ac signal that can be stepped up by the transformer and amplified in an ac amplifier. These circuits are called *diode choppers* or *modulators*. The high-frequency output signal after amplification or transmission over ac circuits is easily converted back to dc or a low frequency by passing the ac through a second chopper that switches the signal back to the original form. In some applications these circuits may be referred to as *lock-in modulators* or *synchronous rectifiers*. Because an input signal that is 90° out-of-phase with the switching signal does not produce a dc output, such rectifiers may be used to distinguish the phase of a signal and then are called *phase-sensitive rectifiers* or *demodulators*.

Modulators use nonlinear components to change the frequency band of

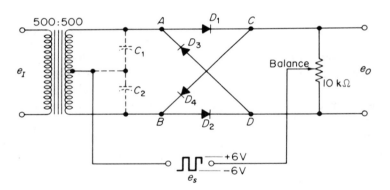

Figure 9.10. *Synchronous modulator or rectifier.*

a signal. If the input signal e_I in Fig. 9.10 is speech, as from a microphone, and the switching signal e_s is a radio frequency, then the output signal includes a radio frequency signal that is modulated—i.e., changed—by the speech and is suitable for radio transmission. Modulators are discussed in greater detail later, but a fact of interest now is the difficulty caused by the voltage drop in the forward biased diodes. Unless the diodes are carefully matched and the circuit components are balanced, the desired output signal is concealed by a signal at the switching frequency caused by the unbalanced diode voltage drops. This difficulty is substantially eliminated by using FETs in place of the diodes.

9.7 SYNCHRONOUS OR LOCK-IN DETECTORS

The FET bridge circuit shown in Fig. 9.11 is selected to explain the characteristics and usefulness of a synchronous detector. Actually, the FETs have brought about a great improvement in the characteristics of choppers, modulators, and synchronous detectors. In replacing diodes, the FETs have eliminated the problems caused by the diode voltage drops because channel conductance is controlled by the gate without superimposing an appreciable current in the bridge. With a 10 V peak gate signal the residual unbalance is less than 10 mV rms, and input signals are usable from 10 mV to 10 V.

A synchronous detector is ordinarily a diode bridge that is switched at a fixed synchronizing frequency. Synchronous detectors have important applications in the detection of weak signals at the synchronous frequency in a background of noise. The characteristics of synchronous detectors can be illustrated by assuming that the switching signal e_s shown in Fig. 9.12 is a 60 Hz square wave. If the input signal e_I is a 60 Hz sine wave, then the gate output signal e_o is the result of the inversion of the input signal every other half cycle, as shown in Fig. 9.12. Observe that the output signal has a dc component when e_I and e_s are in-phase and no dc component when they are 90° out-of-phase.

Suppose now that the input signal is a 59 Hz sine wave while the switching signal remains a 60 Hz square wave. Because the signals differ in frequency, the dc component varies with time and is a low-frequency sine wave having the difference frequency, 1 Hz. If the low-frequency output signal is now observed by connecting a long-period dc meter to the output, the meter follows the output signal as long as the frequency difference is

Figure 9.11. FET synchronous detector.

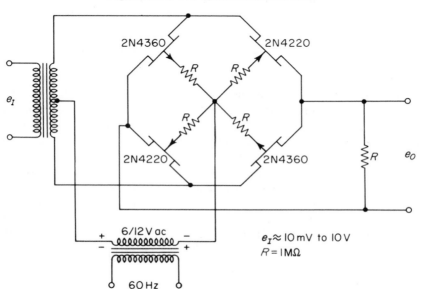

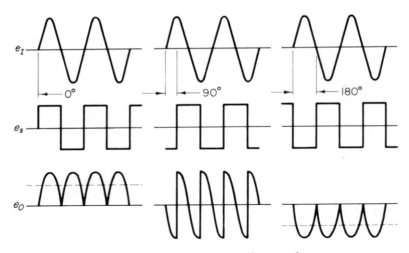

Figure 9.12. *Synchronous rectifier waveforms.*

small. However, if the input frequency and the switching frequency are very different, a long-period meter cannot follow the output signal. For this reason any input signal that deflects the meter must fall in a narrow frequency spectrum centered at the switching frequency. A typical 50 μA ammeter used as the output indicator responds to a 3 Hz bandwidth. If the reference frequency is 60 Hz, then the meter responds as if the system had a Q of $\frac{60}{3}$, or 20. If the reference frequency is raised to 600 Hz and a galvanometer is used that has a 10 times longer period, then the system has an effective Q of 2000. By synchronous detection techniques, signals deeply buried in noise are retrieved and measured with accuracy. Lock-in amplifiers (***correlators***) are used in radio telescopes, in space probe telemetry, and in research laboratories.

9.8 LOGIC SWITCHING CIRCUITS (Ref. 12)

Logic circuits are used in machine controls and computers to produce a desired action without outside help whenever a group of input signals, or controls, attains a prescribed condition. Logic circuits are also important components of much simpler systems. A "3-way" light switch is an example in which a light is operated when both switches A ***and*** B connect to a common wire.

As an elementary example of a logic circuit consider the one shown in Fig. 9.13. The electric solenoid, or a similar device, is operable both by a foot switch S_1 and a manual switch S_2. By connecting the warning lamp and diode as shown, the lamp indicates operation of the manual switch and is not affected by the foot switch. The lamp studies certain facts and reports the

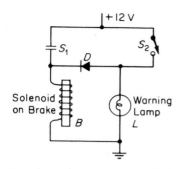

Figure 9.13. *Elementary logic circuit.*

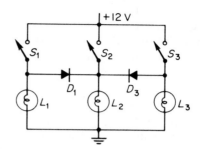

Figure 9.14. *Diode logic circuit.*

logical conclusion. If the lamp is connected in parallel with the solenoid, the lamp reports a different logical conclusion concerning the switches.

An interesting logic circuit using diode switches is shown in Fig. 9.14. The circuit shows 3 lamps that are turned ON by 3 switches. Although lamps are shown, the loads could be relays or a mixed group of electrically-operated devices. Each manual switch operates its associated load, and the side switches S_1 and S_2 also operate the center load. The center switch does not operate either side load. If the polarity of the battery is reversed, the logic is entirely different because the center switch is then able to operate the side loads and each side switch operates only its own load. If the battery is replaced by an ac source, each switch supplies half power to the neighboring load. As an application of this kind of circuit we might imagine that the center load is a blower and the side loads are two different heaters. The logic operation ensures that the blower is ON whenever a heater is ON but the blower can be operated by itself.

Ready-made integrated logic circuits are available in several classes of logic elements:

1. DTL (diode-transistor logic)
2. DCTL (direct-coupled transistor logic)
3. RTL (resistor-transistor logic)
4. TTL (transistor-transistor logic)
5. ECL (emitter-coupled logic)

Each class of logic has advantages and disadvantages which are determined by the operating speed, fan-in and fan-out capability, noise immunity, power requirements, and versatility. The operating speed determines the amount of work a computer can accomplish per unit time. Fan-in represents the number of inputs that can be connected to a logic network that has one output. Fan-out represents the number of networks that can be driven from one output. Noise immunity and versatility are obvious advantages, and the power requirements affect both the operating cost of a large computer and the life of the integrated circuits. Most computers use TTL, but ECL is faster and more difficult to implement.

9.9 GATES

The logic circuits in computers, broadly referred to as *gates*, are fabricated in many different ways. As an example of diode logic circuits two diode gates are shown in Figs. 9.15 and 9.16. The OR gate in Fig. 9.15 has 3 inputs but only 1 output. The switches represent the input or command signals that are either 0 V or +6 V. In a computer the signals are probably obtained by connecting the diodes to the collectors of 3 different flipflops in which 0 V (switch down) represents the no-signal condition. The output at *A* is switched from 0 V to +6 V if any one of the switches is up. Because the gate is opened by any one of the inputs, *A or B or C*, this is called an OR gate. The grounded diodes are back-biased and nonconducting, so the circuit current is limited by the resistor.

In Fig. 9.16 the gate has 3 inputs; the output at *B* is about +0.6 V as long as any one of the switches is down, because the current supplied by the 1 kΩ resistor forward biases the diode(s). However, when all the switches are up, the output is +6 V. The gate is called an AND gate because it is opened only by applying all 3 inputs, *A and B and C*.

Simple OR and AND gates are rarely used because of the need to separate the input circuits, one from another, by introducing a power loss in each input. Moreover, logic operations require inverters and these provide power gain which reestablishes the switching level. Therefore, a gate is always followed by a transistor inverter. A closed OR gate followed by an inverter produces a normally ON signal, +6 V, because the transistor is OFF. Opening the gate produces a *not*-ON output. For this reason the diode-transistor-logic (DTL) gate in Fig. 9.17 is called a NOT-OR or NOR gate and the DTL gate in Fig. 9.18 is called a NOT-AND or NAND gate. In a computer, *not*-ON command signals are as readily available as ON signals. Therefore, there is no problem in using the NOR and NAND logic. However, inverting a command signal converts a NOR gate to an AND gate and inverting the output converts an AND gate to a NAND gate. For this reason a logic system often has only NOR or NAND gates. Discrete circuits generally use NAND logic, while with IC's the kind of logic does not concern the user.

The logic of a circuit may be analyzed by (1) defining the kind of input signal that represents a 1 (ON); (2) determining whether the input switching that produces the 1 input is OR or AND; and (3) determining whether the 1 input produces a 1 output or a NOT 1 output. If the 1 input produces a 1 output, the logic is OR or AND. If the 1 input produces a NOT 1 output, the

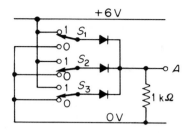

If one input (OR) is 1 (ON) an
output 1 exists at *A*.

Figure 9.15. *OR logic.*

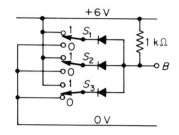

If all (AND) inputs are 1 (ON)
an output 1 exists at *B*.

Figure 9.16. *AND logic.*

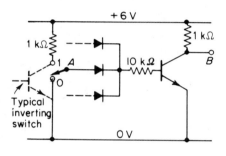

If one input (OR) is 1 (ON) an output 0
(NOT) exists at *B*

Figure 9.17. *NOR gate.*

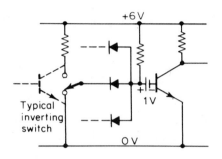

If all inputs (AND) are 1 (ON) an
output 0 (NOT) exists at *B*

Figure 9.18. *NAND gate.*

logic is NOR or NAND. The signal representing the output may be chosen to be quite different from the 1 input, in which case a distinction between OR and NOR, or AND and NAND, may be either arbitrary or unimportant.

The diodes in a logic circuit are devices that separate a group of command signals and change the degree of separation in response to the signal polarity. Logic circuits can be constructed by using resistors or transistors to separate the command circuits, or by using combinations of these elements. Because the integrated circuits of modern computers are fabricated with materials and techniques of the transistor manufacturer, the logic circuits of computers therefore use either diodes or transistors. Transistor logic circuits are more difficult to design and to maintain but have the advantages of fewer components, higher switching speeds, and higher power efficiencies. Insulated gate MOS transistors offer similar advantages and show great promise of replacing transistors in the integrated circuit computers.

Examples of two direct-coupled transistor logic (DCTL) circuits are shown in Fig. 9.19. In Fig. 9.19(a) four transistors are coupled to a common

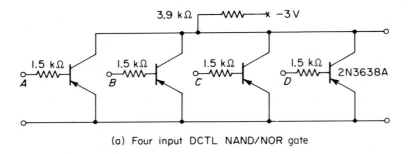

(a) Four input DCTL NAND/NOR gate

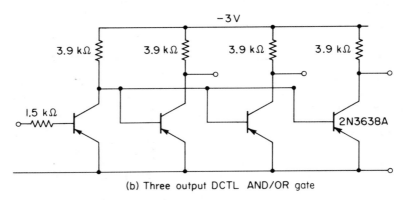

(b) Three output DCTL AND/OR gate

Figure 9.19. *Direct-coupled transistor logic.*

load to provide a fan-in of four. Suppose that all the input terminals *A, B, C,* **and** *D* are grounded. Because none of the transistors is conducting, the output terminal is at the +6 V supply voltage and is **not** grounded. If now *A, B, C,* **or** *D* is connected to +6 V, the output is switched to ground and, therefore, the circuit produces NOT-OR or NOR logic. If *A, B, C,* AND *D* are normally connected to +6 V, then *A, B, C,* **and** *D*—all the inputs—must be grounded to switch the output to +6 V. Therefore, this circuit has NOT-AND, or NAND logic. If the logic circuit of Fig. 9.19(a) is constructed as an integrated circuit on a single silicon chip, the resistors may be fabricated by making the base very thin. The circuit in Fig. 9.19(b) illustrates a single-input, noninverting, direct-coupled gate with a fan-out of three.

9.10 A DIGITAL COMPUTER

Digital computers solve problems in which the information can be represented by digits. The computer performs logical and arithmetical operations according to a program of instructions. Digital computers are distinguished by a speed, accuracy, and versatility that enable them to perform most data-handling operations.

An electronic digital computer represents the numerical values in a calculation by the discrete states of flipflops and switching circuits of the computer. Calculations are carried out with binary numbers by representing the binary numbers 0 and 1 as the two states of a switch—OFF and ON, respectively. A large scale computer employs an incredible number of semiconductor devices and an interesting complex of switching circuits.

The essential parts of a digital computer are:

1. Arithmetic systems that perform numerical operations, mainly by counting and summing. (Multiplication, division, and integration are performed by addition and temporary storage.)
2. Storage units that hold numerical and coded information as binary numbers in flipflops.
3. Inputs for reading the information supplied by the user and for converting data from analog-to-digital (A/D) form.
4. Outputs for presenting information in a usable form that may be printed out or presented as an analog signal by a digital-to-analog (D/A) converter.
5. Control systems that determine the sequence of computer operations. (Control information is in digital form and is processed by the same kind of logic and storage circuits used in computing.)
6. Memory, which usually includes temporary storage in flipflops or magnetic cores, and bulk storage on magnetic drums or tape.

9.11 NUMERICAL CONTROL

Electronic switching, logic, and storage are often used to control machines and industrial processes. The control information is coded in numerical or binary form for conversion into process commands by a computer especially designed to meet the needs of a complex machine or industrial process.

An elementary example of a machine control may be illustrated by a relay which is closed by a microsecond pulse, provided an AND gate sees that certain other switches are properly operated.

The circuit in Fig. 9.20 makes it possible to close a slowly operating relay in response to a microsecond pulse. The transistors operate as a monostable flipflop or as a bi-stable lock-up, depending on the position of the switch S. If the switch is open, an input pulse turns on TR-1 and TR-2, which then closes the relay. The sudden rise of the relay voltage causes a bias current to be supplied through the capacitor C and R_4, which holds the transistor ON after the pulse ends. Eventually the capacitor-charging current falls, the

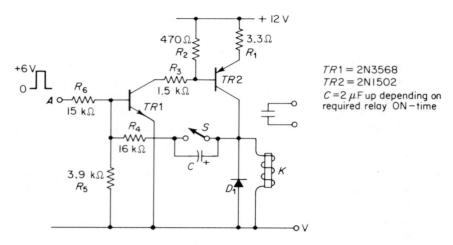

Figure 9.20. *Pulse-operated relay.*

transistor turns OFF, and the circuit returns to the initial OFF state. If the switch S is left closed, a trigger pulse locks up the relay until the switch is opened or a negative-going input pulse turns the transistor OFF. Negative-going pulses can be prevented from turning the relay OFF by replacing the resistor R_6 by a diode.

The addition of logic gates to control the pulse-operated relay is illustrated by the circuit in Fig. 9.21. The switch S_2 and the RC elements connect-

Figure 9.21. *Gated relay amplifier.*

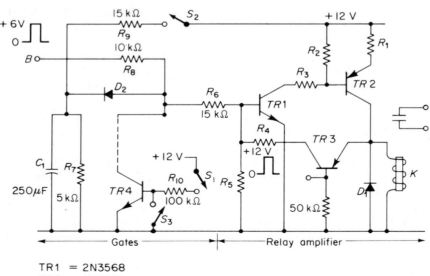

TR1 = 2N3568
TR2 = 2N1502
TR3 = 2N3638
TR4 = 2N3568

ing to diode D_2 are a means of delaying the triggering of the relay circuit until a time delay after S_2 is closed. This lock-out circuit prevents circuit turn-ON transients from operating the relay prematurely. The circuit attenuates the input pulses by shunting C_1 across the pulse input circuit. After S_2 is closed, the capacitor C_1 charges and the diode D_2 is back biased, so that C_1 is removed a second later as a shunt. The transistor TR-4 is included to show that a transistor can be used to short the input pulse until either S_1 is opened or S_3 is closed. Taken together, the resistor R_8, the diode D_2, and the transistor TR-4 comprise an AND gate.

The transistor TR-3 in the relay lock-up circuit is a voltage-operated transistor switch that replaces the mechanical switch S in Fig. 9.20. The transistor turns ON the lock-up bias except that application of a positive-going pulse will switch TR-1 OFF and release the relay.

When the gated relay amplifier is used for machine control, the pulse signals may be supplied by a magnetic tape that is controlling the machine, and the switches may be limit controls to prevent faulty operation or safety devices to protect an operator.

9.12 COMPLEMENTARY MOS LOGIC

The space saving and low-power advantages of CMOS devices have already been described in the previous chapter, and for the same reasons many logic circuits are constructed using CMOS devices. The circuit of a CMOS NAND logic switch is shown in Fig. 9.22. A study of the circuit shows either one or the other of the series connected p-channel FETs or both of the parallel-connected n-channel devices are always open except during the switching instant, which means power is used only for switching. The circuit also shows that FET terminals are shared by two or three devices so there can be space-saving advantages in the manufacture of the circuit by IC techniques.

The high impedance characteristic of the MOS gate gives the designer a

Figure 9.22. CMOS NAND gate.

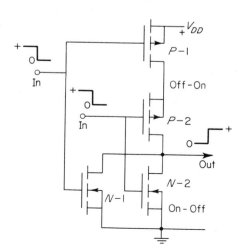

new freedom to fan into several hundred gates, and a similar freedom to design circuits without resistors. Because the CMOS circuits require low-operating power they may be designed with relatively high supply voltages and with MOS devices that need a relatively high switching signal. Thus, CMOS logic may be designed to have a relatively high immunity to noise signals, or, when noise is not a problem they may be designed to operate with only a 1.5 V supply voltage.

9.13 COMPLEMENTARY MOS MEMORY

The circuit of a MOS memory cell is illustrated in Fig. 9.23. An information bit, a 1 or a 0, is stored in the capacitance C of the gate G and the

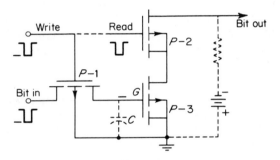

Figure. 9.23. *CMOS memory cell.*

connecting line. The capacitor is charged when P-1 is turned ON by a write signal, and the charge is normally $+0$ V or -4 V, depending on the state of the flip-flop connected to the input. Whether the bit stored in the memory cell is a 0 or 1 is read by applying a read signal to P-2, and the data output line shows whether P-3 is ON or OFF without discharging the gate capacitor. When wiring space is conserved by using a single-read-write line connected to both P-1 and P-3, over 1000 bits may be stored on a single small chip. These memory circuits cost less than one-fourth cent per bit, and the circuit operating power is about 100 μW/bit. CMOS switching and memory circuits have a relatively high noise signal immunity but a slow operating speed that may soon be overcome. CMOS circuits are often operated on a nominal 1.5 V supply and are used in electronic watches, telephone switching, pocket calculators, missiles, satellites, and medical systems.

SUMMARY

Diodes operate as voltage-controlled switches in response to forward and reverse voltages of one-half volt. Diodes are used for switching low-level signals in choppers, modulators, and demodulators. Diode rectifiers, voltage multipliers, and transient suppressors are familiar examples of diode switching circuits.

Logic circuits used in computers and machine controls may use diodes as gates for connecting and isolating a group of circuits according to a logical plan. Because logic circuits need amplifiers—which invert—the basic OR and AND logic is always inverted and operated as NOR and NAND logic.

Transistor logic circuits have the advantage of low operating power requirements and fast switching speeds. Computer logic circuits are manufactured as integrated circuits and the choice between diode logic and transistor logic is generally a decision of the manufacturer.

A digital computer uses switching, logic, and storage circuits to perform logical operations and calculations using binary numbers. Computers for numerical control compare tape instructions with process feedback signals to control a complex machine or an industrial process.

Where slower switching speeds are acceptable, the CMOS circuits are making possible the building of inexpensive computers for consumers and are bringing about important improvements of even the largest industrial computers by reducing the power requirements and increasing the memory capacity of computers.

PROBLEMS

9-1. Specify a forward current rating, a reverse voltage rating, and a power rating to be used for the diode cutout of a 10 A, 12 V battery charger.

9-2. Specify diode ratings, as in Prob. 9-1, if the load shown in Fig. 9.2 is a 100 W soldering iron. What is the rms load voltage, the average voltage, and the power input?

9-3. Explain the operation of the voltage doubler shown in Fig. P-9.3. What is the dc voltage across the input capacitor C_1?

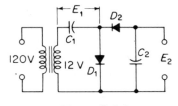

Figure P-9.3.

9-4. Describe the operation of the logic circuit, Fig. 9.13, if the battery is reversed.

9-5. Describe the operation of the logic circuit, Fig. 9.14, if the battery is replaced by a 12 V ac supply.

9-6. Analyze and describe the operation of the logic circuit shown in Fig. P-9.6.

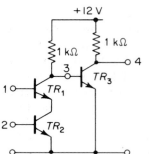

Figure P-9.6.

9-7. Analyze and describe the operation of the logic circuit shown in Fig. P-9.7.

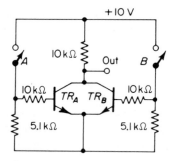

Figure P-9.7.

9-8. Sketch the waveforms obtained in Fig. 9.12 if the switching signal is at triple and double the frequency of the input signal.

9-9. Sketch the waveforms obtained in Fig. 9.12 if the switching signal is at one-half and one-third the frequency of the input signal.

9-10. An inverted command signal changes a NOR gate to an AND gate, while a NAND gate becomes an OR gate. Explain.

9-11. Refer to Fig. 9.3 and give the advantages and disadvantages of using the diode D in series with the 10 Ω base resistor instead of in series with the power supply.

9-12. Repeat problems 9.8 and 9.9 assuming the input signal e_I is a square wave.

CHAPTER TEN

Active Reactances, Filters and Tuned Amplifiers

Active circuits employ transistors to supply a power-assist for changing the gain, impedance, and frequency characteristics of a network. The simplest examples of active circuits are the amplifiers studied in the preceding chapters. In an amplifier, transistors control the flow of power supplied from a battery, thereby assisting the input signal by providing power gain in a network that otherwise would give a power loss. Transistors are also commonly used to modify, in various ways, the impedance and frequency characteristics of networks. For example, the negative feedback of an active circuit flattens the frequency response of an amplifier, and positive feedback increases the selectivity of a tuned circuit or produces an oscillator. In many secondary roles, active circuits change the magnitude or sign of impedances and produce both desirable and undesirable effects. The aim of this chapter is to show how active elements produce useful changes in the reactive impedance of a circuit and to show how these changes are used in active filters and in tuned amplifiers.

10.1 ACTIVE REACTANCES

An important example of an active reactance has been known since the early years of radio—that *the input capacitance of a vacuum tube is approximately proportional to the voltage gain of the stage*. This phenomenon, known as the *Miller effect*, is easily explained. Suppose, as shown in Fig. 10.1, an amplifier has a voltage gain of 100, an input impedance of 10 kΩ, and a capacitance C_{ob} between the collector and the base. Furthermore, assume that the capacitance C_{ob} is only 10 pF, the capacitance between about 12 cm of # 18 lamp cord. Let an ac input signal be applied between the transistor base and the ground, with the polarity of the signal indicated by a + (plus) sign on the input. Suppose also that capacitor has a negligibly high reactance compared with the collector load resistance and that the signal at the collector is 100 times the input signal and is polarized to make the collector — (minus) with respect to ground. The total voltage across the capacitor is 101 times the input signal. Hence, the current in the capacitor is 101 times the current demanded by a 10 pF capacitor connected across the input generator alone. The signal source has to supply both the amplifier input current and the current through the feedback capacitor. In effect, the signal source has to supply the reactive current of a base-to-ground input capacitor that is 101 times the collector-to-base capacitor. The phenomenon of a feedback capacitor loading the input as though the amplifier input capacitance were the feedback capacitor increased by the voltage gain of the stage is known as the *Miller effect.*

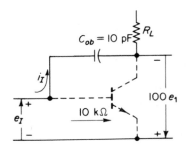

Figure 10.1. *CE amplifier showing Miller effect.*

For the specific example given (Fig. 10.1) the effective input capacitance is 101(10 pF), or 1000 pF. A reactance chart (see Appendix) shows that 1000 pF has a reactance of 10,000 Ω at a frequency of 17 kHz. In other words, the load on the signal source will drop to 7,100 Ω at 17 kHz, and the source must have a low impedance if the input signal is to be maintained above 17 kHz. Because this loading of the signal is caused by the capacitance of only 12 centimeters of wire, we see that the Miller effect is significant, even in low-frequency amplifiers. In broad-band video and RF amplifiers, the Miller effect is an important consideration for the designer and is a source of potential trouble that should be understood by servicing and applications personnel.

10.2　MILLER EFFECT CAPACITANCE

The equivalent Miller effect capacitance C_M is a base-to-ground input capacitor that loads the input in the same way a (smaller) feedback capacitor C_{ob} loads the input. If, as shown in Fig. 10.2, an amplifier has a voltage gain, G_v, then the total voltage across C_{ob} is $(1 + G_v)$ times the input voltage. The input current is $(1 + G_v)$ times larger than the input voltage alone can pro-

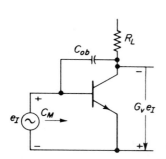

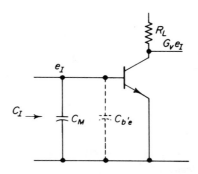

Figure 10.2. *Amplifier with feedback capacitance* C_{ob}.

Figure 10.3. *Miller effect equivalent of* C_{ob}.

duce through C_{ob}. The high input current implies that the input capacitance C_M is:

$$C_M = (1 + G_v)C_{ob} \qquad (10.1)$$

For high-gain stages, where the Miller effect is of greatest importance, C_M is given to a practical accuracy by the simpler form:

$$C_M = G_v C_{ob} \qquad (10.2)$$

The total input capacitance C_I of a stage (Fig. 10.3) is the sum of the Miller effect capacitance C_M and the total base-to-emitter capacitance C_{be}. Therefore:

$$C_I = C_{be} + G_v C_{ob} \qquad (10.3)$$

10.3 SHUNT OR MILLER FEEDBACK (Ref. 1)

The Miller effect, described above as a capacitance feedback, can be generalized to represent the effects of shunt feedback. In fact, the uses and advantages of shunt feedback are a direct result of the input loading, and a generalization of the Miller effect is of considerable usefulness.

If the shunt feedback element is a generalized impedance Z_f, then the equivalent load across the input of the amplifier can be written, following Eq. (10.2), in the form:

$$Z_M = \frac{Z_f}{G_v} \qquad (10.4)$$

When the feedback element is a pure resistance as in Fig. 10.4, the net input resistance (Fig. 10.5) is the input resistance of the stage without shunt feedback in parallel with the resistance:

$$R_M = \frac{R_f}{G_v} \qquad (10.5)$$

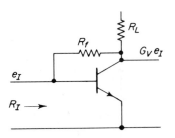

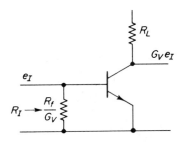

Figure 10.4. *Amplifier with shunt feedback.*

Figure 10.5. *Input impedance with shunt feedback.*

The importance of shunt feedback in determining the input impedance of an amplifier is illustrated by the CE stage shown in Fig. 10.6. The amplifier is biased by connecting a resistor from collector to base, and the capacitor at the midpoint of the bias resistor prevents ac feedback without disturbing the dc feedback. With the switch S closed, the measured voltage gain is 500 and the input impedance is about 1700 Ω. These values substituted in the TG-IR [Eq. (2.4)] indicate an approximate β value of 100. With the switch open, the input is shunted by a resistance:

$$R_M = 200 \ \Omega \tag{10.6}$$

The feedback does not change the voltage gain but does change the stage current gain by preventing the signal current from reaching the base-emitter input. Because the shunt resistance is $\frac{1}{9}$ the transistor input impedance, the

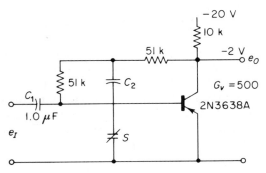

Figure 10.6. *Shunt feedback amplifier.*

effective stage current gain is approximately $\beta/10$. However, by combining the gain-impedance relation [Eq. (2.4)] with Eq. (10.5) and solving for the stage current gain, we find that the dc current gain is:

$$S = \frac{R_f}{R_L} \tag{10.7}$$

Substituting the circuit resistance values in the above equation, we get $S = 10$, which is in reasonable agreement with the effective current gain and the indicated value of β.

In summary, shunt feedback stabilizes the stage current gain by making the product of the input resistance and the voltage gain approximately constant [Eq. (10.4)]. By making the input resistance vary inversely with the voltage gain, feedback tends to make the current gain S a constant that is controlled by the ratio of two fixed resistances. Collector-to-base feedback also stabilizes the Q-point voltage by making the collector voltage so low that a change of the voltage produces an offsetting change of the current gain β and of the bias current. The collector voltage is usually well regulated if set at about 2 V at room temperature.

The gain frequency characteristic of a stage having shunt feedback is illustrated in Fig. 10.7. Curve A is the low-frequency characteristic obtained when the shunt capacitance C_2 is relatively large. The 3 db cutoff is determined mainly by C_1 and the 1700 Ω transistor input impedance, but the curve has a slope of 12 db per octave at low frequencies when both capacitors are determining the cutoff. Curve D is obtained by removing C_2, so the 3 db cutoff at 800 Hz is fixed by C_1 and the 200 Ω input impedance with feedback.

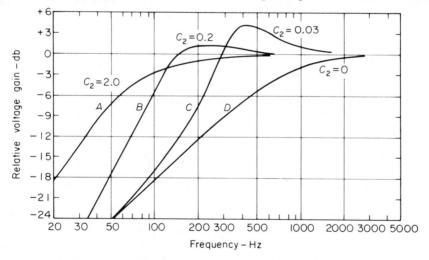

Figure 10.7. *Feedback amplifier frequency response.*

The intermediate curves B and C are particularly interesting because they show a small gain above cutoff and a steep slope in the transition from the pass-band to the cutoff region. Except for the amplifier gain, the curves A and D can be produced by a two-stage cascade of RC elements alone, but curves B and C have the characteristics of an active filter. The capacitor C_2 gives the equivalent input impedance a small inductive component which causes the amplifier gain rise and the steep slope at cutoff. Active filters generally have reactive components in the feedback network which improve the gain-frequency characteristic, just as in the example under discussion. However, because the curves depend on the amplifier gain, a well designed active filter usually has enough negative feedback to stabilize the gain and the cutoff characteristics.

10.4 RESISTANCE-CAPACITANCE FILTERS

The most common way of removing unwanted signal frequencies is by the use of a simple RC resistance-capacitance filter. Low frequencies are removed by a series capacitor and a shunt resistor, as shown in Fig. 10.8. High frequencies are removed by interchanging the resistor and capacitor, as shown in Fig. 10.9. For both filters the cutoff frequency:

$$f_c = \frac{1}{2\pi RC} \tag{10.8}$$

is that for which the reactance of the capacitor equals the resistance, the loss is 3 db, and the network phase shift is 45°. When the loss of the single-stage filter is high, the slope of the frequency characteristic is 6 db per octave, or 20 db per decade.

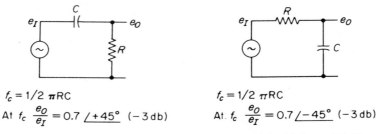

$f_c = 1/2\,\pi RC$

At f_c $\dfrac{e_O}{e_I} = 0.7\,\underline{/+45°}\,(-3\,\text{db})$

Figure 10.8. *Low-cut RC filter.*

$f_c = 1/2\,\pi RC$

At f_c $\dfrac{e_O}{e_I} = 0.7\,\underline{/-45°}\,(-3\,\text{db})$

Figure 10.9. *High-cut RC filter.*

The shunt and series capacitors used between stages of an amplifier must be correctly proportioned because they act as a filter in limiting the frequency response. A typical interstage RC filter is represented in Fig. 10.10. If, as is usually the case, the high-frequency cutoff frequency is at least 10 times the low-frequency cutoff, then the filters can be considered as independent of each other. Assuming independent filters, we find the 3 db low-frequency cutoff of the interstage shown in Fig. 10.10 is the frequency at which the reactance of the coupling capacitor C_C equals the series resistance $R_L + R_B$. Similarly, the 3 db high-frequency cutoff occurs when the reactance of the shunt capacitor C_S is equal to the resistance of R_L in parallel with R_B. In general, the input resistance and capacitance of the second transistor $TR2$

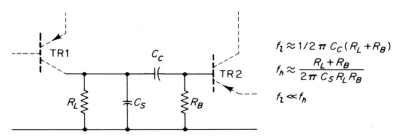

Figure 10.10. *Amplifier interstage RC filter.*

cannot be neglected in calculating the cutoff frequencies, so a calculated cutoff frequency should always be checked by a measurement.

10.5 TWO-STAGE RC FILTERS

Simple RC filters are inadequate for many purposes because of their very gradual transition from the passband to the cutoff region. A more rapid transition at cutoff is obtained by using a two-stage RC filter. If the cutoff frequency is defined again as the frequency at which the reactance of the capacitor equals the resistance, then the two-stage filter has a 90° phase shift at cutoff and the slope is 12 db per octave in the high loss region. RC filters are cascaded either as shown in Fig. 10.11 or by separating a series of single-stage filters with transistor stages.

Figure 10.11. *Two-stage RC filter.*

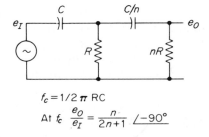

The frequency characteristic of a two-stage RC filter is shown in Fig. 10.12; also shown for comparison is the frequency characteristic of a single-stage RC filter. The curves represent the characteristics of both high-pass and low-pass filters. As the curves show, all the RC filters have a gradual transition from the pass-band to the cutoff region. The principal improvement obtained by cascading is the increased slope a decade removed from the cutoff frequency. Although the impedance level of the second stage is often made 2 to 5 times ($n = 2$ to $n = 5$) that of the first stage, the curves show that little real advantage is obtained.

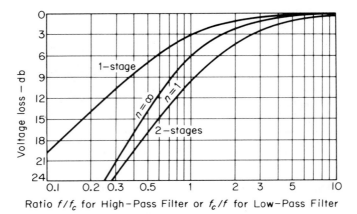

Figure 10.12. *RC filter frequency response.*

In general, passive filters provide a sharp transition at cutoff only if the filters have low-loss, high Q inductors. High Q inductors suitable for audio-frequency designs are bulky, heavy, and expensive. Small inductors and transformers are usually of questionable value and require shielding to minimize 60 Hz pickup by the magnetic cores. However, active filters using simple RC circuits are capable of duplicating the performance of high Q LC filters.

10.6 TUNED CIRCUITS AND THE Q FACTOR

Circuits that transmit a narrow frequency band and reject all other frequencies are called **tuned circuits**. Such circuits are used where a signal must be separated from noise, static, or other signals. Because the rms voltage of electrical noise is usually proportional to the bandwidth of a signal channel, the noise is minimized by limiting the bandwidth as much as possible. On the other hand, a given signal cannot be transmitted in too narrow a band without destroying part of the signal. For these reasons the bandwidth of a tuned circuit is determined by the needs of the signal channel.

A discussion of the relative bandwidth of a tuned circuit is simplified by using the term Q to describe the sharpness of tuning. *Q is defined as a measure of the quality of an inductor or capacitor.* However, the quality of a tuned circuit can be measured by comparing the peak response frequency to the bandwidth at points 3 db down from the peak response. As indicated in Fig. 10.13, the Q of a tuned circuit is given to a close approximation by the relation:

$$Q \simeq \frac{f_0}{BW} \tag{10.9}$$

where f_0 is the center frequency and BW is the bandwidth at the 3 db points.

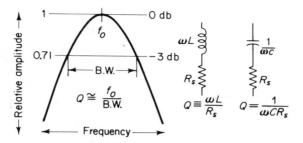

Figure 10.13. *Q curve and formulas.*

If the center frequency is 1000 Hz and the bandwidth is 50 Hz, the tuned circuit Q is 20. The Q factors of tuned circuits are ordinarily between 2 and 100.

The Q of a tuned circuit is usually controlled by the energy loss in the inductance. The ratio of the inductive reactance ωL to the equivalent series resistance R_s of the inductance is known as the Q, or figure of merit, of the coil. Algebraically:

$$Q \equiv \frac{\omega L}{R_s} \qquad (10.10)$$

The Q of an inductance varies with frequency but usually has a broad maximum, relatively independent of frequency, which is a useful design parameter. The Q of inductances varies from 1 to about 400. The Q of capacitors varies from about 100 to 10,000. The Q of an electromechanical resonator may be much higher.

10.7 TUNED AMPLIFIER WITH Q CONTROL

The tuned amplifier shown in Fig. 10.14 is of unusual interest. The Q of the amplifier can be changed from 2 to 10 by the Q control without changing the peak frequency appreciably, and the peak frequency is shifted over only a small range by changing the collector supply voltage. The amplifier behaves as a negative resistance at the peak frequency and produces an output voltage, e_o, that is about Q times the input voltage e_I. Because this amplifier makes e_o larger than e_I, the negative resistance acts as a voltage in series with e_I that supplies power at the terminals x-x. Since any impedance within the generator e_I determines the Q in part, a resistor must be used in series or in parallel with the generator so that the Thevenin equivalent of the source is 5 kΩ.

Removing the generator allows the amplifier to oscillate, and a sine wave output can be obtained by reducing the collector voltage to about 6 V. Surprisingly, an oscillator can be constructed using an emitter follower and RC coupling because the RC network gives a voltage gain of about 1.03. The transistor should have a high current gain at a collector current of 1 mA.

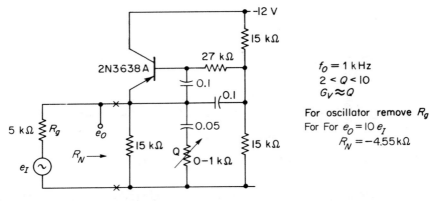

Figure 10.14. *Active tuned amplifier with Q control or oscillator.*

10.8 ACTIVE FILTERS

Because of their small size, high transconductance, and circuit simplicity, transistors and FETs make possible many new types of filter circuits. Impedance levels within a filter are easily transformed, or a small reactor may be made to serve in place of a large reactance, as in a capacitance multiplier. With transistors, the designer has at his disposal negative impedances and negative resistance. The impact of these many possibilities is only just beginning to be realized, partly because the design of these circuits is difficult.

The main reason for using active filters is that transistors either make possible the reduction of the resistance losses of a filter or the obtaining of inductive reactances without using real inductances. The possibility of constructing low-frequency filters and oscillators without bulky inductors is particularly attractive.

Active filters are constructed by using a high-gain operational amplifier with a complicated feedback network or by using a cascade of low-gain stages with simple RC feedback networks. The operational amplifier offers high gain and stability, but the feedback networks must be carefully designed to avoid creating instability problems. Some of the advantages of operational amplifiers can be attained by using FET-transistor feedback pairs to provide more gain than is available from a single transistor. The cascaded stages with simple feedback networks are stable, easily fabricated, and the filter characteristics are independent of the collector supply voltage. The circuits of this chapter usually show a single transistor as the active device. With minor changes in the circuit FETs may often be substituted for the transistor and bring the advantages of simpler bias circuits, higher impedance levels, and smaller capacitors.

10.9 ACTIVE HIGH-PASS FILTER

The active high-pass filter shown in Fig. 10.15 employs a two-stage cascaded RC filter between the input and the base of an emitter follower. By connecting the resistor of the input stage to the emitter output terminals, the filter uses feedback to augment the input signal and thereby to increase the output signal at the effective cutoff frequency. In fact, curve *A* shown in Fig. 10.16 shows that the filter and the emitter follower together have a voltage gain of about 1.3. A study of the circuit shows that the active filter has the correct phase relations for an oscillator but lacks the required loop gain. In other words, the filter characteristics are obtained by the use of a limited amount of positive feedback. Because the transistor gain cannot exceed 1,

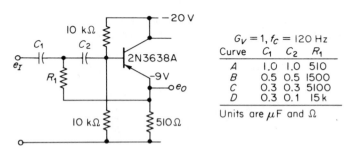

$G_V = 1, f_c = 120$ Hz

Curve	C_1	C_2	R_1
A	1.0	1.0	510
B	0.5	0.5	1500
C	0.3	0.3	5100
D	0.3	0.1	15 k

Units are μF and Ω

Figure 10.15. *Active high-pass filter.*

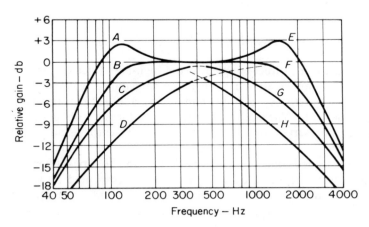

Figure 10.16. *Active filter frequency characteristics: high-pass A through D; low-pass, E through H.*

these filters are insensitive to the collector voltage, even though there is positive feedback.

The filter represented by curve *B* provides a monotonic response having the sharpest break at cutoff. Curve C is obtained from a filter that has very little effective feedback and so is much like a two-stage RC filter. The differences between the curves are caused mainly by differences in the amount of feedback.

10.10 ACTIVE LOW-PASS FILTER

The low-pass filter shown in Fig. 10.17 has a two-stage cascaded low-pass RC filter in the input but is otherwise similar to the high-pass filter described in the previous section. The curves of both filters have broadly similar characteristics, and an explanation of the low-pass filter is substantially the same as that of the high-pass filter. An important difference between the two filters comes from the fact that the low-pass filter has a dc path between the input terminal and the transistor base. The filter, as shown, requires a dc bias voltage at the input terminal that is about one-half the collector supply

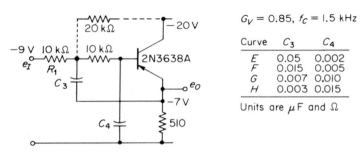

$G_V = 0.85$, $f_c = 1.5$ kHz

Curve	C_3	C_4
E	0.05	0.002
F	0.015	0.005
G	0.007	0.010
H	0.003	0.015

Units are μF and Ω

Figure 10.17. *Active low-pass filter.*

voltage and can be direct-coupled to the high-pass filter of Fig. 10.15. The lowpass filter can be driven from a low-impedance, grounded source by changing the resistor R_1 to 20 kΩ and adding a 20 kΩ bias resistor, as indicated in Fig. 10.17. The suggested change introduces a 6 db loss but provides the bias and preserves the 10 kΩ Thevenin equivalent resistance of the input stage.

The cascaded connection of the low-pass and high-pass filters provides a band-pass response curve by using the filters represented by the curves B and F. A tuned circuit characteristic is obtained by adjusting filters A and E to peak at the same frequency, but similar characteristics are obtained more simply by a single-stage, active, tuned circuit.

10.11 CE STAGE FILTERS

The single-stage filters shown in Figs. 10.18, 10.19, and 10.20 use CE stages that have a no-load voltage gain of about 20. Depending on the RC

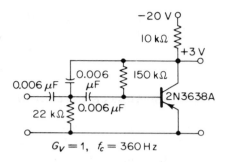

Figure 10.18. *Active high-pass filter.*

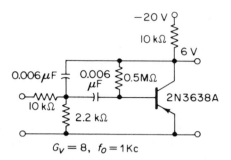

Figure 10.19. *Active tuned amplifier.*

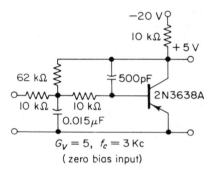

Figure 10.20. *Active low-pass filter.*

feedback networks, the filters provide a high-pass, low-pass, or tuned-amplifier gain characteristic. The curves in Fig. 10.21 show the normalized gain-frequency characteristics of each filter, with 0 db representing the overall gain in the pass-band, as given in the circuit diagrams. A band-pass characteristic is obtained by cascading the low-pass and the high-pass filters.

The cutoff frequency of all the active filters may be changed from 2 to

Figure 10.21. *Active filter response curves.*

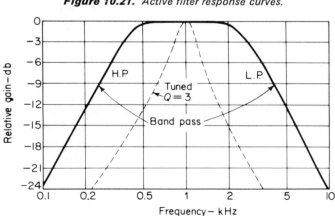

3 decades by changing the capacitors by a common factor. Changes of the cutoff frequency up to about a 3 : 1 change may be obtained, without seriously changing the shape of the curves, by changing the resistors of the input network by a common factor.

10.12 IMPEDANCE INVERSION

Either by design or by accident, transistors often cause a capacitance to appear as an inductance in another part of a circuit. An impedance that is the reciprocal of a pure capacitance is a pure inductance, and vice versa. Impedance inversions are produced by a combination of high gain and feedback.

As an illustration of an impedance inversion, consider the amplifier in Fig. 10.22. It has a collector-to-base feedback resistor and a collector load that is almost a pure reactance. If the collector load is the reactance X, the voltage gain is $G_v = g_m X$. The collector-to-base feedback resistor causes the input impedance, by Eq. (10.4), to be:

$$Z_{in} = \frac{R_f}{g_m X} \tag{10.11}$$

Now, if the load is a pure reactance, Eq. (10.11) states that, except for a constant multiplier, the input impedance is an inversion of the collector load impedance.

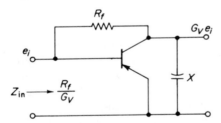

Figure 10.22. Reactance inverter.

The magnitude of an inverted impedance is more easily estimated by changing Eq. (10.11) to a polar form that expresses the voltage gain of the stage as a magnitude with the impedance angle stated. In polar form the input impedance of Eq. (10.11) becomes:

$$Z_{in} = \frac{R_f}{G_v \angle -90°} = \frac{R_f \angle +90°}{G_v} \tag{10.12}$$

Equation (10.12) states that shunt feedback reflects a capacitance load back to the input of a stage as a shunt inductive reactance. When this inductive input impedance is combined with the shunt input capacitance, an amplifier may have a very high input impedance and may exhibit transient ringing or a tendency to oscillate. In an amplifier having two or more CE stages and a high voltage gain, the cumulative effect of feedback in each stage, or of overall

feedback, may be sufficient to produce positive feedback and sufficient negative resistance to convert an amplifier to a persistent oscillator.

An impedance inversion is illustrated by the amplifier shown in Fig.
10.23. The amplifier has collector-to-base feedback and is driven from a high
impedance, or current, source. The collector load is the 0.1 μF capacitor
which, at the frequency of interest, has a reactance of 1500 Ω. This reactance
is about $\frac{1}{10}$ the collector circuit resistance, so that the load is almost a pure
capacitance. A measurement of the signal voltages shows that the amplifier
responds as if a tuned circuit shunts the base. The frequency of resonance is
1100 Hz, and the base-to-collector voltage gain is 120 at resonance. The
estimated input resistance of the transistor alone is 1000 Ω.

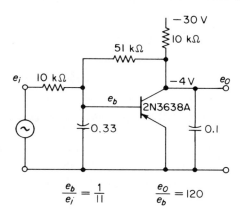

Figure 10.23. *Impedance-inverting amplifier.*

An equivalent circuit which represents the tuned circuit across the
amplifier input is shown in Fig. 10.24. By Eq. (10.12) the equivalent shunt
inductance has a reactance of 425 Ω. A reactance chart shows that the
equivalent inductance and the 0.33 μF capacitor should resonate at a frequency very close to the observed 1100 Hz. The transistor input resistance
limits the circuit Q to about 2.

Figure 10.24. *Equivalent tuned circuit.*

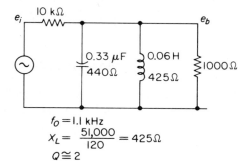

10.13 NEGATIVE RESISTANCE

Feedback that returns power in phase with an input signal reduces the input power and is equivalent to a negative resistance placed across the input. The amplifier in Fig. 10.25 shows how a negative resistance may be produced by collector-to-base feedback in a single-stage amplifier. We assume the amplifier load is small compared with the collector resistance and that the reactance X_C of the feedback capacitance is high compared with the reactance X_L and the input resistance R_I. With these assumptions the current in X_L is

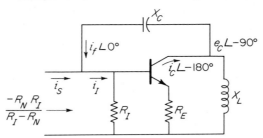

Figure 10.25. *Single stage with negative resistance.*

90° out of phase with the input current i, and the feedback current i_f is in phase with the input current. When the loop gain is 1, $i_f = i$, and the voltage e is maintained by i_f with $i_s = 0$. Thus, the feedback acts as a negative resistance $R_- = -R_I$ when

$$S\frac{X_L}{X_C} = 1 \tag{10.13}$$

Solving Eq. (10.13) for R_- we find

$$R_- = -\frac{X_C}{X_L}R_E \tag{10.14}$$

Low values of negative resistance are the most troublesome in the circuit under study, and Eq. (10.14) shows that R_- varies inversely with the frequency squared as long as the initial assumptions are valid. The equations suggest that the negative resistance effects in amplifiers may be controlled by reducing the feedback capacitance and the amplifier gain until the loop gain is significantly less than 1. Carefully designed negative resistances are sometimes used to reduce the resistance loss in a circuit or to increase the signal in a transmission line.

10.14 OPERATIONAL AMPLIFIERS

Operational amplifiers are named from early applications in which they were used to perform mathematical operations (addition, multiplication, integration, etc.). An **operational amplifier** is a high-gain, direct-coupled, differential amplifier that is designed to provide a high degree of stability and considerable freedom in the application of overall feedback. Operational amplifiers are commonly used as components of analog computers, but the availability of inexpensive integrated circuit amplifiers has made the packaged operational amplifier useful as a replacement for almost any low-frequency amplifier. Operational amplifiers are now used in instruments, data processing equipment, and in the electronic portions of closed-loop control systems or servo mechanisms. Operational amplifiers are commonly known as OP amps.

Operational amplifiers are fabricated as a cascade of differential stages having common mode rejection. Hence, they require both positive and negative power supplies. A differential amplifier has two inputs and two outputs. Therefore, the amplifier provides phase inversion for degenerative feedback and can be connected to provide in-phase or phase-inverted amplification. Because high-gain broad-band amplifiers oscillate when feedback is applied, an operational amplifier is usually stabilized by a built-in RC cutoff. With a single shunt capacitor the high-frequency cutoff is made a simple 6 db/octave, limiting the internal phase shift to 90°. As long as the phase shift of the external feedback network does not bring the total loop phase shift too close to 180°, stable operation is assured. An amplifier which has an open-loop gain of 10^6 and is rated to provide a gain bandwidth product of 1 MHz has a built-in high-frequency cutoff beginning at 1 Hz. If, as illustrated in Fig. 10.26, the gain of this amplifier is limited to 10,000 by external feedback, then it is useful up to a frequency of only 100 Hz. Operational amplifiers are available as integrated circuits having open-loop gains up to 10^8 with stabi-

Figure 10.26. *Operational amplifier gain-frequency curves.*

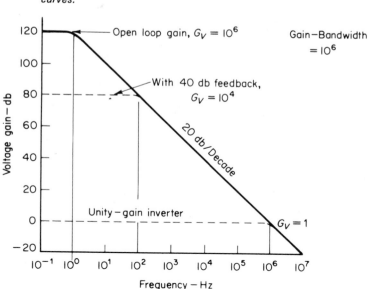

lized gain-bandwidth products of 100 MHz. Such amplifiers will provide a gain of 10,000 up to 10 kHz.

10.15 FEEDBACK CONTROL OF OPERATIONAL AMPLIFIERS (Ref. 2)

The use of external feedback elements to set the voltage gain of an operational amplifier is explained by reference to the circuit in Fig. 10.27. In this example the input signal e_I is applied to the phase-inverting input designated by the minus sign. The resistor R_O connects the output back to the summing point A, and the input signal is connected to the summing point by R_I. Assuming that the amplifier has a very high gain, the voltage e_S at the summing point is 5 to 7 orders of magnitude smaller than the output voltage, and, for practical values of the resistors e_S, is at least 2 or 3 orders of magnitude smaller than the input voltage e_I. Because e_S is negligible, we can assume that the summing point is at ground potential and that the amplifier input current is negligible compared to the current in the resistors. The current i_I is, therefore, the same in both resistors.

Ohm's law and the summing point conditions provide the following equations for the input and output voltages:

$$e_O = -i_I R_O \qquad (10.15)$$

and

$$e_I = i_I R_I \qquad (10.16)$$

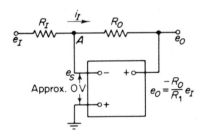

Figure 10.27. Operational amplifier with feedback.

The voltage gain is found by dividing the first equation by the second, whence:

$$G_v = \frac{e_O}{e_I} = -\frac{R_O}{R_I} \qquad (10.17)$$

Equation (10.17) implies that, insofar as the summing point voltage and current can be neglected, the gain of the amplifier is determined only by the ratio of the two resistors. By making the amplifier open-loop gain high, the gain with feedback is accurately controlled by the external resistors. If the resistors R_I and R_O are made equal, the amplifier is a unity gain inverter. The term *inverter* is used to indicate the sign reversal. If the resistors are replaced by impedances Z_I and Z_O, the amplifier output is functionally related to the input by the equation:

$$e_O = -e_I \frac{Z_O}{Z_I} \tag{10.18}$$

Equation (10.18) states that the transfer equation of a high-gain amplifier is accurately determined by the feedback impedances and is independent of the amplifier itself. The usefulness of operational amplifiers for engineering studies and for applications comes from the fact that the transfer function represented by Eq. (10.18) is easily determined by readily available circuit elements, both real and imaginary.

10.16 INTEGRATING AMPLIFIER

The use of an OP amp as the analogue of a mathematical relation is illustrated by the circuit in Fig. 10.28. Replacing the feedback resistor in Fig. 10.27 by a capacitor makes the operational amplifier an integrating amplifier. For such an amplifier (Fig. 10.28) the voltage across the capacitor e_O is inversely proportional to C_O and is proportional to the integral of the summing point current. Written algebraically:

$$e_O = -\frac{1}{C_O} \int i_I \, dt \tag{10.19}$$

However, the input current by Ohm's law is:

$$i_I = \frac{e_I}{R_I} \tag{10.20}$$

Substituting Eq. (10.20) into Eq. (10.19) gives:

$$e_O = -\frac{1}{R_I C_O} \int e_I \, dt \tag{10.21}$$

Equation (10.21) states that the output voltage of the integrating amplifier is

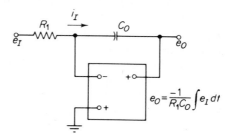

Figure 10.28. *Integrating amplifier.*

proportional to the time integral of the input voltage. In a practical application the capacitor C_O may be 1 μF and R_I may be 10 kΩ. Therefore, the time constant is 0.01 second. The time constant of the integrator is the time required for the output to change by an amount equal to the average value of the input. For this example 1 V input applied for 0.1 second produces an output voltage change of 10 V.

10.17 DIFFERENTIATING AMPLIFIER

A differentiating amplifier is formed by interchanging the resistor and the capacitor of an integrating amplifier. The input voltage is a function of the input current, as follows:

$$e_I = \frac{1}{C_I} \int i_I \, dt \tag{10.22}$$

Differentiating both sides of the input relation gives:

$$i_I = C_I \frac{de_I}{dt} \tag{10.23}$$

The output voltage is $-i_I R_o$, whence:

$$e_o = -R_o C_I \frac{de_I}{dt} \tag{10.24}$$

For a sinusoidal input voltage $e_I = E \sin \omega t$ the output is the cosine wave:

$$e_o = -R_o C_I \omega E \cos \omega t \tag{10.25}$$

Differentiating amplifiers are rarely used because the increasing response with increasing frequency ω makes a system sensitive to stray noise and aggravates the instability problems.

10.18 ANALOGUE COMPUTERS

An analog computer uses an electrical network to simultate a physical system, or the equations that represent a system. The response of the network may be used to study the performance of a servo system, the vibrations or control of an aircraft, or the performance of a complex power network. Studies of this kind usually require an engineer capable of writing the equations believed to represent the system. Given the equations, a technician can set up and operate the analogue computer.

An analog computer uses operational amplifiers to represent the terms of a differential equation. The solution of the equation is obtained by observing the response of the analog circuit when the computer is driven by a signal representing the independent variable. To avoid using differentiating amplifiers, the equation under study is successively integrated until the computer can be constructed as a series of integrating amplifiers. An analog computer also uses summing amplifiers inverters, and scale factors (potentiometers).

An elementary form of an analog computer is illustrated in Fig. 10.29.

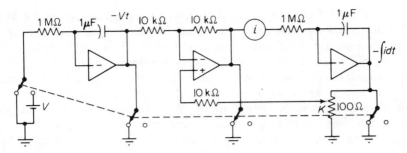

Figure 10.29. *Analog computer for $i = -2K \int idt + Vt$.*

The equation to be solved describes the rise of current in an inductance, for which the differential equation is:

$$L\frac{di}{dt} + Ri = V' \qquad (10.26)$$

Integrating both sides of the equation to eliminate the need for a differentiating amplifier, and substituting $V = V'/L$ we obtain the equivalent equation:

$$i = -\frac{R}{L} \int i\, dt + Vt \qquad (10.27)$$

When a voltage V is applied at the input of the first integrating amplifier, the computer in Fig. 10.29 gives the required solution for i as the output current of a summing amplifier. The output of the summing amplifier is Vt plus the integral $i\, dt$ inverted and multiplied by a scale factor $2K$. For any time function of V that is applied to the input, the computer provides a solution of Eq. (10.27). The switches are needed to ensure that the amplifiers are zeroed at the start of a cycle. Note that the gain of the amplifier + side is 2.

SUMMARY

Many new forms of filter circuits are made by using active circuits to modify the transmission characteristics of networks. With controlled positive feedback an active circuit may reduce the resistance loss of a network or inprove the cutoff characteristics. In this way active RC filters may have the response characteristics of LC filters or tuned circuits.

Negative feedback applied through a reactance network modifies the input characteristics of an amplifier with effects similar to those produced by positive feedback. For example, an inverting amplifier with a capacitance load and feedback has a reactive input impedance. By similar means, active circuits make practically useful inductive reactances, negative resistances, and low capacitive reactances.

The high effective input capacitance of a CE stage, known as the **Miller effect**, is an active reactance which is produced by a small feedback capacitance. In a similar effect, any impedance used for shunt feedback in a high-gain amplifier effectively loads the amplifier input. The resulting input impedance is the feedback impedance divided by the voltage gain of the stage.

A negative resistance is produced when feedback returns a current or voltage in phase with an input signal. With shunt feedback the negative resistance offsets the equivalent input resistance when the loop gain is 1. The negative resistance produced by collector-to-base feedback capacitance is a common cause of instability and oscillation in high-frequency amplifiers.

An operational amplifier is a high-gain amplifier having a high-frequency roll-off that allows great flexibility in the use of external feedback. Without feedback these amplifiers have gains of 10^6 and bandwidths of 10–100 Hz.

An analog computer uses operational amplifiers to represent the terms of differential equations. Operational amplifiers invert, differentiate, or integrate depending on the resistance or reactance used for the feedback element. With multicomponent feedback networks an operational amplifier may be used as an active filter, or may provide a nearly ideal transfer characteristic for an instrument or for an industrial control system.

PROBLEMS

10-1. Find the high-frequency cutoff and the voltage gain-bandwidth product of the amplifier shown in Fig. P-10.1.

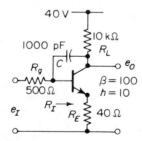

Figure P-10.1.

10-2. (a) Review the Miller effect and write out your own proof of Eq. (10.3). (b) Illustrate with a practical example.

10-3. (a) Specify capacitor values for the amplifier in Fig. 10.6 that will have the characteristic B of Fig. 10.7 with a 0 db loss at 350 Hz instead of 150 Hz as shown. (b) Explain qualitatively what effect you would expect if the gain of the amplifier having characteristic C were reduced to 200 without changing the filter components.

10-4. Show that the values given in the text are reasonable for the input impedance and for the cutoff frequencies of the stage shown in Fig. 10.6.

10-5. By calculations show that all values given in Fig. 10.24 are at least approximately correct.

10-6. Draw the circuit with component values of a filter, using the circuits in Fig. 10.15 and Fig. 10.17, which will approximate the band-pass characteristic shown in Fig. 10.21.

10-7. Construct a phasor diagram that confirms curve C in Fig. 10.7 at 400 Hz and at 200 Hz. (Difficult)

10-8. Calculate one point on each curve in Fig. 10.12 to check the curves at $f = f_c$ and at $f = \frac{1}{2} f_c$. (Difficult)

10-9. Show how to modify the analog computer in Fig. 10.29 so that the response of a series R-L-C circuit can be studied.

10-10. Refer to Fig. 10.29 and assume that the input integrator is disconnected and the input V is applied to the input of the summing amplifier. (a) What is the differential equation of the new computer? (b) If $i = \sin t$ and $k = 1$, what is the input V?

10-11. (a) Plot curves showing the resistance characteristic produced when a variable negative resistance is in parallel with a positive resistance. (b) Repeat for a negative resistance in series with a positive resistance.

10-12. Show that the emitter current for the tuned amplifier in Fig. 10.14 is 1.3 mA.

10-13. For the tuned amplifier shown in Fig. 10.14, show when $e_o = Qe_I$ that $R_N = QR_g/(Q-1)$.

10-14. Find the 3 dB cutoff frequencies for the interstage coupling network used in Fig. 5.1, and assume C_S (Fig. 10.10) is 0.005 μF. Ans: 17 Hz, 130 kHz.

CHAPTER ELEVEN

Transistor Oscillators

An oscillator is a circuit that is capable of sustaining an ac output signal by converting dc power to ac power. Almost any high-gain amplifier will break into oscillation if the output is coupled back to the input. Oscillators can be constructed to generate a variety of signal waveforms and are widely used as convenient sources of sinusoidal ac signals for testing, control, and frequency conversion. As a source of square wave, ramp, or pulse signals oscillators are commonly used in switching, signalling, and control circuits. Because of the importance of oscillators as parts of circuits and as everyday tools, the student should know the more common forms and understand how their frequency and waveform are determined. This chapter is an introduction to the theory of oscillators and a review of the common forms of semiconductor oscillators.

Oscillators are generally either sine wave (a simple oscillator), square wave, or sawtooth (probably called a multivibrator), or a form of a pulse or blocking oscillator. The design of each of these types is quite different and will depend on whether or not the oscillator is required to have good frequency stability. In general, designing oscillators is a specialty, and most of what has been written about vacuum tube oscillators can be applied to transistor oscillators.

11.1 CONDITIONS REQUIRED FOR OSCILLATIONS (Ref. 3)

The circuit requirements that are necessary in a sine wave oscillator can be easily described, and a simple calculation will usually determine whether a given circuit meets these conditions. The circuit of the Franklin type of oscillator in Fig. 11.1 will be used to explain the theory of oscillators. This oscillator is simply a high-gain amplifier which has its output fed back to

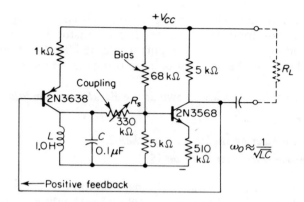

Figure 11.1. *Franklin type oscillator, 500 cps.*

the input. Each stage of the amplifier, a CE amplifier, inverts the signal so that the signal returned to the input is in phase with and augments the assumed input signal. We must recognize, however, that the signal is in phase at the resonant frequency of the tuned circuit, where the latter is equivalent to a high resistance, only if the spurious phase shifts in the circuit are small.

One condition required for oscillation is that the round trip gain, or *loop gain*, must be at least 1 or slightly more. A loop gain of 1 is enough, theoretically, to maintain an oscillating condition, but a gain of slightly more than 1 is required to ensure easy starting and a reasonably stable signal amplitude. The voltage gain of the second stage of the Franklin oscillator is about 10. The voltage gain of the first stage depends on the resistance of the resonant circuit at resonance. This is $Q\omega L$, where Q is the Q-factor of the inductance, ω is 2π times the resonant frequency, and L is the inductance of the coil. In the circuit shown the iron core coil had an inductance of 1 H and a Q of 6 (by a bridge measurement), so that the resonant resistance at 500 Hz was

$$R = Q\omega L = 6(6.28 \times 500) \times 1 = 20,000 \ \Omega \qquad (11.1)$$

The voltage gain of the first stage is the ratio of this load resistance divided by the 1000 Ω emitter resistance, or about 20. The two-stage voltage gain is 200, except for the loss introduced by the 330 kΩ resistor that couples the first stage to the second stage. The input impedance of the second stage is slightly less than 5000 Ω, so the coupling loss is a little more than 5 ÷ 335. Therefore, the calculated loop gain is 200 × 5 ÷ 335, or about 3. We are assured by this gain factor that the circuit should oscillate if the phase relations are correct.

The second condition required for oscillation is that the loop phase shift must be 0° [or 360°]. Direct-coupling is used in this circuit and there are no spurious shifts to be expected at the low audio frequencies for which the oscillator is intended. The coupling resistor R_S is large enough so that the tuned circuit is not appreciably loaded. If this resistor were replaced by a small capacitor, considerable phase shift would be introduced, and the frequency would shift off the resonant point to bring the net phase shift back to $0°$.

These two conditions are known as the **Barkhausen conditions** for oscillation, namely:

1. *The loop gain must be slightly greater than 1.*
2. *The loop phase shift must be $0°$ or $360°$.*

A quality oscillator meets several additional conditions.

An oscillator required to supply a sine wave should be biased so that the amplifier stages are operated at linear Class A Q-points. The student will observe that both stages of the Franklin oscillator are biased just as they would be for a linear amplifier. The coupling resistor is so large that it does not affect the Q-point of the second stage. Otherwise, a change in the bias resistor of that stage would be required. However, because excess loop gain must be provided, a sine-wave oscillator usually requires a means of reducing the gain at high signal levels so that the amplitude is limited without causing excess waveform distortion.

A rapid phase change is necessary so that a small variation in the circuit phase shift (caused by a change in the circuit or the transistors) can be compensated by a small change in the operating frequency. Moreover, the frequency determining network should have a stable phase characteristic. We note that a rapidly changing phase characteristic implies that the frequency determining network is sharply resonant (high Q) or that it has a rapidly changing frequency response. In the Franklin oscillator the tuned circuit operates at its unloaded Q as long as the resonant resistance is less than the load imposed by the collector and R_S combined. However, most tuned-circuit oscillators using a single transistor have reactances in the feedback loop that make the oscillator frequency considerably different from a calculated L-C resonant frequency.

A requirement for a stable oscillator is that the external load should not affect the adjustments of the oscillator. In the example under discussion we can assume that this requirement is met by readjusting the coupling resistor whenever the load is changed, but for greater stability of frequency and amplitude the load should be separated by adding a buffer stage.

11.2 PARASITIC OSCILLATIONS AND FEEDBACK
INSTABILITY (Refs. 3, 10)

Anyone working with high-gain circuits knows too well the difficulties encountered in preventing spurious oscillations. At high frequencies a short length of wire has considerable inductive reactance, and the small capacitances between components may cause considerable coupling between circuits. Together, these reactances are the source of most parasitic oscillations in high-gain circuits, especially at high frequencies. At low frequencies unwanted oscillations are more likely to be caused by mutual coupling in the power supply filters. A helpful solution of spurious feedback problems is to isolate the circuit responsible for the oscillation and then find which specific elements are responsible for the circuit's meeting the Barkhausen conditions. The prevention of oscillations, or ringing, in negative feedback amplifiers is an interesting and related problem.

The amount of feedback that can be used in a high-gain amplifier depends on how well the loop gain and phase shift can be controlled to avoid the Barkhausen conditions, particularly at high frequencies. An adequate discussion of this problem is beyond the scope of this book, but we can outline some of the techniques used to reduce instability in a feedback amplifier.

The amount of feedback is generally controlled by a resistor connected from output to input. If the resistor is too small, the amplifier usually oscillates at a frequency well above the frequency range for which the amplifier is intended. High-frequency oscillations of this kind may be stopped by increasing the feedback resistor or decreasing the amplifier gain. Either of these changes should reduce the loop gain to less than 1 at the frequency of oscillation. However, the feedback need not be reduced if the phase shift at the critical frequency can be reduced to make the feedback negative (degenerative) instead of positive (regenerative).

The phase shift in the feedback loop is determined by the rate at which the amplitude-frequency response falls off at high frequencies. Since the phase shift with a single high cutoff is only 90°, the loop phase shift may be reduced by making one cutoff occur at a much lower frequency than all other cutoffs. This technique makes an OP amp relatively stable with considerable feedback.

Another technique reduces the cutoff rate at frequencies just above the critical frequency. For this purpose a small capacitor is used in parallel with the feedback resistor, and the optimum value may be found by trial. Because an uncomplicated circuit usually has less phase shift, a feedback amplifier tends to be more stable when simply designed and carefully built to avoid unnecessary and stray capacitance.

11.3 RESONANT CIRCUIT OSCILLATORS

Resonant circuit, sine wave oscillators are commonly used in radio receivers and transmitters. The resonant circuit for determining the operating frequency is usually a simple coil and capacitor or an equivalent electromechanical resonator. With care in the design, such oscillators can be made to have a good power efficiency, a reasonable waveform, and very good frequency stability. The best known resonant circuit oscillators are named after the radio pioneers Hartley, Colpitts, and Clapp.

One circuit used in broadcast receivers is the tuned emitter oscillator shown in Fig. 11.2. The circuit found in a receiver may have additional circuit elements, either to permit using the oscillator for a dual purpose or to provide coupling. In the tuned emitter oscillator a resonant circuit is coupled to the emitter, the base is bypassed to ground, and regenerative feedback is obtained by closely coupling a coil in the collector circuit to the tuned circuit coil. The two coils are phased as indicated by the dots on the coils in the figure. Output power is usually obtained by coupling a third coil. In this oscillator the transistor must supply the power demand of the load and enough power to the emitter to maintain the oscillations. Because the load is closely coupled to the frequency determining circuit, this type of oscillator has poor frequency stability and a poor waveform; however, a receiver load is small and

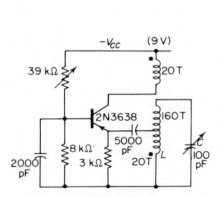

Figure 11.2. *Tuned emitter oscillator: typical broadcast, 0.5 to 1.5 MHz.*

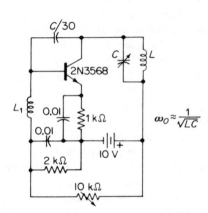

Figure 11.3. *Capacitor feedback oscillator.*

constant, and high Q-factors are easily obtained at radio frequencies. The resistors shown in the circuit are selected to bias the transistor and to stabilize the operating point in much the same way that resistors are selected for a linear amplifier.

A capacitor feedback oscillator is shown in Fig. 11.3. Inductances are used in both the base and the collector, but they are not coupled, and the feedback signal is provided by the capacitor connected between the collector and base. Usually it is sufficient to tune either the collector or the base circuit, and the correct phase turnover occurs at a frequency slightly below the resonant frequency. If both circuits are tuned, the oscillator operates more readily, and the internal capacitance of the transistor generally provides enough feedback. For these reasons the capacitor feedback oscillator is often used at high frequencies where the internal capacities of the transistor naturally become a part of the circuit. Similarly, almost any amplifier having a tuned base circuit and a tuned collector circuit tends to oscillate unless the stage voltage gain is limited enough to ensure stable operation.

A crystal oscillator known as a **Pierce oscillator** is illustrated in Fig. 11.4. A quartz crystal is a very high Q resonant element that behaves as a series resonant circuit in parallel with a capacitor, as shown in Fig. 11.4. Because the series resonant frequency and the parallel resonant frequency of the crystal are close together (within 1 percent), the reactance of a crystal changes rapidly for a small change in frequency near resonance. For this reason and because a crystal is time and temperature stable, a quartz crystal is capable of holding an oscillator frequency to better than ± 0.01 percent of the nominal frequency.

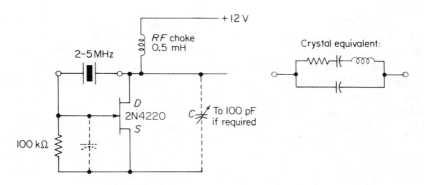

Figure 11.4. *Pierce-type FET crystal oscillator.*

The use of an FET for the active element in a crystal oscillator has two advantages: first, the high impedance gate does not adversely load the crystal and lower the Q; second, the low drain voltage protects the crystal from damage by over-excitation or high-voltage breakdown. The Pierce oscillator requires a capacitive reactance across both the gate and the drain, and the crystal provides an inductive reactance at the operating frequency. This reactance π (pi) configuration is necessary to establish the 180° phase turnover. If the choke is inductive at the crystal frequency, a shunt capacitance across the drain may be needed, but the drain circuit is not tuned in the ordinary sense.

If the crystal is connected across the gate resistor, the oscillator operates as a capacitance feedback oscillator and the drain circuit must have an inductive reactance. An inductive reactance is obtained by tuning the drain circuit at a frequency above the crystal frequency. Because capacities exist naturally across the gate and drain circuits, a crystal oscillator usually operates better with the crystal connected as the feedback element.

11.4 HARTLEY AND COLPITTS OSCILLATORS (Refs. 7, 10)

The well-known Hartley oscillator, shown in Fig. 11.5, uses a tapped coil and a single tuning capacitor that couples the collector and base circuits. The emitter and the tap on the coil are at ground potential. The collector and the base circuits share a single resonant circuit, and the amount of the base drive is determined by the position of the tap on the coil. The coupling between the two parts of the coil is not important. A good waveform is obtained if the Q of the coil exceeds about 20 and the output loading is relatively light. An output signal for low-impedance loads is obtained by coupling a third coil, and, for high-impedance loads, by capacitor-coupling to the collector.

The *Hartley oscillator* in Fig. 11.5 is designed to operate in the supersonic frequency range from about 8 kHz to 64 kHz. Higher or lower frequencies may be obtained by changing the reactance components so that the impedance values at the desired frequency are the same as those illustrated in the figure. For example, the Hartley oscillator operates at a 10 times higher frequency by reducing all capacitors by a factor of 10 and by reducing the inductance of the coil by a factor of 10. The Q of the coil should be at least 10 at the lowest operating frequency for good stability and waveform.

A *Colpitts oscillator*, shown in Fig. 11.6, is similar to the Hartley oscillator except that the emitter is tapped to a point on the capacitor side of the resonant circuit. Usually the capacitor at the base end of the tuned circuit is fixed, and the oscillator frequency is changed by varying the capacitor on the collector end. Because the fixed base capacitor tends to reduce the base

Figure 11.5. *Hartley oscillator, 8 kHz to 64 kHz.*

Figure 11.6. *Colpitts oscillator, 8 kHz to 64 kHz.*

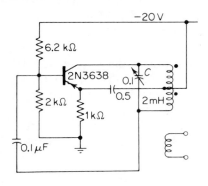

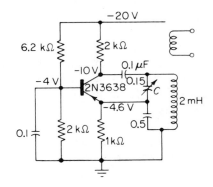

drive at high frequencies, the waveform of the Colpitts oscillator tends to improve at high frequencies. For such reasons, the choice between the Hartley or the Colpitts oscillator depends on minor differences between their performance characteristics or on the ease with which one or the other oscillator can be adjusted to meet the requirements of a design.

The circuits of the Hartley and Colpitts oscillators are proportioned to exhibit the similarities between them. Each of the oscillators has emitter and base resistors that fix the operating Q-point in essentially the same way bias resistors are used in an amplifier. Each oscillator has three points coupled to the transistor and each has the collector and the base at opposite ends of the tuned circuit. Both oscillators can be tuned up to an 8 times higher frequency by reducing the tuning capacitor from the value indicated. However, the effect of the resistors in these oscillator circuits is to make the actual operating frequency as much as twice the frequency calculated from the equivalent L and C values of the resonant circuit.

11.5   CLAPP OSCILLATOR

The **Clapp oscillator** is used where a high degree of frequency stability is required in a variable-frequency oscillator. The Clapp oscillator in Fig. 11.7

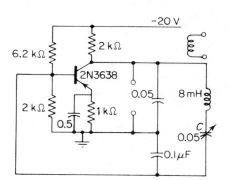

Figure 11.7. Clapp oscillator, 12 kHz to 25 kHz.

is similar to the Colpitts oscillator except that the inductance is replaced by a series resonant circuit. By making the collector and base coupling capacitors large—i.e., low reactances—the transistor is effectively removed from the frequency determining elements which are the series resonant LC circuit. The Clapp oscillator operates as a variable frequency oscillator over only a limited frequency range. Hence, except for fixed frequency applications or in frequency standards where complicated range switches are acceptable, it is not often used. In the circuit shown a 2 to 1 increase in frequency is produced by a 4 to 1 reduction of the tuning capacitor. A greater frequency change requires a change of the coil and the coupling capacitors.

11.6 POWER OSCILLATORS

An oscillator required to deliver considerable power may be designed to operate Class B or Class C, especially at radio frequencies. For a fixed size transistor, Class B or Class C biasing gives a higher power efficiency and permits higher output powers. Single-sided—i.e., "single-ended"—Class C operation is used in radio frequency circuits because tuned circuits have a high Q at these frequencies, and the flywheel effect ensures an acceptable waveform. Push-pull Class C operation may be used where a better wave-form or higher output powers are required.

A common difficulty in the design of Class C oscillators comes from the need to bias the oscillator approximately Class A for starting and to provide some method for increasing the bias as the oscillations build up. A variable bias is usually obtained by rectifying a part of the oscillator signal and filtering the bias, using a large capacitor. In a simple oscillator the base-emitter diode may serve as the rectifier with the base coupling capacitor acting as the bias filter capacitor. Because the bias changes with the amplitude of the oscillator signal, the capacitor charge must also change. If the capacitor is too large, the charge changes too slowly, and the oscillator operates inter-mittently as a blocking oscillator. If the capacitor is too small, the oscillator will not start easily, or its waveform will show considerable second harmonic distortion.

Any oscillator must be over-driven enough to ensure starting under the most unfavorable operating conditions of supply voltage, temperature, etc. If an oscillator is required to have an excellent waveform also, the tran-sistors are usually operated Class A and the signal amplitude is limited by an automatic gain control (AGC). In a typical AGC circuit a fraction of the output signal is rectified, filtered, and used to control the resistance of an FET or transistor which acts to limit the loop gain of the oscillator. The AGC action not only ensures a good waveform but also holds the output amplitude independent of the operating frequency and improves the frequency stability.

Additional information concerning tuned oscillators is found in the descriptions of typical radio frequency circuits given in the transistor hand-books. The problems of oscillator design are much the same, whether tran-sistor or tube driven, and considerable information concerning oscillator problems and oscillator design is found in the vacuum tube literature.

11.7 RESISTANCE-CAPACITANCE OSCILLATORS (Ref. 3)

Audio frequency oscillators are usually a form of a resistance-capaci-tance feedback oscillator. A sine wave RC oscillator is generally a low dis-tortion Class A amplifier, followed by an RC network that shifts the signal phase until the in-phase condition is satisfied. The RC network introduces a considerable power loss that must be made up in the amplifier, but the RC circuits avoid the use of inductances, which are unsatisfactory at audio frequencies.

The RC oscillators can be classified according to whether the amplifier has an odd or even number of stages. If the collector of a single-stage amplifier is returned to the base through a simple ladder network, the ladder must provide the 180° phase shift. The phase-shift oscillator of Fig. 11.8 accomplishes the phase turnover by providing a 60° phase shift in each section of the three identical RC sections. The resistance of the last section is provided in part by the input resistance of the transistor. With three identical sections and an external collector resistor equal to the other resistors, the 180° phase shift is obtained when the input-to-output current ratio is 56. To bring the loop current gain up to 1, the transistor must have a minimum β of 56. When the collector resistor value is between two and three times the value of the resistors in the ladder, a minimum current gain of 46 will suffice. Lower current gains can be used by adding a fourth section in the RC ladder and by decreasing the impedance level of each section progressing toward the base. With these changes and the base section resistor shorted, the required current gain may be as low as 30.

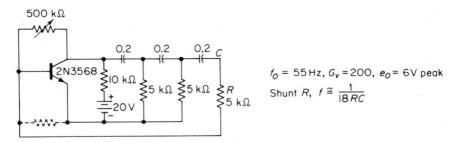

Figure 11.8. *RC phase-shift oscillator.*

For the RC oscillator shown in Fig. 11.8 the waveform is a sine wave, and the frequency is most stable when the circuit current gain is just sufficient to maintain the oscillations. A circuit current gain of 1 is established by adjusting the supply voltage or by shunting the base with a resistor that lowers the current gain. Best results are secured when the transistor current gain is not above the required minimum.

Phase-shift oscillators are constructed by using field-effect or insulated-gate devices. A three-section ladder of alike sections requires a voltage gain of nearly 30, which is more than is obtainable from many FET devices. In the circuit illustrated in Fig. 11.9 the ladder has each section progressively at a three times higher impedance level. The 180° phase turnover is obtained in this ladder when the overall voltage loss is about 18. With the output of

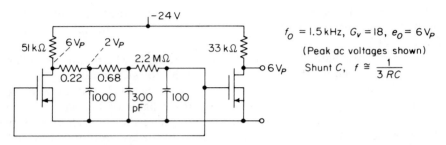

Figure 11.9. MOS FET low distortion phase-shift oscillator.

the ladder returned to the FET gate, oscillations are sustained if the FET has a voltage gain exceeding 18. By increasing the impedance step-up to 10 times, the loading of each section on the previous section is almost negligible and the circuit oscillates if the voltage gain exceeds 12. However, because the FET output impedance is not zero, as is usually assumed, the circuit must have about 3 db more voltage gain than the theoretical minimum.

The frequency commonly given for a phase-shift oscillator is derived by assuming impractical values for the active element load resistor, and the calculated frequency may be in error by a factor of two. In round numbers, the frequency of a three-section shunt R oscillator is given by the equation:

$$f \cong \frac{1}{18RC} \tag{11.2}$$

and the frequency of a three-section shunt C oscillator is given by the equation:

$$f \cong \frac{1}{3RC} \tag{11.3}$$

where RC is the time constant of a typical section. With four sections, the frequency of the shunt R oscillator is a factor of 2 higher and the frequency of a shunt C oscillator is a factor of 2 lower. A required frequency is easily obtained by adjusting one or two of the resistors.

Phase-shift oscillators usually produce 5 to 10 percent distortion. The oscillator of Fig. 11.9 is unusual because it will supply a very low distortion signal. The second FET, on the right side, is driven by the oscillator gate signal, which is quite free of harmonics because the ladder is a low-pass filter. The enhancement-mode MOS transistors eliminate the need for a series blocking capacitor and allow a high impedance level in the ladder.

In general, phase-shift oscillators are best adapted for essentially fixed frequency operation. The large capacitors required in transistor circuits have to be changed in steps. Small frequency changes can be made by adjusting one or more of the resistors, but changing the resistors changes the required current gain, and the waveform will be distorted at the extreme frequency adjustments. The important advantages of the phase-shift oscillators are their simplicity, frequency stability, and reasonably good waveform.

11.8 TWIN-T AND WIEN BRIDGE OSCILLATORS

A more constant operating frequency can be obtained in RC oscillators by employing a feedback network that shifts the phase rapidly for a given small change in the signal frequency. Simple RC circuits that shift the phase rapidly are high-loss networks like the twin-T and the Wien bridge. A high loss in the feedback network makes a two-stage amplifier necessary. A two-stage amplifier provides an output signal that is in phase with the input. Hence, the feedback network should have a 0° net phase shift at the operating frequency. A Wien bridge and a twin-T meet the conditions required of the feedback network, and, when the circuit loss is high, both circuits provide a rapid rate of phase shift with frequency. Either circuit makes an RC oscillator with excellent frequency stability.

The Wien bridge oscillator is used as an audio frequency test oscillator and has been extensively described in the vacuum tube literature. When the FET became available, a number of oscillator circuits were published which were essentially a tube-type circuit with FETs replacing the vacuum tubes. The Wien bridge oscillator shown in Fig. 11.10 uses a high-gain integrated circuit differential amplifier with the + and − inputs connected to the output of the bridge. The output of the amplifier is connected to the input of the bridge.

The left side of the bridge is made up of the series RC arm and the parallel RC arm. The right side of the bridge is made up of a fixed resistor and a voltage controlled FET resistor. The voltage that controls the FET resistor is obtained by rectifying and filtering part of the output signal peaks.

Figure 11.10. *Wien bridge oscillator with integrated circuit differential amplifier and FET gain control.*

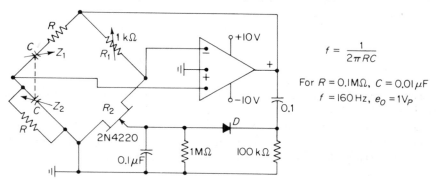

$$f = \frac{1}{2\pi RC}$$

For $R = 0.1\,M\Omega$, $C = 0.01\,\mu F$
$f = 160\,Hz$, $e_O = 1V_P$

With no output signal the FET resistance is low, the bridge is off balance, and the signal input to the differential amplifier is largest on the + (in-phase) side of the amplifier. When the output signal builds up to the desired level, the AGC signal biases the FET toward pinchoff and reduces the differential input signal.

The frequency of the Wien bridge oscillator is the frequency at which the phase angle of the arm Z_1 is the same as the phase angle of the arm Z_2, namely,

$$f = \frac{1}{2\pi RC} \tag{11.4}$$

Because the differential amplifier has a high gain, the bridge is almost balanced at this frequency, and the signal at the input to the amplifier is in phase with the signal input to the bridge. The AGC circuit maintains just enough unbalance in the bridge to keep the loop gain slightly more than 1. The frequency is varied over a 10 to 1 frequency range by varying the ganged variable air capacitors C. The frequency decade is changed by a 10 to 1 change of the resistors R. The resistor R_1 is used only to make an initial adjustment of the bridge operating point and is about twice the resistance of the FET.

The twin-T oscillator of Fig. 11.11 is an interesting example of a simple, high-stability transistor oscillator. A twin-T or a bridge-T null filter behaves like a high-Q series-resonant circuit across the feedback path. At the frequency $f = 1/2\pi RC$ the twin-T has a sharp null, or balance. By decreasing

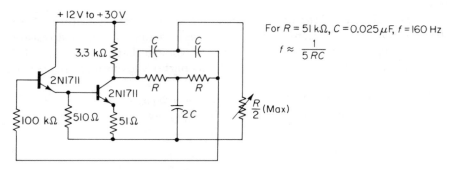

Figure 11.11. Twin-T RC oscillator.

the shunt resistor slightly, the twin-T transmits a small in-phase signal that is rapidly changing in phase at the null frequency. When the twin-T is off-balance as in Fig. 11.11 the frequency is given by $f = 1/5RC$. With a high-gain CC-CE amplifier, this circuit produces a fixed frequency oscillator which has a remarkably high stability of operating frequency.

11.9 NONSINUSOIDAL OSCILATORS

Oscillators that produce a nonsinusoidal waveform have many important applications. Usually they are a form of *relaxation oscillator*—an oscillator having an excess of loop gain so that the active element is driven well into cutoff. For a part of the cycle, energy is stored rapidly in one of the reactive elements of the circuit, and at a later part of the cycle this energy is discharged more slowly. These circuits are characterized by a tightly closed high-gain loop, one or more energy storage elements, and an amplifier that operates as an ON-OFF switch. Examples of these oscillators are known as multivibrators, blocking oscillators, and relaxation oscillators.

11.10 MULTIVIBRATORS (Ref. 12)

Multivibrators are used for the generation of square waves which have a fairly accurate and stable operating frequency. A multivibrator has a stable frequency if the half period of the cycle is determined mostly by the resistors and the capacitors of the circuit and only to a small degree by the transistors.

The multivibrator described in Chap. 8, Fig. 8.10, is a low-power type that is easily synchronized to a frequency standard by capacitor-coupling the reference signal to the base or the collector. If this multivibrator is not synchronized, the frequency varies with the collector supply voltage and the circuit makes a simple voltage-to-frequency converter for transmitting voltage information.

The multivibrator shown in Fig. 11.12 is a high-current type that produces a very stable output frequency. The stability is caused by the fact that the transistors are switched ON and OFF so well that the capacitor charge

Figure 11.12. Multivibrator, stable frequency type.

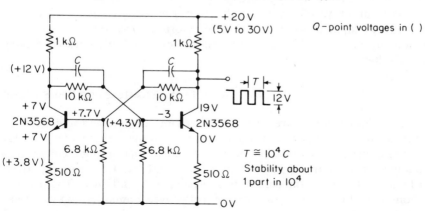

and discharge times are determined by the resistors of the circuit. In contrast
to the multivibrator discussed in Chap. 8, the frequency of the one shown in
Fig. 11.12 is practically independent of the collector voltage for supply voltages of 10–30 V. The frequency of oscillation depends on the coupling
capacitors and an equivalent timing resistance that is best determined by
measuring the operating frequency. The multivibrator frequency is of the
form:

$$f = \frac{K}{R(C_1 + C_2)} \tag{11.5}$$

where C_1 and C_2 are the values of the coupling capacitors, R is the equivalent
timing resistance, and K is approximately 2, a constant. When the capacitors
are equal, the waveform is symmetrical and the frequency is given by:

$$f = \frac{1}{RC} \tag{11.6}$$

For the multivibrator shown in Fig. 11.12, the equivalent resistance R is
about 10 kΩ. If $C = 0.1$ μF, Eq. (11.5) gives:

$$f = \frac{1}{10^4 \times 10^{-7}} = 1 \text{ kHz} \tag{11.7}$$

Usually, the transistor Q-points should fall in the active region if the
capacitors are removed. Whenever the circuit resistances are symmetrical,
the Q-points may be observed by connecting both bases or both collectors.
The Q-point of the multivibrator in Fig. 11.12 is indicated on the circuit
diagram. This Q-point check provides an easy way of locating circuit troubles.
Because of the inherent symmetry of the circuit a multivibrator may not
always start. Sometimes, to ensure easy starting, a diode is connected in
series with one of the feedback resistors, so that the circuit is asymmetrical
during the instant when the collector supply voltage is turned ON. When
operating, the symmetry is restored.

The output waveform of a multivibrator is nearly a square wave when
the feedback capacitors are alike. The waveform can be converted to a sawtooth by driving an RC integrator or can be converted to a pulse by driving
an RC differentiator.

11.11 TRANSISTOR INVERTERS

Transistor inverters are a form of multivibrator designed to convert dc
power at high efficiency into a square wave ac. In these circuits (Fig. 11.13)
a pair of transistors operates in a push-pull transformer output configuration
and a part of the output is returned to each input base. The bases are driven
hard enough to switch the collectors alternately ON and OFF. The result is
that one side of the output transformer is effectively connected to the battery
through the emitter resistor until the magnetizing current builds up and

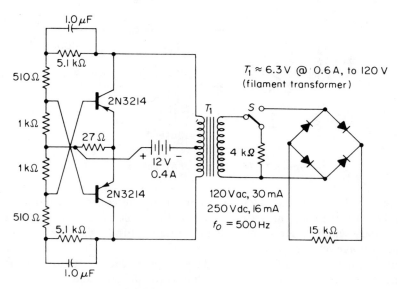

Figure 11.13. DC to ac inverter.

saturates the core. At this point the ON transistor is switched OFF, the feedback voltage reverses, and the opposite transistor is turned ON for the second half of the cycle. The half cycle is determined by the time the collector current requires to saturate the core. Hence, the operating frequency is determined by the core and winding design and is proportional to the applied voltage. Power conversion efficiencies of 80–95 percent are obtained at switching frequencies in the audio range upward to at least 100 kHz. The circuit shown here is interesting because it uses an ordinary filament transformer and operates with an efficiency of about 75 percent. The load is adjusted so that the power input is about the same as the volt-ampere rating of the transformer. The ac waveform is approximately a square wave.

11.12 RINGING CONVERTER OR FLYBACK OSCILLATOR

Low-power, high-voltage, dc-to-dc converters and ignition systems sometimes use a single transistor in a circuit called a *ringing converter*. The converter illustrated in Fig. 11.14 converts 12 Vdc to 1000 Vdc at 1 mA with an efficiency of 50 percent. Similar circuits used in TV receivers are called *flyback oscillators*.

The converter operates by storing energy in the transformer core when the transistor is ON and by transferring the energy to the load when the transistor is turned OFF. During the conduction period the transistor is driven as a switch by the regenerative feedback of the base winding. As shown

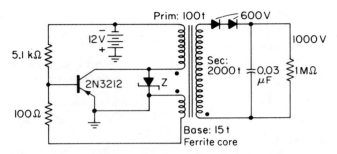

Figure 11.14. *Ringing or flyback oscillator.*

by the waveforms in Fig. 11.15, the current in the collector winding rises linearly, with the time rate of rise fixed by the inductance and resistance of the primary winding. However, when the rate of current rise falls, the transformer voltage reverses in a regenerative action that drives the transistor into cutoff. Similarly, the induced secondary voltage reverses, thereby causing the diodes to conduct and supply energy to the 0.2 μF capacitor and load. The reversed base voltage holds the transistor OFF until the energy in the transformer is released and the cycle repeats.

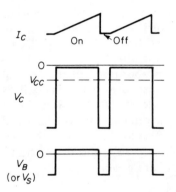

Figure 11.15. *Flyback waveforms.*

If the energy-release time is less than the storage time, the primary voltage is larger than the input voltage by the ratio of these times. Consequently, the total voltage step-up is proportional to the transformer turns ratio multiplied by the ratio of ON time to OFF time. As an approximation, the peak voltage e_p is given by the relation:

$$e_p \cong V_{cc} \frac{n_2}{n_1} \cdot \frac{t_{\text{ON}}}{t_{\text{OFF}}} \tag{11.8}$$

A long ON time requires a high Q winding, and a short OFF time requires a high resistance load. In the example illustrated the turns ratio accounts for a 20-to-1 step-up. Hence, the 80-to-1 voltage step-up implies an ON to OFF ratio of over 4-to-1. The transistor in a ringing converter must withstand somewhat more than the inductive voltage rise and is ordinarily protected by a Zener diode if the load is removable, as in an ignition system. This problem of absorbing the no-load energy tends to limit the power capacity of these converters. The term *ringing* comes from the fact that the lightly loaded transformer tends to oscillate when shock-excited. The term *flyback* comes from the short OFF time.

11.13 BLOCKING OSCILLATORS

A blocking oscillator generates a short high-energy pulse, followed by a long period in which the active element is cut off, or blocked. Blocking oscillators are used either as free-running or synchronized oscillators, as sources of steep wavefront pulses, and as simple dc-to-ac inverters. For some applications a blocking oscillator is made to generate only a single pulse in response to each trigger pulse by biasing the transistor at cutoff so that a trigger signal moves the operating point into the active region just long enough to start a pulse cycle. These are called *driven oscillators* and are used to amplify and reshape the trigger pulse into a particular form. Blocking oscillators have the advantage of very low quiescent current between pulses. Hence, high pulse currents can be drawn from the output for short durations without exceeding the power capability of the transistor. Only a single low-power transistor is required in a blocking oscillator.

A blocking oscillator is similar to a ringing converter except that the transformer and the circuit are designed to produce a short ON period and a long OFF period. By this change of the switching times, the ON pulse is given high energy and a wave shape that is controlled by the transformer. The transistor in a blocking oscillator is resting more than half the cycle, whereas the transistor in a ringing converter is working most of the cycle. A push-pull power inverter loads each transistor alternately half time.

The circuit of a typical blocking oscillator is shown in Fig. 11.16, and typical waveforms are given in Fig. 11.17. The transformer coupling the

Figure 11.16. *Blocking oscillator.*

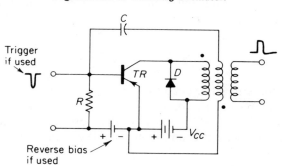

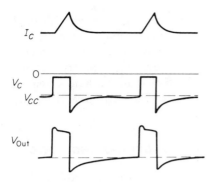

Figure 11.17. *Blocking oscillator waveforms.*

collector to the base usually has a step-down ratio of about 5-to-1, and the series capacitor, like the capacitor of a multivibrator, is charged rapidly and discharged slowly. When the transistor is caused to conduct, the closely coupled transformer drives the base so hard in the forward direction that the capacitor is rapidly charged through the forward-biased diode. The output pulse is generated by the rapid turn-ON of the collector current, and the steepness of the pulse wavefront is limited only by the leakage inductance of the transformer. The top of the pulse is flattened by collector current saturation (β fall-off), and the flat top of the pulse terminates when the transformer saturates. With saturation, the feedback signal drops and is no longer able to forward bias the transistor. The capacitor, in discharging, reverse biases the transistor and is caused to discharge slowly through the high valued base circuit resistor. The cycle is restarted either by forward bias or a trigger signal. The time constant RC of the base resistor and capacitor determines the quiescent time between pulses. Observe that the transformer is polarized to aid regeneratively both the turn-ON and the turn-OFF parts of the cycle.

Because the pulse depends in a complicated way on the circuit components, pulse transformers are designed to give a particular pulse shape at a particular operating frequency, and the intended circuit is usually supplied with the transformer. Blocking oscillators can be designed to produce pulses having rise times of less than 0.1 μS and the pulse can be shaped to be either a single pulse or a damped wave train. Notice that the diode D_1 protects the collector junction from high voltages produced by the transformer at the termination of the output pulse.

SUMMARY

An oscillator is an active circuit with regenerative feedback. Oscillations occur when the loop gain exceeds 1 and the loop phase shift is zero degrees. A quality oscillator needs AGC control of the loop gain, a rapidly changing phase-frequency characteristic, and highly stable circuit components.

The stability and transient characteristics of a feedback system are improved by shaping the feedback loop gain and phase response so that the conditions for oscillation are not too closely approached.

Most radio frequency oscillators are modified forms of the Hartley or Colpitts LC oscillators. Low-frequency oscillators are usually RC oscillators with the Wien bridge circuit preferred for laboratory signal sources.

Power inverters employ a multivibrator-type oscillator for square wave switching of the dc input power. Flyback oscillators store energy in a magnetic core for rapid discharge and a voltage step-up. Blocking oscillators produce a short-duration, high-current pulse followed by a long OFF time. Because of the short duty cycle, small transistors are able to carry the high pulse currents required for signalling and switching.

PROBLEMS

11-1. Refer to the tuned emitter oscillator circuit, Fig. 11.2, and (a) explain the purpose of the 2000 pF base-to-ground capacitor. (b) If the oscillator frequency is 700 kHz when C is 100 pF, what is the value indicated for the inductance L? (c) Show that the loop phase relations are correct.

11-2. (a) Find values of the capacitors that make 800 kHz the low-frequency limit of the Hartley oscillator (Fig. 11.5). (b) What inductance value does your circuit require? (c) If the minimum value of the tuning capacitor is 100 pF, what is the upper frequency at which your oscillator will operate?

11-3. Suppose a shunt-R phase-shift oscillator A is built, using the R and C values formerly used in a shunt-C oscillator B. The frequency of oscillator A is about $\frac{1}{6}$ that of B. What characteristic of the ladder network accounts for the large frequency ratio?

11-4. Assume that each section of a 3-section shunt-R phase-shift oscillator can be considered alone without correcting for the loading of the adjacent sections. (a) What is the phase shift required of each section and (b) what impedance relations produce this phase shift? (c) Compare the voltage loss in three such sections with the loss values given in the text.

11-5. (a) In the Wien bridge oscillator shown in Fig. 11.10, why is R_1 given as about twice R_2? (b) Suppose the FET in Fig. 11.10 was a p-channel device, would the oscillator fail to oscillate, or would there be some other indication of a change? (c) Would reversing the diode D correct the difficulty? Explain your conclusions.

11-6. (a) Refer to the ringing oscillator circuit shown in Fig. 11.14 and explain why the back voltage cannot be limited by connecting the

Zener diode across the transformer primary. (b) How would the output waveform be changed if a Zener diode were used in the blocking oscillator circuit (Fig. 11.16)?

11-7. In the MOS FET circuit shown in Fig. P-11.8, assume that R_1 is small compared with R_3 but that the loop gain is greater than 1. (a) Explain why this circuit will or will not oscillate. (b) If this circuit does oscillate, what is the frequency of oscillation if $C_1 = C_2$ and $R_2 = R_3$? (c) What is the loss in each of the R-C circuits at the frequency of oscillation? (d) Why is R_1 shown as variable?

11-8. Discuss the important similarities and differences between the circuit shown in Fig. P-11.8 and the Franklin oscillator shown in Fig. 11.1.

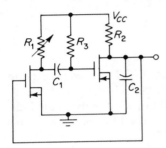

Figure P-11.8.

11-9. (a) Submit a sketch of an oscillator circuit copied from a transistor radio circuit and explain how the oscillator is coupled into the RF and IF sections of the radio. (b) Reduce the oscillator to a simple form, identify the bias circuit, and show that the resistors are approximately correct for Class A biasing.

11-10. A 4-section phase-shift oscillator uses four 10 kΩ shunt resistors and four 0.6 μF capacitors. (a) What is the approximate frequency of oscillation? (b) What is the frequency of oscillation if the resistors are in series and the capacitors are in shunt? (c) Why might a 4-section oscillator be used in preference to a 3-section oscillator?

CHAPTER TWELVE

Frequency Changing—Modulation, Demodulation and Distortion

In most of today's communication systems high-frequency waves are employed as carriers. The information signal, or voice, is transmitted by slowly changing the amplitude or the frequency of the carrier. Subsequently, the carrier may be translated many times from one place in the frequency spectrum to another, as is exemplified in a simple radio transmitting and receiving system. The typical AM radio transmitter and superheterodyne receiver, represented by the block diagram in Fig. 12.1, has three frequency translations between the transmitting microphone and the receiving loud-speaker. In the transmitter the voice frequencies are impressed on a radio frequency carrier by modulating the carrier. In the receiver the incoming radio frequency (RF) is reduced by the mixer to an intermediate carrier frequency (IF) and is finally converted to an audio frequency in the detector. All these frequency changes employ controlled distortion and filtering of signals by means of which an information signal is preserved substantially free of distortion.

12.1 MODULATION (Refs. 9, 10)

The process of impressing a signal on the carrier is called *modulation*. A modulated carrier has a group of sideband frequencies above and below the carrier frequency. Because the information bearing sidebands are close to the carrier frequency, a modulated carrier can be transmitted by way of sharply tuned, high-frequency channels. In the receiver the incoming signal is amplified in a tunable amplifier to eliminate spurious signals or noise and is combined in a frequency converter or mixer with the output of a tunable oscillator. By making the oscillator track the frequency of the tuned amplifier, the received frequency is changed to a fixed IF, usually about 455 kilohertz.

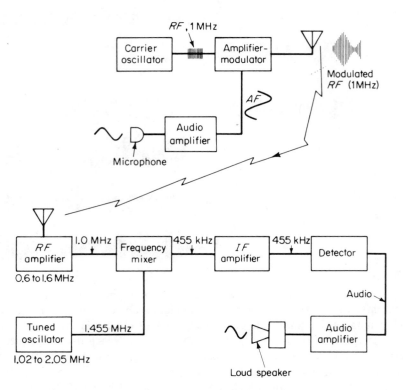

Figure 12.1. *Broadcast band radio transmitter-receiver block diagram.*

This frequency conversion permits amplification at the low IF, where selectivity and high gain are easily attained, because the amplifier frequency is fixed. Because the IF still carries the sidebands, the low-frequency audio signal can be retrieved by rectifying the IF carrier and filtering to remove all but the audio frequencies. In a radio system this final demodulation and filtering process is called **detection**.

The most common types of carrier modulation are amplitude modulation (AM) and frequency modulation (FM). AM is generally used either for its simplicity or where restricting the sideband frequencies to a relatively narrow band is desirable. FM is used where a wider frequency range may be transmitted as a means of reducing interfering noise and distortion. Our concern in this chapter is· mainly with AM because of its everyday use, particularly in radio, television, and recording systems.

12.2 AMPLITUDE MODULATION

Amplitude modulation is produced by passing two or more signals through a nonlinear device. As shown in Fig. 12.2, the simple superposition

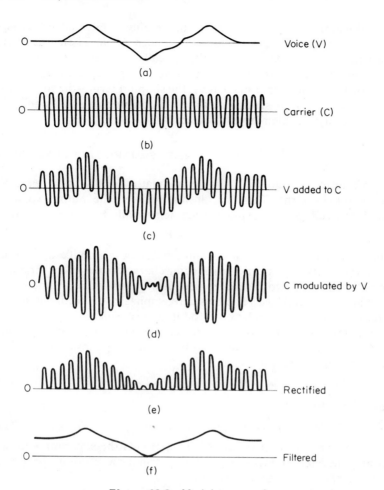

Figure 12.2. *Modulator waveforms.*

of a voice signal and a carrier in a linear system produces a waveform [Fig. 12.2(c)] which shows that the separate identity of the signals is preserved. When the carrier is modulated by the voice, the amplitude of the carrier has the waveform of the voice signal, as illustrated in Fig. 12.2(d). In effect, modulation causes the carrier envelope to vary as a replica of the signal, as if the carrier were multiplied by 1 plus the signal amplitude, with the signal amplitude always less than 1. The distinction between superposition and modulation is much the same as the distinction between addition and multiplication. Signals add in a linear system and are multiplied in a nonlinear system. By examining what happens to two signals in a nonlinear system, the frequency changes produced by amplitude modulation and demodulation are easily explained.

Modulation is usually produced either in diodes operated at high signal levels as switches, or in an amplifier biased near cutoff. The effect of these nonlinearities can be examined mathematically by using square-law and higher order terms of the transfer equation which represents the input-output relations of the modulator. Fortunately, the simplest term—the square law—leads to an accurate representation of most modulators at small signal levels and of well-designed, high-level modulators.

12.3 SQUARE-LAW MODULATION

The transfer equation of a simple modulator usually has a linear term and a square-law term. This relationship between the output voltage e_O and the input voltage e_I is expressed mathematically by the equation:

$$e_O = K_1 e_I + K_2 e_I^2 \tag{12.1}$$

The linear, or first, term represents the modulator as an amplifier. The square-law term represents the distortion-producing characteristics of the modulator. An amplifier is biased to maximize the linear term, while a modulator is operated at large signal levels and biased to emphasize the second and higher order terms of the transfer relation. Modulation is produced when two input signals are impressed on a nonlinear device. For example, if an information signal $A \sin \omega_s t$ and a carrier signal $B \sin \omega_c t$ are superimposed and introduced as the input signal e_I, then:

$$e_I = A \sin \omega_s t + B \sin \omega_c t \tag{12.2}$$

and the modulator output is:

$$e_O = K_1(A \sin \omega_s t + B \sin \omega_c t) + K_2(A \sin \omega_s t + B \sin \omega_c t)^2 \tag{12.3}$$

In Eq. (12.3) the first term on the right represents the superimposed input signals amplified by the factor K_1. When put in a more useful form by expanding the square, the second term becomes:

$$K_2(A^2 \sin \omega_s t + B^2 \sin^2 \omega_c t + 2AB \sin \omega_s t \sin \omega_c t) \tag{12.4}$$

Then, using the trigonometric relation $2 \sin^2 x = 1 - \cos 2x$, and dropping the factor $K_2/2$, the second term becomes:

$$(A^2 + B^2) - A^2 \cos 2\omega_s t - B^2 \cos 2\omega_c t + 4A \sin \omega_s t(B \sin \omega_c t) \tag{12.5}$$

The first term $(A^2 + B^2)$ of Eq. (12.5) represents a dc modulation component that is not of interest at this time. The second and third terms of Eq. (12.5) have double the frequency of the audio and of the carrier, respectively, and are eliminated in a tuned RF amplifier. The last term is the amplitude modulated carrier, which is changing in amplitude as if the carrier amplitude B were multiplied by the audio signal $(A \sin \omega_s t)$. However, in a tuned RF amplifier the modulated signal will include both the unmodulated linear term $B \sin \omega_c t$ and the modulated term $A \sin \omega_s t(B \sin \omega_c t)$. The ratio

$(A/B)100$ is the percent modulation, and, to avoid distortion in demodulating the signal by simple detection, the percent modulation must be less than 100. Therefore, for A less than B, the envelope of an amplitude-modulated carrier is of the form represented in Fig. 12.2(d).

To show that a modulated carrier can be represented as a fixed carrier and a pair of sidebands, we use the trigonometric relation:

$$2 \sin x \sin y = \cos(x - y) - \cos(x + y) \qquad (12.6)$$

and change the modulated term of Eq. (12.4) to the form:

$$2AB \sin \omega_s t(\sin \omega_c t) = AB \cos(\omega_c - \omega_s)t - AB \cos(\omega_c + \omega_s)t \qquad (12.7)$$

In other words, a carrier which varies in amplitude is exactly equivalent to a pair of sidebands at frequencies just above and below the carrier frequency. The changing part of the carrier is equivalent to a pair of sideband signals in a frequency band just twice the frequency band of the information signal. For this reason a radio frequency channel must be capable of transmitting sidebands if it transmits intelligence or is required to change amplitude rapidly.

12.4 CLASS C MODULATOR

One of the best known modulators is the Class C modulated amplifier shown in Fig. 12.3. Although called an amplifier, the transitor operates more like a high-speed switch and connects the collector battery to the tuned circuit for about one-third of each cycle. When the switch is closed, the battery stores energy in the tuned circuit. When the switch is open, the tuned circuit releases energy to the amplifier load. By making the reactive elements store considerable excess energy, the tuned circuit is given a flywheel effect that maintains the output waveform acceptably free of harmonic distortion.

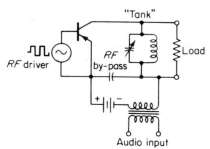

Figure 12.3. *Class C modulated amplifier.*

The Class C amplifier has an efficiency of 60 to 80 percent, because the collector current flows only when the voltage drop across the transistor is low. For efficient switching the transistor is driven hard at the carrier frequency and is biased OFF except at the peak of the cycle. Class C amplifiers are sometimes operated push-pull as a way of increasing the power output. Each side of the push-pull amplifier operates independently of the other, so the power output is doubled.

Class C amplifiers are easily modulated by varying the collector supply voltage, because the carrier amplitude is linearly proportional to the collector supply voltage. When the power supply includes the output of an audio frequency amplifier, the carrier amplitude varies with the audio signal. The circuit shown in Fig. 12.4 is easily set up to exhibit the characteristics of Class C modulators.

Normally the output of a Class C amplifier passes through several tuned amplifiers which remove any trace of the audio frequency signal. In examining the output waveform of the modulator shown in Fig. 12.4, by using an oscilloscope, a capacitor C must be inserted that is small enough to remove the audio frequency components. When the audio frequency approaches the carrier frequency, the signal component is not easily removed, and the modulated waveform takes on an unfamiliar distorted appearance. Furthermore, the signal recovered after demodulation will be distorted if the signal frequency is higher than about $\frac{1}{30}$ of the carrier frequency. The difficulty is that at high modulating frequencies the envelope of the carrier does not have enough points to represent the signal waveform.

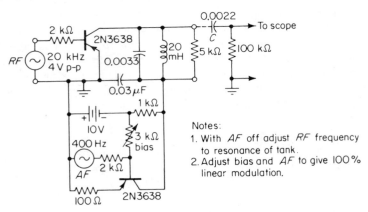

Figure 12.4. *Class C modulated amplifier example.*

12.5 AN EXAMPLE OF A TRANSISTOR MODULATOR

The modulated amplifier shown in Fig. 12.4 is designed to operate at low frequencies and exhibits the characteristics of a modulated Class C amplifier. The tank circuit is comprised of a capacitor and a ferrite cored inductor tuned to about 20 kHz. Operating as a simple switch, the transistor is driven by a laboratory oscillator at the tank frequency. The audio amplifier is resistance-coupled for simplicity, but this causes a power loss in the collector resistor. Hence, the amplifier cannot exhibit the high efficiency typical of radio frequency Class C amplifiers. However, this modulator is useful for low-power applications where simplicity is more important than power efficiency.

The power in a modulated wave is the sum of the powers of the separate frequency components. When a carrier is 100 percent modulated, the audio peaks reduce the carrier to zero, and each sideband has an amplitude of one-half that of the carrier and a power of one-fourth the carrier power. The carrier power is supplied by the Class C amplifier; the sideband power, which is 50 percent of the carrier power, is supplied by the audio amplifier. In order to maintain high signal-to-noise ratios—that is, intelligibility in the presence of background noise—the highest practical percentage of modulation must be maintained without undue distortion of the high-level audio signals. The important point is that intelligibility is achieved in communications only by maintaining the correct power levels of the carrier and audio signals impressed on the modulator.

12.6 MODULATED OSCILLATORS

Circuit simplicity can be achieved by modulating a carrier frequency oscillator, as shown in Fig. 12.5. An RF Class C oscillator can be modulated either by varying the collector supply voltage at the signal frequency or by introducing ("injecting") the signal frequency at the transistor base. The component values shown in Fig. 12.5 are suggested for an experimental study of a modulated oscillator. A resistance-coupled amplifier can be used to vary the collector supply, or a low-power oscillator—even a microphone—can be used to modulate the base.

Modulated oscillators, called **converters**, are often found in the input stages of radio receivers. These modulators are used to convert the incoming radio frequency to a lower fixed carrier frequency for easier amplification and

Figure 12.5. *Modulated Colpitts oscillator, f_o = 20 kc.*

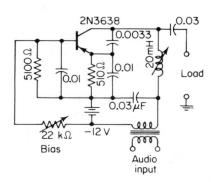

filtering. Because the incoming radio signal is weak, base injection is usually employed. In a broadcast band receiver the incoming signal frequencies range from 540 kHz to about 1600 kHz. The oscillator frequency is tuned to a frequency 455 kHz above the incoming radio frequency, and the modulator output is the difference, a fixed 455 kHz. Where the performance requirements are more demanding, the oscillator transistor may be separated from the transistor that converts frequency, in which case the latter transistor is called a *mixer*.

12.7 FREQUENCY MULTIPLIERS (Ref. 2)

By taking advantage of the fact that the pulses of the collector current have appreciable harmonic content, a Class C amplifier signal can be used to generate output power that is a harmonic of the exciting signal applied to the base. Power at the harmonic frequency is obtained by tuning the collector tank circuit to the desired harmonic of the exciting voltage and adjusting the collector current on-time. Harmonic generators of this kind are used in radio transmitters when the frequency is too high to be generated directly by a crystal, or where the frequency is so high that the power gain of a stage is impractically low.

Frequency multipliers are particularly useful at microwave frequencies (above 1000 MHz) mainly because of the difficulties involved in controlling these very high frequencies. With diode frequency multipliers a crystal controlled frequency may be multiplied several hundred times to serve as a crystal-stabilized microwave signal.

Near the upper frequency limit of an oscillator, more power may be obtained by operating the oscillator at a lower frequency and using a 4 or 6 times frequency multiplier. The diodes used in microwave frequency multipliers produce a very steep wavefront when switched from forward to reverse. Both step-recovery and varactor diodes are used in frequency multipliers.

12.8 SUPPRESSED CARRIER MODULATION (Ref. 5)

The carrier component of a modulated wave is not affected by modulation, and so contains none of the information being transmitted. By eliminating the carrier, the power requirements of the transmitter are reduced by a factor of 3, and the modulator uses only diodes as the nonlinear element, whereas ordinary amplitude modulation is not so easily attained. The simplicity of diode modulators and the lower power level are important considerations in the design of complex communication systems. Modulators having a negligibly small carrier output are known as *suppressed carrier* or *balanced modulators*. One channel of stereo FM is converted to a 39 kHz suppressed carrier signal, which is then transmitted as part of the FM signal.

The diode modulator shown in Fig. 12.6 uses four diodes in a lattice configuration that is driven by a high-frequency carrier. The low-level signal is introduced by way of the iron-cored transformer on the left, but the signal path to the transformer on the right is determined by the carrier. On alternate

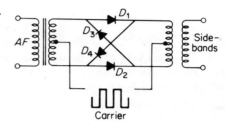

Figure 12.6. *Lattice modulator.*

half cycles either diodes D_1 and D_2 are turned ON, or diodes D_3 and D_4 are ON, and, therefore, the signal conduction path is reversed every half cycle of the carrier. The effect of the modulation is to produce a negligibly small signal in the transformer on the right when there is no low-frequency input and to produce alternate positive and negative square waves at the carrier frequency when the low-frequency signal is present. As indicated in Fig. 12.7(c), a single cycle of the signal produces a two-cycle beat at the carrier frequency. The double frequency signal is the beat produced by interference of the sidebands, which have a frequency difference of twice the signal frequency.

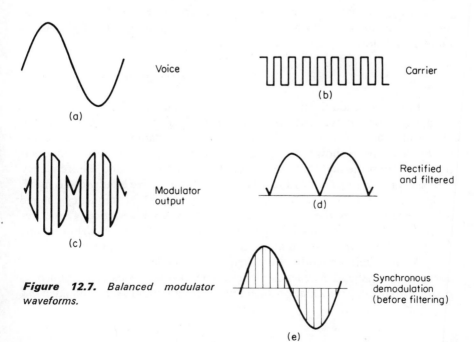

Figure 12.7. *Balanced modulator waveforms.*

Demodulation of a suppressed carrier wave cannot be effected by simple rectification because the filtered output signal follows the wave envelope [Fig. 12.7(d)] and has double the frequency of the information signal. Demodulation is achieved either by restoring the carrier signal before rectification or by synchronous detection, as described in Chap. 9.

12.9 DEMODULATION—DETECTION

The recovery of an information signal from a modulated wave is known as demodulation, although the basic process is the same as modulation. When a carrier and its sidebands are introduced as inputs to a modulator, the difference frequency ω_D

$$\omega_D = (\omega_c + \omega_s) - \omega_c = \omega_s \qquad (12.8)$$

is the desired information signal ω_s. Because a carrier-frequency signal is required in demodulation, the carrier must be restored at the receiver in suppressed carrier systems. For the same reason distortion-free detection requires an input signal that has an unmodulated carrier amplitude at least as large as the peak modulation amplitude.

Rectification by removing one side of an AM wave produces an average voltage [Fig. 12.2(e)] that follows the carrier envelope and is, therefore, the desired information signal. As in a power rectifier, the average is obtained by passing the rectifier output through a low-pass filter to remove the carrier ripple. However, problems with the control of distortion force compromises and careful design of the filter, especially if the upper signal frequency approaches $\frac{1}{20}$ the carrier frequency. As illustrated by Fig. 12.2(f), the filtered audio signal does not have the high-frequency components of the audio input signal shown by Fig. 12.2(a).

12.10 FREQUENCY MODULATION

Frequency-modulated systems vary the frequency of a carrier wave of fixed amplitude above and below a central frequency in response to the amplitude variations of the signal voltage. An example of a frequency-modulated wave is shown in Fig. 12.8. The advantages of frequency modulation stem from the fact that the carrier amplitude is fixed and the frequency can be changed over a wide range. Therefore, the effective bandwidth is much higher than with amplitude modulation. The amount of frequency change that is produced by the signal is called the *frequency deviation*. Large frequency deviations provide high signal-to-noise ratios, but practical limitations usually limit the deviation to about five times the signal bandwidth.

As in amplitude modulation, frequency modulation produces a carrier along with sideband frequencies above and below the carrier frequency. A single-frequency information signal produces an infinite series of sideband

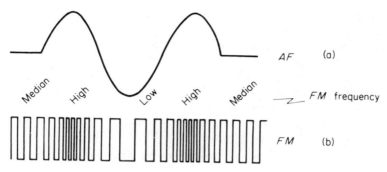

Figure 12.8. *Frequency-modulated waveform.*

frequencies, but practice shows that a total bandwidth of four to five times the frequency deviation is adequate for high-fidelity transmissions.

The circuits employed for frequency modulating an oscillator depend on the carrier frequency. Most types of low-frequency oscillators can be modulated by varying the collector supply voltage, but a linear relation between the voltage and frequency may be difficult to obtain except for small deviations. At radio frequencies the capacitor that determines the oscillator frequency may be varied electronically by the low-frequency modulating signal. At moderately high frequencies the reactance of a capacitance multiplier can be varied by a change of the Q-point conditions which change the gain and, therefore, the capacitance. At high frequencies the capacitance of a semiconductor diode is varied by changing the reverse voltage impressed on the diode.

Some of the advantages of frequency modulation arise from the fact that FM signals can be amplified in simple switching or Class C amplifiers. The fact that the signal peaks are clipped in an FM amplifier is an advantage because clipping tends to eliminate impulse noise (ignition or static noise) and does not change the zero crossings of the signal, which are used in demodulation as a measure of the FM frequency. In fact, an FM receiver usually has an RF stage and a mixer or converter, followed by a high-gain IF amplifier that is designed to limit the signal amplitude, thus eliminating noise peaks and amplitude variations. The zero crossings are then detected in a frequency discriminator or a ratio detector. Both detector circuits convert frequency changes to a signal having an amplitude that varies with the frequency. A resonant circuit tuned off-frequency is a simple form of discriminator because the signal amplitude varies with signal frequency, and the resulting AM wave is then detected in an ordinary AM detector. An off-tune, resonant circuit discriminator is nonlinear and critically dependent on the amount of detuning, whereas a frequency discriminator is especially designed to provide linear conversion.

FM is used in high-frequency communication systems when the need for a high quality of signal transmission cannot be met by an AM system. At high carrier frequencies the frequency spectrum is available for high-frequency deviations, background noise is quite low, and a large frequency deviation is easily produced. AM communication is used at low frequencies where the limited available frequency spectrum makes impractical the use of a significant FM deviation ratio. Magnetic tape recorders use a modulated carrier, usually FM, in recording dc and low frequencies that are below the useful frequency range of the tape pickup heads.

12.11 FREQUENCY AND PHASE DISTORTION

An ideal amplifier or communication system produces an output that exactly duplicates the input in all respects except magnitude. Actual systems fall short of the ideal by failing to amplify the different frequency components of the input voltage equally well, by giving an output that is not proportional to the amplitude of the input, or by making the relative phases of the different frequency components in the output differ from the relative phases in the input signal. These distortion effects are referred to as *frequency*, *amplitude*, and *phase distortion*, respectively.

Frequency distortion is caused by the inductances and capacitors of a circuit that limit the range of frequencies transmitted by a system or amplifier. Frequency distortion is sometimes introduced deliberately in order to remove noise or reduce pickup when the noise power is concentrated in a frequency spectrum that is different from the main signal spectrum. The reduction of noise by filtering is of value only when the ratio of the signal power to the noise power, the S/N ratio, is improved and the resulting frequency distortion is tolerable.

Phase distortion occurs if the phase relations between the different frequency components of a wave are changed so that the shape of the signal waveform is altered. Phase distortion is important in the transmission of pulses or pictures, but is relatively unimportant in the reproduction of sound because the human ear is not able to detect the relative phase of sounds. In general, the phase characteristics of a circuit are exactly determined by the frequency characteristics, and the necessary low-frequency and high-frequency cutoffs in circuits must introduce phase distortion. Usually, the simplest way to reduce phase distortion is to eliminate or compensate the frequency distortion that is the cause. The effects of frequency and phase distortion on signal waveforms are discussed in Chap. 13. The methods of compensating for the loss in amplifiers at low and at high frequencies are usually considered in descriptions of wide-band or video amplifiers (Refs. 3, 10).

12.12 RINGING

Ringing, or *transient distortion*, is a form of phase and frequency distortion that is caused by spurious high-Q resonant circuits. In feedback amplifiers the ringing appears as a high-frequency damped wave caused by a tendency of the system to oscillate when shock-excited. In electromechanical systems ringing is caused by undamped, mechanically resonant elements that are coupled to the electrical system. Ringing is usually eliminated by removing the resonant elements or by reducing the effective Q of the resonant system. The control of ringing in a feedback amplifier is discussed in Sec. 11.2.

12.13 AMPLITUDE DISTORTION

Amplitude distortion comes from a nonlinear relation between the output signal and the input signal and is generally caused by nonlinearity in the power stage transistors. Amplitude distortion produces frequencies in the amplifier output that are not in the input. The distortion process is essentially the same as modulation. In a simple amplifier the most disturbing distortion is the second harmonic of the signal frequency and the associated dc, the modulation components of the square-law modulator. We hear the second harmonics as "distortion," and the dc causes an increase of the collector current at high-signal levels. In a push-pull output stage the second harmonic and dc distortion terms are balanced out, and the main distortion components are the third and higher order harmonics. Intermodulation distortion is produced by the modulation of one signal by another and is particularly disturbing in the reproduction of sound because intermodulation frequencies are not harmonics of the signals which produce the intermodulation. Amplitude distortion is prevented only by constructing or adjusting an amplifier so that it is capable of producing high output power without distortion. Amplifiers are rated to produce a given power output with a specified total harmonic distortion.

12.14 APPLICATIONS OF FREQUENCY CHANGING— A RADIO (Refs. 5, 9, 10)

A broadcast band radio receiver is an assembly of amplifiers, oscillators, and frequency-changing circuits. We have examined these circuits in detail as separate elements, and the student is now prepared to understand how an assembly of circuits produces a working radio. A broadcast band radio is selected for study because radios are available for experimental study, and the student of electronics finds an abundant opportunity to apply his knowedge in the repair of radios.

Almost all modern radio receivers use a superheterodyne circuit that is represented by the block diagram shown in Fig. 12.1. Such receivers consist of a radio frequency section or amplifier, a mixer or converter, an intermediate frequency amplifier, a detector, an audio amplifier, a loudspeaker, and an ac or dc power supply. The circuit diagram of a typical transistor broadcast receiver is illustrated in Fig. 12.9, except for minor details that have been omitted to simplify the description.

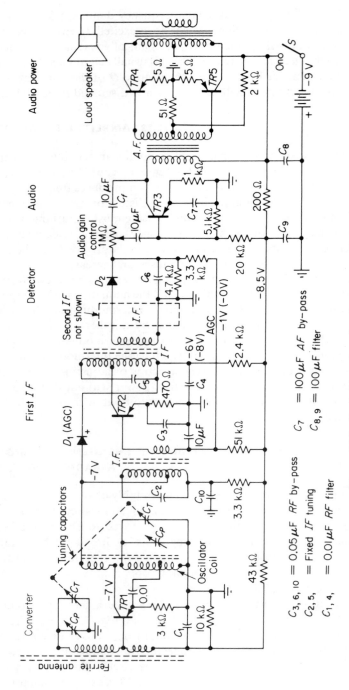

Figure 12.9. Typical broadcast band transistor radio receiver.

232

The antenna of a portable receiver is usually a coil of wire about 3 in. long wound on a magnetic ferrite core. The core serves to enlarge the effective pickup volume of the coil and increases the efficiency. A variable capacitor C_T across the coil tunes the circuit to the incoming radio frequency (RF), thereby increasing the effective signal voltage and selecting the signal from those of different frequencies. The coil and the capacitor are designed to minimize the cost and the physical size of the tuned circuit, so the impedance at resonance cannot be matched directly into the transistor. For this reason, the transistor is connected to the resonant circuit on a tap that optimizes the coupled signal. The input stage of the better receivers is a radio frequency amplifier which requires a tuned circuit in both the base and the collector circuits. Each circuit must be tuned by varying the capacitors or the ferrite cores. The less expensive receivers omit the radio frequency amplifier and couple the antenna circuit to the base of the modulator, sometimes called the *first detector*. The first detector is either a pair of transistors making up an oscillator and mixer or a single transistor (converter) that serves both purposes.

The receiver shown in Fig. 12.9 uses a tuned emitter-type oscillator as the converter and is tuned to a frequency 455 kHz higher than the incoming signal. The oscillator is tuned by a common control simultaneously with the signal selecting tuner so that the frequency of the mixer output will always be the same. Accordingly, the intermediate frequency (IF) amplifier requires no adjustment and uses circuits tuned to a single constant frequency. Tube-type receivers are usually designed for an IF value of 455 kHz, but the transistor receivers are more likely to employ a 225 kHz intermediate frequency. The choice of the 225 kHz frequency is made by compromising a number of conflicting requirements, a typical problem in design. If the receiver is tuned by using capacitors, then small, adjustable "padding" capacitors C_P are placed in parallel with the main capacitors C_T to make the RF tuning and the oscillator "track." *Tracking* means that the oscillator frequency differs from the received radio frequency by a fixed amount that always equals the intermediate frequency.

Customarily, the IF amplifier is a series of two or three identical stages, although only one such stage is shown in Fig. 12.9. Because of difficulties with interaction between stages and the tendency for high-gain stages to oscillate, the voltage gain of each transistor stage in a series is limited to about 10 (20 db). The stage gains are limited either by connecting the transistors at a low point on the tuned circuit or by shunting the circuit with a resistor. In some instances a combination of both methods is used because too low a shunt resistor may produce too wide an IF passband.

The output of the IF amplifier is coupled to a diode D_2, which rectifies the signal. The rectified signal serves two purposes: the envelope of the IF signal provides the audio signal for the output amplifier, and the long period average of the IF carrier is used as a dc bias to control the gain of the IF amplifier. This dc bias is referred to as the *automatic gain control* (AGC) signal.

12.15 AUTOMATIC GAIN CONTROL

The purpose of AGC is to maintain the output of the IF amplifier at a constant level that is independent of the received signal strength. As a radio receiver is tuned from one station to another, the incoming RF signal may vary from a few microvolts to several hundred millivolts. This change of RF signal level necessitates an offsetting change in the IF and RF amplifier gains in order to avoid distortion and to present both strong and weak stations at about equal volume levels when the receiver is tuned.

High-frequency amplifiers are designed with the emitter resistor by-passed to eliminate ac feedback. With the feedback eliminated, the gain of a transistor stage is approximately proportional to the product of the transistor current gain and the dc emitter current. The current gain β varies somewhat more slowly than the transconductance at low emitter currents but falls off rapidly at high emitter currents. The net effect of a Q-point change can be represented as a dependence of the stage power gain on the emitter current, as shown in Fig. 12.10. By operating the RF and IF stages at the peak of the gain curve, the gain can be reduced either by decreasing or increasing the dc emitter current. Because the impedance matching changes with a change of Q-point, the AGC gain reduction is more than is indicated by the curve.

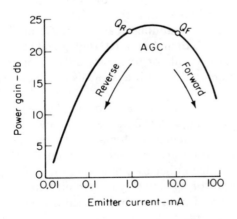

Figure 12.10. IF amplifier power gain vs. emitter current.

The most common form of AGC amplifier, as shown in Fig. 12.11, operates the variable gain stage with a constant collector voltage, and the AGC bias is polarized to reduce the power gain by decreasing the emitter current. A less frequently used system has the AGC bias polarized to increase

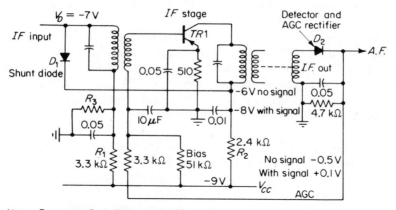

Note: R_3 or preceding collector make $V_0 = -7$ V
TR1 is any germanium *IF pnp* transistor (2N1524)

Figure 12.11. *Reverse AGC circuit with shunt diode.*

the bias current. By supplying the collector voltage through a series resistor, the increasing emitter current is accompanied by a decrease in the collector voltage.

AGC systems that decrease the gain by increasing the emitter current are called *forward AGC*. Those that decrease the emitter current are called *reverse AGC*. In both AGC systems the Q-point change produces an impedance mismatch, and a considerable change in the power gain can be obtained with a small change in the AGC bias. The reverse AGC requires few components and few control stages. The forward AGC has the advantage of accepting higher signal levels as the gain is reduced. Since most systems of AGC tend to overload at high signal levels, the auxiliary diode D_1, shown in Fig. 12.11, is nearly always used as a variable resistor across the input. The collector resistors R_1 and R_2 are proportioned to back bias the diode at low signal levels and to forward bias the diode when the conventional AGC changes the collector current and the voltage drop in R_2. Because the auxiliary diode shunt eliminates the large signal problem, most radios now use reverse AGC. Occasionally an auxiliary dc amplifier stage is used in the AGC line as an AGC voltage amplifier or as a voltage variable emitter resistor to change the AGC gain. The choice between the different types of AGC systems is determined by weighing the cost of a complex system against the limited performance characteristics of a simpler circuit. The only observable difference between a forward and a reverse AGC may be the polarity of the diode D_2 and that of the diode D_1, if the latter is used.

12.16 AUDIO VOLUME CONTROL

The audio signal developed by the diode detector is about 1 V across the 4.7 kΩ diode load resistor. A 4 Ω loudspeaker with $\frac{1}{4}$ W input requires

1 V, but the high ratio of the impedance levels makes it necessary to provide a power gain of at least 30 (30 db). A typical small radio has a single audio driver stage that is transformer-coupled to a push-pull Class B output stage. An excess of audio gain is necessary to receive weak stations and to permit a choice of the sound level by the volume control. When receiving strong stations, the gain in the driver stage (20 db) is probably just offset by the loss in the volume control.

A simple potentiometer, tapered logarithmically, is a satisfactory volume control if the following stage does not appreciably load the output tap. A potentiometer is used when followed by a vacuum tube or an FET stage, but transistors require either a specially tapered potentiometer or a variable series resistor. The correct value of the series resistor is calculated by multiplying the attenuation ratio desired by the total base circuit resistance. The total resistance is the sum of the impedance looking into the base and the impedance of the source that drives the base. For the radio shown in Fig. 12.9 the source resistance is 2000 Ω and the base input resistance is negligible. Disregarding the collector-to-base feedback, we find that the 1 MΩ gain control gives an attenuation ratio of $10^6/2000$, or 500 (54 db). The control taper should be the reverse of the taper used in a vacuum tube amplifier because large changes of resistance are required as the control nears the high-loss end.

The volume control used in the radio (Fig. 12.9) is arranged to provide shunt feedback at the low-gain settings. The feedback capacitor C_f is selected to reduce the feedback at low frequencies so that the base frequencies become relatively louder at low-gain settings of the control. This type of frequency adjustment is commonly employed in radio and hi-fi equipment in order to offset the tendency of the human ear to hear mainly the high-frequency components of low-level sounds.

The audio power stage of a typical radio is a transformer-coupled Class B amplifier similar to the amplifier shown in Fig. 12.9. The Class B amplifier has the important advantage that the collector current is low when the audio level is low, and the transistors can be mounted on relatively small heat sinks. The loudspeaker resistance is 4, 8, or 16 Ω, with the higher values preferred for transistor applications.

12.17 RADIO TRANSMITTERS

The circuit of the radio transmitter illustrated in Fig. 12.12 reveals the fact that an AM transmitter is a comparatively simple electronic system. The upper half of the circuit represents the radio frequency oscillator, crystal stabilized to ± 0.005 percent, and a Class C radio frequency power amplifier. The lower half of the circuit represents a dynamic microphone (sometimes a loudspeaker), a Class A audio amplifier, and a Class B audio amplifier.

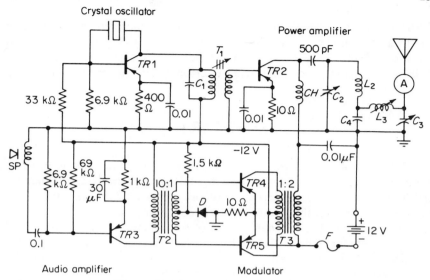

Figure 12.12. *Transistor radio transmitter.*

Part of a radio receiver is sometimes used as the audio section of the transmitter. The audio power amplifier is popularly known as the modulator, although the radio frequency Class C amplifier is the nonlinear device that mixes the audio and the radio frequency signals. The transmitter circuit is presented to show how circuits that have been studied as separate entities combine to make a radio transmitter, and there is no reason to explain details of the circuit, except for the antenna-coupling network, C_4, L_3, and C_3.

An efficient transmitting antenna has an impedance that is approximately 50 Ω. Antennas less than a quarter wavelength are inefficient and present an impedance that has only a few ohms in series with a high reactance. Short antennas should be avoided because the power output of a transmitter is limited, and, to be effective, the available power must be efficiently coupled to the antenna. The capacitor C_4 provides a low impedance connection to the tank circuit, while C_3 and L_3 make two adjustments available by which the coupling circuit is tuned to the transmitter frequency and the antenna resistance is transformed to load the tank. The reactances must be well designed to minimize power losses in the coupling network. The ammeter or current indicator shows when the antenna current is a maximum or that the transmitter is properly adjusted. Anyone making adjustments to a radio transmitter is required by law to have a license.

The importance of communications today and the demands for space in the radio frequency spectrum have given rise to the development of much more complicated radio frequency and microwave wave systems than those illustrated above. A small improvement in performance generally requires a complicated complex of modulators, filters, frequency multipliers, and amplifiers. These circuits require much skill and understanding and offer challenging opportunities for trained electronics technicians and engineers.

SUMMARY

Information signals are impressed on a high-frequency carrier wave by modulation and are recovered to the original signal form by demodulation. Amplitude modulation processes may be represented by the frequency changes that occur in the square-law term of a nonlinear system. The transmission of information requires a band of frequencies which is represented in an AM wave by the sideband frequencies.

Distortion in a linear system is caused by amplitude distortion (modulation), frequency response changes, and phase distortion. Amplitude distortion is corrected by reducing signal levels, by feedback, or by designing more linear amplifiers. Frequency response distortion is corrected by changing the reactance elements that cause amplitude variations with frequency, by equalization, or by feedback. Phase distortion in a simple lumped-element system is determined by the frequency response and is corrected by correcting the frequency response.

An AM transmitter and receiver, as used in broadcasting, is an example of the use of modulation and demodulation in the transmission of information. Today's communication systems require many frequency changes and a variety of modulation schemes that depend on the requirements for speed and accuracy in transmitting information through the system.

PROBLEMS

12-1. Explain how to distinguish between the waveform of a signal that is modulated by 60 Hz and a signal that has 60 Hz pickup superimposed, as in an amplifier that has inadequate shielding.

12-2. In a simple square-law modulator the input frequencies are 1 kHz and 10 kHz. List all the output frequencies.

12-3. A radio receiver has an IF amplifier tuned to 250 kHz. The incoming RF is 600 kHz AM modulated by an audio signal which has a 100 Hz to 6 kHz frequency spectrum. What are the upper and lower frequency limits of the signal that must be transmitted by the IF amplifier?

12-4. Compare the requirements of a demodulator receiving information signals having suppressed carrier modulation with the requirements of a demodulator receiving AM signals.

12-5. An amplifier has a 200 Hz, a 5 kHz, and a 6 kHz signal input (a) List eight distortion frequencies, (b) several intermodulation frequencies, and (c) several harmonic frequencies.

12-6. Suppose a detector is receiving a 1 MHz signal modulated by 1 kHz. What are the audio output frequencies of an ideal AM detector and what audio frequencies does a square-law detector produce?

12-7. If the audio input to the bridge modulator shown in Fig. P-12.7 is a sine wave, sketch the chopped RF signal that is observed at AB, assuming transformer T does not transmit the audio signal.   (*Hint:* Subtract the average value of the chopped signal.)

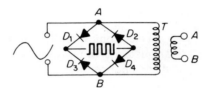

Figure P-12.7.

12-8. A radio is transmitting a 1 MHz carrier modulated by 5 kHz, and the signal passes through a single tuned circuit. (a) If a 3 db loss of the sidebands is just tolerable, what is the maximum or minimum tolerable Q? (b) Is this Q a maximum or minimum? Why? (c) What kinds of distortion are produced by passing the radio signal through this tuned circuit?

12-9. (a) If broadcast stations are spaced at 10 kHz intervals, what is the upper frequency limit that can be allowed for the audio modulation? (b) Submit a diagram to explain your answer. (c) What would you give as a more practical upper limit?

CHAPTER THIRTEEN

Wave Shaping and Nonsinusoidal Waves

In electronic circuits and their many applications extensive use is made of various kinds of nonsinusoidal waves, such as square and sawtooth waves, gated and chopped waves, and pulses. Some of these waveforms are generated directly by switching circuits or by relaxation oscillators such as have been described in Chap. 11. Other waveforms are produced by using diodes or saturating amplifiers to modify a sine wave. Still other waveforms are reshaped by a filter or equalizer. The processes most commonly used to shape waveforms involve clipping, clamping, gating, frequency discrimination, and integration or differentiation. This chapter is concerned with the use of non-linear devices for signal shaping and presents an introduction to the characteristics of nonsinusoidal waves.

13.1 CLIPPING (Refs. 4, 10)

Clipping flattens a portion of a wave by limiting the peak amplitude to an arbitrary level that is lower than the amplitude of the original signal. Clippers are classified as *peak clippers*, *base clippers*, or *slicers*, depending on the way they operate on the wave.

A *peak limiter* operates by preventing either the positive or the negative, or both, amplitudes of an input wave from exceeding a value set by the clipper. Examples of positive peak clippers are shown in Fig. 13.1. The common diode clipper shown in Fig. 13.1(a) uses a battery to determine the positive peak output signal. If, as shown, the battery has 4.5 V and the forward voltage drop of the diode is about 0.5 V, then the diode is effectively open for input signals less than $+4.5$ V or for negative signals, but the diode closes to prevent a positive-going output from exceeding $+5$ V. A Zener diode limiter that produces essentially the same result is illustrated in Fig. 13.1(b).

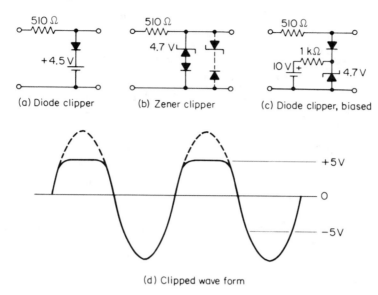

(a) Diode clipper   (b) Zener clipper   (c) Diode clipper, biased

(d) Clipped wave form

Figure 13.1. *Diode clipping circuits.*

Here, because two diodes are operating in series, the peaks are not so sharply limited as with a diode limiter, but the Zener limiter does not require a battery. Another form of clipper, illustrated in Fig. 13.1(c), has the advantage of the diode clipper—i.e., more abrupt limiting—and uses a Zener diode to set the limiting voltage. A current for forward biasing the Zener diode may be supplied from any convenient supply (10 V in the example), and the clipping level is easily adjusted by changing the Zener diode.

The clippers in Figs. 13.1(a), 13.1(b), and 13.1(c) produce essentially the same output waveform as shown in Fig. 13.1(d). Because only one diode is involved, the clippers shown in Figs. 13.1(a) and 13.1(c) produce more abrupt clipping than does a circuit having several diodes in series. Furthermore, Zener diodes do not have a sharp breakdown in the very low voltage range. Therefore, a diode clipper should be used for clipping peaks in the 1 V to 4 V range.

A Zener diode used without the series diode makes a positive and negative peak clipper that limits the peaks of one polarity at the voltage rating of the Zener diode and limits the peaks of the opposite polarity to about 0.5 V. This clipper, as shown in Fig. 13.2(a) and 13.2(b), is used in triggering applications where a limited trigger signal of fixed polarity must be assured.

Both positive and negative peaks are clipped when a second Zener diode and a plain diode are connected, both in reverse, as indicated by the dotted connection in Fig. 13.1(b). By similar modifications of the circuits shown in Fig. 13.1(a) or 13.1(c), these also become double peak clippers.

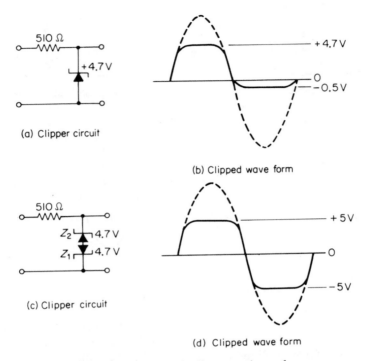

(a) Clipper circuit

(b) Clipped wave form

(c) Clipper circuit

(d) Clipped wave form

Figure 13.2. *Zener diode clippers and waveforms.*

However, as shown in Fig. 13.2(c), only two back-to-back Zener diodes are required to produce a double peak clipper. When the input voltage is positive, the Zener diode Z_1 is forward biased and conducts, but diode Z_2 is open for any voltage below the breakdown value. When the input voltage exceeds the breakdown voltage of Z_2 plus about 0.5 V in Z_1, the output signal is clipped. With negative input voltages the diodes exchange roles and the negative peaks are determined by the breakdown value of Z_1. The waveform of a double peak clipper is shown in Fig. 13.2(d).

13.2 BASE CLIPPER (NOISE SUPPRESSOR)

In many circuits, particularly for counting and switching, a circuit must be made immune to small noise signals and permitted to operate only when the input signal exceeds a particular value. A *noise suppressor*, also called a *base clipper*, is shown in Fig. 13.3(a). With input signals less than $+4.5$ V or negative, the diode is forward biased, and the output terminal is held at the 4.5 V base line, as shown in Fig. 13.3(b). Positive signals having amplitudes exceeding the battery voltage open the diode and are passed without

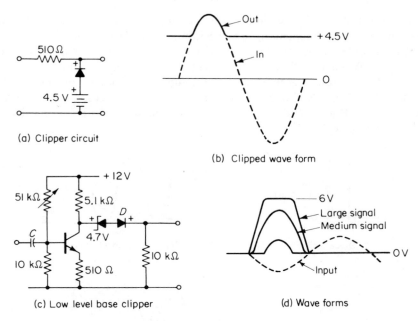

(a) Clipper circuit

(b) Clipped wave form

(c) Low level base clipper

(d) Wave forms

Figure 13.3. *Base clippers.*

attenuation. Noise signals less than 4.5 V cannot open the diode and are attenuated. If the series resistor is 10 kΩ, the noise is attenuated by at least 40 dB.

A noise suppression circuit using the Zener diode, as shown in Fig. 13.3(c), transmits the positive peaks but has the advantage of leaving the reference base line at 0 V. Signals too low to effect Zener diode breakdown are attenuated. The Zener circuit is useful in direct coupling an amplifier to a trigger circuit, provided the Zener diode rating is high enough to block the Q-point voltage of the amplifier with an additional allowance to cover the noise signals. For some applications the Zener diode is preferable to a capacitor because the latter stores charge and, in discharging makes the base line vary. With the Zener circuit, the positive-going signal is easily limited by making the supply voltage only a few volts above the Zener diode rating.

Base clipping circuits are sometimes made by biasing a transistor or FET amplifier well into cutoff. The characteristics of such circuits are discussed in a later section of this chapter.

13.3   LOW-LEVEL DIODE CLIPPER AND NOISE SUPPRESSOR

For low-level signal applications a pair of back-to-back silicon diodes make a simple and effective peak limiter. Using ordinary low-power silicon diodes, the circuit in Fig. 13.4(a) limits the output signal to about 0.4 to 0.6 peak-to-peak volts, depending on the input level. The limited signal has rounded shoulders, but the low-level signals are transmitted with negligible distortion. When the diodes are connected as a series element as in Fig. 1.34(c), the circuit makes an effective noise suppressor that attenuates signals less than 0.1 V rms by about 40 dB and passes signals exceeding 1 V rms with less than a 3 dB loss. With large input signals the diode switching transients are practically negligible.

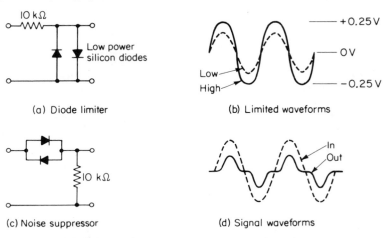

(a) Diode limiter

(b) Limited waveforms

(c) Noise suppressor

(d) Signal waveforms

Figure 13.4. Diode limiter, (a) and (b); noise suppressor, (c) and (d).

By substituting gold-doped diodes in Fig. 13.4(a) and 13.4(c), and by increasing the resistor to 100 kΩ, the clipping level and the noise suppression level of the diode circuits may be lowered about 5 times. Even lower switching levels can be reached by using the back-to-back diodes as elements in the feedback path of an operational amplifier.

13.4 CLIPPING AMPLIFIERS

Signals are sometimes clipped by overdriving an amplifier. Signals that are less than 0.1 V cannot be shaped by diodes alone, but an amplifier can raise the signal to a level that is limited by exceeding the capabilities of the output stage. A transistor stage, as in Fig. 13.5(a), exhibits a peak signal limit if the transistor is either cut OFF or is turned full ON. A transistor limiter may be biased to produce either cutoff limiting, "bottoming," or both. On the load line represented in Fig. 13.5(b), a transistor is OFF at point D and ON at B. If the amplifier Q-point is at D, the output signal may have a positive-going value, but negative signal excursions are limited by collector current cutoff at D. If the amplifier is biased at the Q-point B, the output

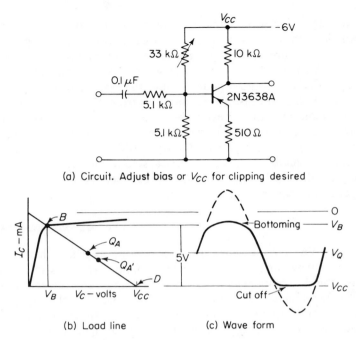

(a) Circuit. Adjust bias or V_{CC} for clipping desired

(b) Load line   (c) Wave form

Figure 13.5. *Clipping amplifier.*

may have negative-going excursions, but positive signals are limited at B because the collector voltage bottoms at V_B. The collector voltage bottoms when the transistor is turned full ON and V_B is the emitter voltage—calculated by assuming the collector is shorted to the emitter. For the stage illustrated in Fig. 13.5(a), V_B is 0.3 V.

If the amplifier is biased to operate at the center of the load line, then both positive-going and negative signals are limited at the points B and D, respectively. An overdriven transistor amplifier is capable of producing reasonably sharp limiting. The peak-to-peak voltage limits are controlled by the choice of the collector supply voltage.

Junction-type FETs possess gradual cutoff and bottoming characteristics that make them generally unsatisfactory for signal limiting. However, the gate diode of an FET makes a satisfactory limiter for replacing the diode of a shunt limiter. The enhancement-mode MOS FETs have a sharp turn-ON characteristic that makes them useful as base clippers, particularly for the suppression of noise in switching circuits.

13.5 SQUARE WAVE CLIPPING

Many applications, especially in the testing of circuits and components, require very precise square waves. A clipping amplifier would seem useful for converting a sine wave signal to a square wave, especially if the signal is passed through a series of stages with each stage improving the signal produced by the previous stage. However, a series of capacitor-coupled stages is unsatisfactory because the Q-point of an overdriven amplifier is shifted by signal rectification in an amount that varies with the signal level. Usually the Q-point shift is toward cutoff, as from Q_A to $Q_{A'}$ in Fig. 13.5(b). This shift of the Q-point is easily observed with a dc voltmeter and gives an approximate measure of the amount of second harmonic distortion in an amplifier. Q-point shift produces unequal times of the zero crossings and an unbalanced square wave. A symmetrical square wave is produced in a system using a high-gain dc amplifier and feedback, because the feedback reduces the Q-point shift to a negligible amount.

13.6 CLAMPING CIRCUITS

A clamping circuit generally uses a diode as a switch to hold one extreme of a wave at fixed potential or ground. When, as in a television amplifier, the dc component of the picture signal must be reinserted, the clamp is called a *dc restorer*. As shown in Fig. 13.6(a), a clamping circuit is formed by combining a diode and a long time constant RC coupling circuit. The diode rectifies—i.e., clips—a small part of the signal peaks and charges

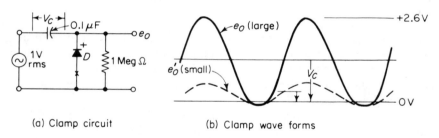

(a) Clamp circuit   (b) Clamp wave forms

Figure 13.6. *Diode clamp.*

the capacitor until the output signal moves over and is clamped to the negative peaks. The drive circuit should have a low resistance to enable easy charging of the capacitor. The shunt resistor should have a high resistance so that the capacitor does not require frequent recharging.

In effect, the diode connects the output terminal to ground whenever the input is negative, and the long time constant maintains the clamped state without requiring too much distortion of the signal peaks. By inserting a battery between the diode and ground [at X in Fig. 13.6(a)], the peaks can be clamped to a voltage off ground. By reversing the diode polarity, the positive peaks of the signal are clamped.

Almost any resistance-coupled amplifier tends to clamp large signal peaks by signal rectification. In a similar manner signal rectification in the collector or drain circuit may produce a dc component that shifts the operating Q-point. Clamping circuits are used whenever the bias point of a circuit must be made to follow the amplitude of an input signal. Often the grid or gate diode of an amplifier is used as the clamp diode, and the fact that a circuit is expected to clamp the input signal may not be obvious.

13.7   SELF-BIASED AMPLIFIERS

The self-biasing amplifier illustrated in Fig. 13.7(a) uses the gate diode as a rectifier to develop bias whenever the input signal is large. A self-biasing amplifier tends to clamp the large signals to ground. Hence, a zero bias Q-point is moved toward cutoff when the input signal becomes large enough to develop a gate bias. A self-biasing amplifier is useful when an amplifier must handle large signals and a bias circuit is impractical. A self-biasing amplifier also offers the advantage that FETs which have differing characteristics can be used without separately adjusting the bias for each FET.

The characteristics of the self-biasing amplifier shown in Fig. 13.7(a) are easily observed. The load resistor R_L is adjusted so that with zero bias the static drain voltage is about 6 V. With the switch S open [Fig. 13.7(a)], the output signal is increased to about 17 V peak-to-peak. Because the gate clamps the input signal to zero bias, the Q-point moves from Q_A to $Q_{A'}$, which keeps the output signal bottoming, thus preventing serious distortion. If the capacitor is shorted so that the gate rectifier cannot change the bias, the drain Q-point remains at A and large signals are clipped by the gate diode.

The diode characteristic of a gate (or grid) also has disadvantages. When an RC-coupled amplifier is subjected to input signals that are much higher than normal, the gate rectifier develops excessive bias and the stage is blocked —i.e., cut OFF—until the capacitor discharges. For some applications blocking forces the use of short time-constant coupling circuits or direct-coupled amplifiers. An important advantage of the insulated gate FETs

Figure 13.7. *Self-biased FET amplifier.*

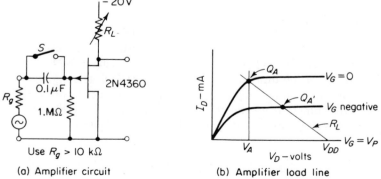

(a) Amplifier circuit   (b) Amplifier load line

comes from the fact that the gate cannot rectify and an RC-coupled MOS FET amplifier is nonblocking.

13.8 GATING, GATES, AND CHOPPERS

At times a waveform is needed that can be formed by turning a sine wave ON for a fraction of a cycle or for one or more complete cycles. A circuit that opens and closes a linear channel is called a *gate, chopper,* or *inhibit circuit*. A gate is simply a fast electronic switch that is operated by an electrical signal. Unfortunately, the gates available until recently have introduced a part of the control signal into the signal channel and have produced a disturbing transient. The FET now makes it possible to open and close a channel with a much smaller transient effect.

13.9 DIODE GATES

The circuit in Fig. 13.8(a) shows one form of a diode gate. The signal channel is from A to B, and the gate control signal is applied at G. The resistors in the channel ensure that the channel impedance is high. The resistors are not required if the source and load have equivalent or higher resistance. The gate diode D_1 is closed when the control at G is $+12$ V, and any signals that are negative or positive up to $+12$ V are inhibited. Depending on the impedance of the gating source, the channel loss is about 40 dB. When the gate signal is OFF, the diode is open for any positive-going pulse or half-wave, as shown in Fig. 13.8(b). Full-wave signals are transmitted by applying -12 V at the gate G.

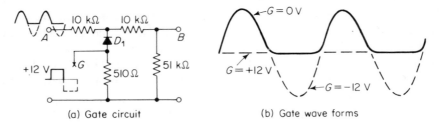

(a) Gate circuit   (b) Gate wave forms

Figure 13.8. *Diode gate.*

The diode gate may be operated by switching the gate from $+12$ V to -12 V, or the gate may be closed by returning the 500 Ω resistor through $+12$ V to ground. The gate is then opened by applying a -24 V pulse at G. Diode gates have the disadvantage that the turn-ON gate signal enters the signal channel as an unwanted transient. Transistor gates have the advantage that the ON voltage from the collector to ground is several times smaller than the ON voltage of a diode gate. The transistor gate shown in Fig. 13.9

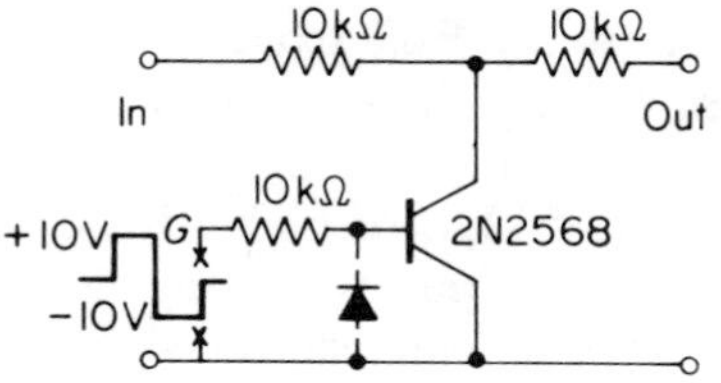

Figure 13.9. *Transistor gate.*

is an effective switch for signals between 10 mV and 10 V, with the lower signal limit set by the fact that the ON voltage of the transistor is 10 mV. If the gating signal is a steady high-frequency square wave and the channel frequencies are low or dc, the transistor is called a **chopper**. Choppers are used to convert dc and low-frequency signals to a chopped high-frequency signal that can be amplified in an ordinary ac amplifier. After amplification, the chopped signal may be reconverted by rectification or synchronous chopping to the original low-frequency signal.

13.10 FET GATE OR CHOPPER

The field-effect devices offer an important advantage for gate and chopper applications: the gate control induces only a very small transient in the signal channel. With no voltage applied to the gate, the channel behaves as a resistor. With the gate biased to pinchoff, the channel resistance is at least 1000 times higher. For low-frequency gate and chopper applications the effect of the gate signal on the channel is usually negligible. If the chopping frequency exceeds a few hundred Hz, then the small capacitance between the gate and the signal channel has a low enough reactance to couple transient spikes from the gate drive voltage into the signal channel. With more complicated circuits which have means for reducing the transient feedthrough, choppers are used at input signal levels of less than 1 μV.

The FET circuit shown in Fig. 13.10 may be used either as a gate or as a chopper. With a negative gate voltage that exceeds the pinchoff voltage, the FET has a dynamic resistance exceeding 100 MΩ. With zero gate voltage the FET has an ON resistance of about 100 Ω. For chopper service the gate should be driven by a square wave which has a peak voltage of about twice the pinchoff voltage of the typical device. The diode D_1 in the gate circuit

Figure 13.10. *FET gate or chopper.*

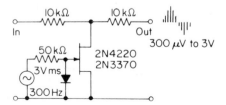

protects the FET from excessive gate current in the forward direction and also reduces the gate signal spikes coupled into the signal channel.

    The chopper circuit shown in Fig. 13.10 is usable at chopper frequencies of up to 5 kHz. At higher frequencies the transients at the chopper frequency set a lower limit to the usable signal. A chopper as simple as shown in the figure is satisfactory for signals from 10 mV to 10 V, a dynamic range of 60 dB.

    The waveforms that are observed in a chopper system are represented by the series of waveforms shown in Fig. 13.11. The waveform *a* represents a low-frequency input signal that has both dc and ac components. The chopper merely eliminates sections of the input signal, as shown in *b*, and the chopped signal still has an average component, dc. Amplification of the chopped signal in an ac amplifier removes the low-frequency signal, and the out-

Figure 13.11. *Chopper waveforms.*

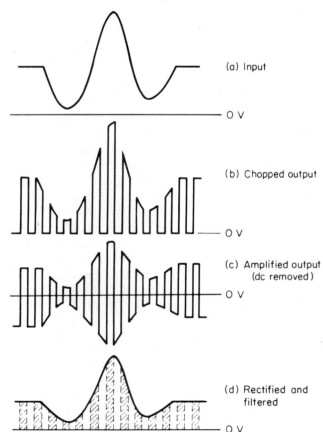

put, shown at c, has approximately equal positive and negative signal excursions. Observe, however, that the dc component of the signal is preserved in the waveform of the output signal because the signal envelope still does not cross the zero axis. When the amplified signal is rectified and smoothed (filtered) to remove the chopper frequency, the final signal is a reproduction of the input signal and includes the dc component, as in d. In effect, a dc chopper amplifier is a modulator-amplifier-demodulator system that is using a square wave, suppressed carrier which operates at the chopper frequency. If the input signal crosses the axis and reverses sign, the output signal must be recovered by a synchronous chopper, because a diode rectifier cannot reverse the sign of the output signal.

13.11   WAVE SHAPING BY FREQUENCY AND PHASE DISCRIMINATION

Almost any practical circuit includes filter networks or has reactance elements that act as filters for removing high-frequency or low-frequency components of the signal. In some cases a filter is inserted to remove unwanted noise, and in other cases a filter is used to exclude very high and very low frequencies from the signal channel. Unfortunately, any network having a nonuniform frequency or phase characteristic tends to change a complex signal waveform. The effect simple filters have on signal waveforms must therefore be considered.

An interesting application of this change of a complex waveform is found in the use of square waves for testing or evaluating the frequency response of amplifiers and networks. In other applications such as television, circuits that change the signal waveform must be avoided, or the distorted waveform must be restored by equalizing networks. In this text we examine briefly the waveform changes commonly observed in circuits which have a single shunt, or series, reactance.

While a pure sine wave always remains a sine wave in linear circuits, the reactance components of a circuit make the amplitude and time phase of a sine wave depend on the frequency. For this reason, the characteristic waveform of a signal that has more than one frequency component is changed when the relative amplitude and phase of the components are changed. Because a square wave is equivalent to an infinite number of superimposed sine waves, one would expect that square waves are considerably changed by circuits which have frequency and phase discrimination.

The effects of frequency and phase discrimination on approximately sinusoidal waveforms are usually analyzed by considering the amplitude and phase changes experienced by the harmonically related sinusoidal components of the given waveform. The modification of square waves and repeated waveforms is more easily analyzed by examining the transient response of the network when the given wave is the input. In most practical applications the transient waveforms are produced by the charge and discharge of energy stored in a single capacitor or inductor; therefore the waveforms are simple exponentials.

13.12 THE STEP FUNCTION

A useful concept for the analysis of transient effects is the *step function*, the signal produced when a voltage is turned ON by a simple switch. As shown in Fig. 13.12, the voltage of a step function is 0 up to $t = 0$, and is 1 thereafter. If the switch is turned ON and OFF alternately, the alternating series of step functions is called a *square wave*. Many signal circuits transmit waves that can be considered as square waves. They are often used for test purposes because they give information concerning the transient response of a circuit which cannot be easily secured by using a sine wave.

time ⟶ *Figure 13.12.* Step function.

When a unit step is applied as the input to an RC low-pass filter, such as is shown in Fig. 13.13(a), the output voltage e_C rises exponentially at a rate that is determined by the time constant $\tau = RC$ of the filter. The output

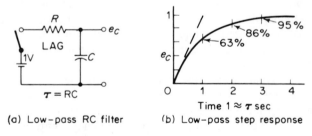

(a) Low-pass RC filter   (b) Low-pass step response

Figure 13.13. Shunt C filter and step response.

voltage e_C is 63 percent of the final output voltage when $t = RC$, 86 percent of the final voltage when $t = 2RC$, and within 5 percent of the final voltage after $t = 3RC$.

When a unit step is applied as the input to a high-pass filter, as shown in Fig. 13.14(a), the output voltage is the instantaneous difference between the unit voltage input and the voltage across the capacitor. Hence, from Fig. 13.13(b) the output voltage just after $t = 0$ is equal to the input voltage and falls exponentially to 0. The output voltage is 37 percent of the input voltage when $t = RC$, 14 percent of the input voltage when $t = 2RC$, and less than 5 percent of the input voltage after $t = 3RC$. The curve which shows the pulse output of a high-pass filter is shown in Fig. 13.14(b).

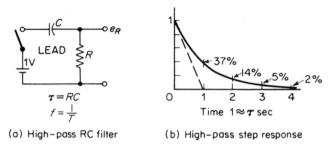

(a) High–pass RC filter   (b) High–pass step response

Figure 13.14. *Series C filter and step response.*

We remark in passing that the output response of a filter that results from a unit step input can be used to determine the sine wave frequency response of the filter, and vice versa. Therefore, a step signal is often used as a quick way of evaluating both the transient and the frequency response of a system.

13.13 SQUARE WAVES

The repeated square wave represented in Fig. 13.15 may be viewed as a series of step functions applied alternately positive and negative at regular intervals. A square wave is said to have a fundamental frequency that is the reciprocal of the time from one turn-ON to the next turn-ON. If the square wave is represented as a Fourier series of sinusoidal waves, the square wave is found to consist of fundamental and odd harmonics which have amplitudes that decrease in inverse proportion to the order of the harmonic. The peak-to-peak amplitude of the fundamental wave is $4/\pi$ times the amplitude of the square wave. The Fourier series of sine waves implies that a square wave is transmitted by a filter as if the signal were a combination of an infinite number of sine waves.

When a square wave is applied as the input to an RC low-pass filter (Fig. 13.16), the output voltage rises and falls exponentially at a rate that is determined by the RC time constant of the filter. If the filter has a time constant that is short compared with the fundamental period of the square wave, the capacitor charges and discharges fully during each half-cycle. Depending on the fundamental frequency, as compared with the 3 dB cut-off frequency, the output response will be of the form represented in Fig. 13.15(a), 13.15(b), or 13.15(c). When the half-cycle time of the signal exceeds

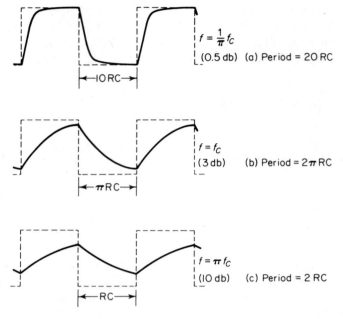

$f = \frac{1}{\pi} f_c$

(0.5 db)   (a) Period = 20 RC

←10 RC→

$f = f_c$

(3 db)   (b) Period = 2π RC

←πRC→

$f = \pi f_c$

(10 db)   (c) Period = 2 RC

←RC→

Figure 13.15. *Square wave low-pass filter response.*

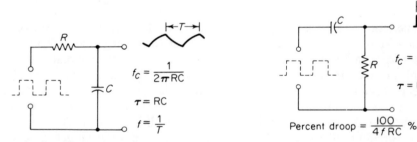

$f_c = \frac{1}{2\pi RC}$

$\tau = RC$

$f = \frac{1}{T}$

$f_c = \frac{1}{2\pi RC}$

$\tau = RC$

Percent droop = $\frac{100}{4fRC}$ %

Figure 13.16. *RC low-pass filter.*   **Figure 13.17.** *RC high-pass filter.*

three times the filter time constant, the output signal has the shape of an almost completed exponential rise or fall, as shown in Fig. 13.15(a) and 13.15(b). If the time constant of the filter is long compared with the half-period of the square wave, the capacitor partially charges each half-cycle, and the resulting output signal is a repeated series of only the initial portion of a charge cycle.

When a square wave is applied as the input to an RC high-pass filter, as in Fig. 13.17, the output voltage is a series of spikes or square waves having sloping tops, as shown in Fig. 13.18. If the time constant of the filter is short

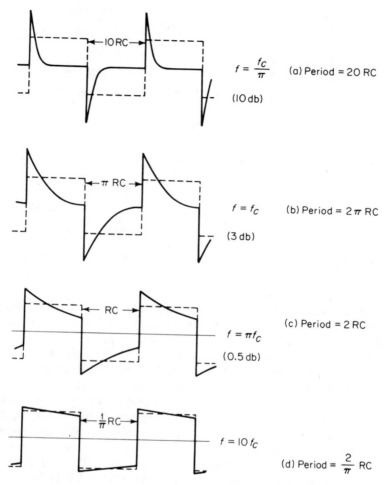

Figure 13.18. *Square wave high-pass filter response.*

compared with the half-period of the square wave, the capacitor discharges so rapidly that the output signal is a series of exponential spikes. If the time constant of the capacitor is large compared with the half-period of the square wave, the capacitor discharges each cycle partially, and the output pulse is a series of square waves which have sloping tops, as in Fig. 13.18(d).

An interesting and important result of the foregoing analysis is shown by the sloping tops of the square waves illustrated in Fig. 13.18(c) and 13.18(d). Although the loss in the high-pass filter at the lowest sinusoidal frequency component is only 0.5 dB, the top of the square wave droops almost 50 percent. As a rule-of-thumb, the droop (see Fig. 13.19), as a percentage

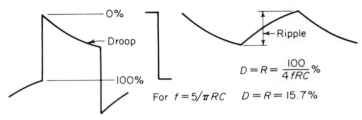

$$D = R = \frac{100}{4 fRC}\%$$

For $f = 5/\pi RC$ $D = R = 15.7\%$

Figure 13.19. *Definitions of droop and ripple.*

Figure 13.20. *High-pass filter example.*

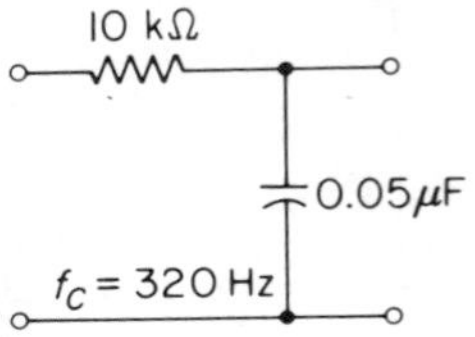

Figure 13.21. *Low-pass filter example.*

of the peak-to-peak input, is $100/4fRC$ for droop values up to about 50 percent. Using the component values given in Fig. 13.20 as an example, the droop with a 3.2 kHz square wave input is 15 percent. This percentage is about twice the tolerable droop of a television video amplifier and shows that the 3 dB low-frequency cutoff of a video amplifier must be more than a decade below the fundamental frequency of the video signal.

Because of the complementary relation between the instantaneous voltages across the capacitor and the resistor, the rule-of-thumb for the droop also gives the peak-to-peak voltage across the capacitor as a percent of the square wave input. Therefore, a low-pass RC filter attenuates the signal by the factor $1/4fRC$, provided the attenuation exceeds 6 dB. Taking Fig. 13.21 as an example, the rule-of-thumb indicates that with a 1 kHz fundamental frequency, $f = 1/2RC$, the square wave loss is 6 dB. At this frequency the sine wave loss factor is about 0.3, or 10 dB. In other words, a low-pass RC filter converts a square wave to a pyramidal wave and attenuates the peak amplitude by about 4 dB less than the attenuation experienced by a sine wave input which has the same fundamental frequency.

13.14 DISTORTED HARMONIC WAVEFORMS

Waveforms that are approximately sinusoidal are usually sine waves that have been distorted in passing through a nonlinear system. Such waves can be represented as sine waves with one or more superimposed harmonics of the fundamental wave. If the amplitude distortion is only 5 to 10 percent, the principal harmonics are usually either second or third, or a combination of both. Higher order harmonics are invariably present, but the principal characteristics of a distorted wave can be understood by neglecting the higher orders. Second harmonic, even-order distortion is produced by nonlinearities that make the positive, or up side, of a wave different from the negative, or down side, of the wave. Third harmonic, odd-order distortion changes both the up side and the down side peaks in the same way.

The distorted wave shown in Fig. 13.22 was made by superimposing a fundamental sine wave—the larger—and a smaller sine wave that is almost, but not exactly, the second harmonic of the fundamental. By this means the succession of waves shows how a distorted wave is changed when the harmonic is shifted in phase relative to the fundamental. When the higher frequency signal is an exact harmonic, the distorted wave is similar to one of the cycles in the figure, but all cycles of the signal are alike. If the distorted wave is passed through a filter, both the relative amplitude and the relative phase of the signal components are changed. Hence, the new waveform may be very different.

The waveforms shown in the upper part of Fig. 13.22 represent a signal which has about 30 percent second harmonic. When the peaks of the signal components coincide, as at A, the up side peak is quite different from the down side peak. Whenever such a wave is observed as the output of an amplifier, we conclude the output stage Q-point is not at the center of the load line because the signal appears to be clipped more on one side than on the other. If the signal has the sawtooth waveform, as at B, the signal probably

Figure 13.22. *Approximately sinusoidal waves fundamental with 30 percent distortion.*

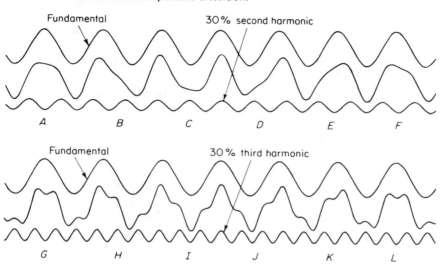

has experienced phase shift also. An amplifier that is overdriven at a funda-mental frequency below the low-frequency cutoff of the output transformer can be expected to have a signal waveform like that at B or D in Fig. 13.22.

Odd harmonic distortion is usually produced by nonlinearities that cause equal limiting of the positive and negative peaks of the signal. The waveform shown in the lower part of Fig. 13.22 represents a signal which appears to have 30 percent third harmonic distortion, but the high frequency is not exactly the third harmonic. When the peaks of the signal components coincide but are 180° out of phase, as at G, the up side and the down side peaks are alike. The "distorted wave" appears flattened and begins to have the characteristics of a square wave. Whenever such a wave is observed at the output of an amplifier, the Q-point is probably centered or the amplifier is balanced, and there is little even harmonic distortion. If the third harmonic is shifted in phase so that the peaks at both frequencies coincide, as at I, the amplifier is probably being overdriven at a frequency below the low-frequency cutoff of the output transformer.

Waveforms of the kind illustrated in Fig. 13.22 are commonly observed in testing amplifiers. A sine wave that is modified by nonlinear distortion of the peaks is a familiar and easily recognized form. However, if the distorted wave is also modified by amplitude-frequency and phase-frequency changes as the result of filtering, the issuing waveform may not be familiar. In most cases, the observed wave can be interpreted by making a comparison with the waveforms given in Fig. 13.22. In other cases, the harmonic components can be identified by a trial and error reconstruction of the observed wave with the use of estimates of the harmonic components.

13.15 WAVEFORM AND SQUARE WAVE TESTS

The reactance elements of circuits and filters change the relative ampli-tude and phase of the harmonic components of a wave. In general, there cannot be an amplitude-frequency change without a corresponding phase-frequency change, and, if one characteristic is given, the other is precisely determined. Therefore, filtering or bandwidth limits in a circuit have their effect in changing the shape of signal waveforms. Where the preservation of waveform is desired, the use of reactance elements should be avoided. On the other hand, if a signal waveform is changed by a frequency characteristic that cannot be removed, an equalizing network may sometimes be construct-ed that will restore the signal to almost the original waveform.

Square waves are used for adjusting equalizing networks. The droop of the square wave is used for adjusting the low-frequency equalization, and the fast rise of the square wave is used for high-frequency equalization. Most oscilloscope probes are adjusted by making a square wave test of the high-frequency equalization. An additional advantage of square wave testing is that the fast rise excites undamped resonances that are made evident as a characteristic ringing waveform.

13.16 WAVE SHAPING BY INTEGRATION AND DIFFERENTIATION (Ref. 4)

From a somewhat different point of view, RC filters are viewed as signal integrators or differentiators. An integrator is a device that produces an output voltage proportional to the algebraic sum of the instantaneous values of the input voltage. A voltage input to the low-pass RC circuit shown in Fig. 13.13 causes the capacitor to receive or lose charge, depending at each instant on the instantaneous voltage input. If the capacitor voltage is always small compared with the voltage across the resistor, the voltage output is proportional to the sum of all the preceding instantaneous input voltages. The magnitude of the output voltage is inversely proportional to the product RC. Hence:

$$e_C = \frac{1}{RC} \int_0^t e_I \, dt \tag{13.1}$$

The high-frequency cutoff of an amplifier integrates signals which have frequencies that are above the cutoff frequency. A sine wave having a frequency ten times the cutoff frequency is converted to a negative cosine wave. The latter is equivalent to a sine wave that lags the input wave by 90°.

If the input to an integrator is a unit step, the output is a linearly increasing voltage called a **ramp function**. The voltage rise across the capacitor of an RC integrator is represented by the exponential curve in Fig. 13.13(b). The curve shows that a linear ramp is produced only for a time that is shorter than about one-half the time constant RC. This result means that a practical circuit integrates an input signal for only a finite time, determined by the RC product. With quality capacitors effective integrating times of several minutes may be obtained.

The foregoing statements are illustrated by the square wave response curves shown in Fig. 13.15. The integral of a square wave is a series of increasing and decreasing ramps. As Fig. 13.15(c) shows, a fairly good integral of the square wave is obtained if the integrating time is less than the time constant of the resistor and capacitor. In fact, the error in integrating the square wave is about 20 percent when the integration is carried up to one-half the RC time constant. For accurate integration the integrating time is usually not more than RC/10.

A differentiator is a device that produces an output voltage proportional to the time rate of change of the input voltage. A voltage applied to the input of a high-pass RC filter (Fig. 13.14) causes the current in the resistor to be proportional to the time rate of change of the voltage across the capacitor. If the voltage across the capacitor follows the input voltage—i.e., if the voltage across the resistor is negligible—the voltage output is proportional to the time rate of change of the input voltage, and the magnitude of the output voltage is proportional to the product RC. Hence:

$$e_R = RC\frac{de_I}{dt} \tag{13.2}$$

If the input to a differentiator is a unit step, the output should be a very narrow pulse or spike. The pulse in Fig. 13.18(a) shows that a differentiator which has a time constant RC can separate—i.e., resolve—abrupt changes of the input signal if separated by five to ten time constants. However, the square wave is differentiated into narrow spikes only by making the time constant of the differentiator approximately equal to the desired time duration of the pulse.

The low-frequency cutoff of an amplifier differentiates signals which have frequencies below the cutoff frequency. A sine wave which has a frequency that is one-tenth the cutoff frequency is converted into a cosine wave. Hence, the output signal leads the input signal by 90°. Because a differentiator transmits high frequencies and attenuates low frequencies, amplifier noise and the pickup of high-frequency interference tend to be aggravated. Therefore, a differentiator is generally avoided or the frequency range is limited.

SUMMARY

The shape of sine waves and similar signals may be changed by clipping circuits using diodes as voltage-sensitive switches. Gates remove or select particular groups of waves or pulses by switching diodes or transistor switches. A gate is generally controlled by an external signal.

A self-biased amplifier operates by clamping the signal peaks to a reference potential and by this means moves the Q-point in response to the signal amplitude.

Nonsinusoidal waves, like square waves, may be shaped by differentiating circuits that accentuate the high-frequency components of the waveform. Similarly, waves are shaped by integrating circuits that attenuate the high-frequency components of the wave.

Waveform distortion occurs when nonsinusoidal signals are transmitted through frequency discriminating networks. Such networks modify the waveform by changing the relative phase and amplitude of the signal components. Sine waves are shifted in phase and changed in amplitude but are not distorted by frequency discriminating networks.

A nonlinear amplitude response distorts waveforms by clipping the waveform asymmetrically or symmetrically. Asymmetrical clipping produces second and higher harmonic distortion, while symmetrical clipping produces only odd harmonic distortion. The resulting waveforms are modified by transmission through a frequency and phase-discriminating network.

Equalizers are networks that restore waveforms by compensating for frequency and phase distortion. Distortion caused in lumped element circuits may be compensated for by simple equalizers, but that produced in transmission lines requires a more complicated type of equalizer.

PROBLEMS

13-1. (a) Describe the performance characteristics of the circuit shown in Fig. P-13.1. (b) Suggest reasons for using this bridge-type of circuit. (c) What happens if the Zener diode is reversed?

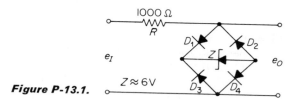

Figure P-13.1.

13-2. The circuit shown in Fig. P-13.2 is called a *slicer*. (a) Sketch the output waveform, assuming ideal diodes and a 30 V rms sine wave input. (b) Repeat, assuming a 2 V rms sine wave input. Neglect the loss in the resistors.

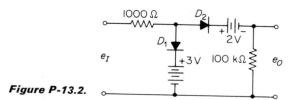

Figure P-13.2.

13-3. (a) Assuming a 5 V rms input signal (Fig. P-13.3), sketch the output waveform and show expected magnitudes. (b) Repeat, assuming R is 5 kΩ. (c) Sketch the waveform expected at A.

Figure P-13.3.

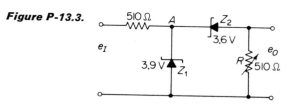

13-4. The circuit shown in Fig. P-13.4 is used as the input section of a peak-to-peak voltmeter. Explain the purpose of D_1 and explain why e_o is proportional to the peak-to-peak input voltage.

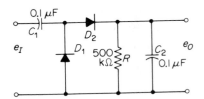

Figure P-13.4.

13-5. Refer to Fig. 13.20 in the text and assume a sine wave input at the frequency $f = 1/2\pi RC$. (a) Calculate the ac voltage loss factor and convert to dB. (b) Why does an integrator reduce the sine wave more than a square wave?

13-6. (a) Confirm the statement that a 3.2 kHz square wave is given a 15 percent droop when passed through the high-pass filter shown in Fig. 13.20. (b) Sketch, with approximate magnitudes indicated, the output obtained when a 3.2 kHz square wave is transmitted through the low-pass filter shown in Fig. 13.21.

13-7. (a) Assume that a 1 V step is applied as the input to the filter shown in Fig. 13.20 and sketch the output wave with magnitudes indicated for the first 100 μS. (b) Repeat, using Fig. 13.21. (c) What happens if the wave is passed through both filters?

13-8. Derive Equation (13.1).

13-9. Derive Equation (13.2).

CHAPTER FOURTEEN

High-Frequency Circuits and UHF Applications

The changes we find in circuits used for high-frequency applications are usually brought about by the need to offset the effects of capacitance. In tuned amplifiers the transistor input capacitance is offset by making the capacitance a part of a resonant coupling network. The instability that makes the adjustment of tuned amplifiers difficult is reduced by neutralizing the capacitance feedback or by mismatching to limit the transistor voltage gain. In broadband amplifiers the high-frequency cutoff produced by capacitance is moved to a higher frequency by using low-impedance circuits and by applying feedback to reduce the amplifier gain in exchange for increased bandwidth.

At UHF and higher frequencies the simple diode has many interesting applications. Diodes are particularly useful for changing signal frequencies to frequencies at which transistors can be used as amplifiers or oscillators. Diodes are used as voltage-variable capacitors for tuning, frequency control, and frequency multiplication. Tunnel and other diodes are used as amplifiers and oscillators at the microwave and higher frequencies.

This chapter is mainly concerned with the techniques and circuits that make transistors and diodes so useful in high-frequency applications. Circuits for the UHF spectrum are emphasized because these circuits illustrate the techniques used at lower frequencies and exhibit some of the problems encountered at microwave frequencies.

14.1 TRANSISTORS AT HIGH FREQUENCIES

The gain of a high-frequency transistor amplifier is limited by capacitance in the transistors and circuit components. As the design frequency is increased, the capacitance effects decrease the attainable power gain of each stage inversely with frequency, and a high-gain amplifier may need many stages. Because the effects of capacitance must be carefully controlled, the circuits used for RF and UHF amplifiers are very different from those of low-frequency amplifiers, and the transistors must have both low input capacitance and low feedback capacitance.

The base-emitter capacitance of a transistor bypasses a part of the input current and begins to reduce the transistor current gain at a relatively low frequency, called the *beta cutoff frequency* f_β. As is shown in Fig. 14.1, the transistor current gain β decreases with increasing frequency and is reduced to a current gain of 1 at a frequency f_T, called the transistor *current gain-bandwidth product*. The transistor data sheets generally give the current gain of a typical transistor and a frequency at which the current gain is between 5 and 10. The product of the given frequency and the current gain is the transistor gain-bandwidth product f_T.

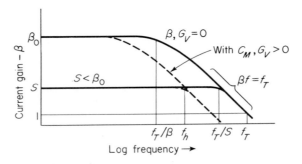

Figure 14.1. *Beta-frequency curve and stage f_n.*

Additional input capacitance is produced by the Miller effect capacitance C_M when a stage has voltage gain. Thus, the combined effect of the transistor input capacitance and the Miller-effect capacitance is to reduce the usable gain bandwidth product to a relatively low value that is represented by the broken line in Fig. 14.1. However, when a stage is designed with feedback reducing the current gain to S, the frequency response may be made relatively flat up to the frequency f_h shown in Fig. 14.1. Thus, each stage of a wide-band amplifier is generally designed with relatively low values of current gain and voltage gain, and a high-gain amplifier may need many stages with only 5 to 10 dB gain per stage.

The transistors of a low-level RF amplifier are generally biased as if they were an audio stage with emitter feedback biasing, and the emitter resistor is bypassed at radio frequencies by a small emitter capacitor. Amplifiers for high-level FM signals can be used without bias because only the zero crossings of the signal are important, and these stages are generally designed to limit the signal by peak clipping. Amplifiers for pulse signals are designed as wide-band amplifiers and are often operated without bias in order to shape the signal.

The small physical dimensions of RF transistors and the effects of capacitance make these devices significantly different from audio transistors. RF transistors have lower breakdown ratings because the junctions are relatively thin. Therefore, the operating Q-points are usually nearer the maximum voltage and power ratings of the transistors.

14.2 BROAD-BAND (VIDEO) AMPLIFIERS (Ref. 2)

A TV video amplifier is a broad-band amplifier designed to transmit all frequencies between 10 Hz and 4 MHz Such a wide band is required in order to reproduce faithfully the fine detail of a TV picture. A simplified form of a video amplifier is shown in Fig. 14.2 to illustrate the limitations produced by capacitance and the techniques used to obtain a wide frequency response.

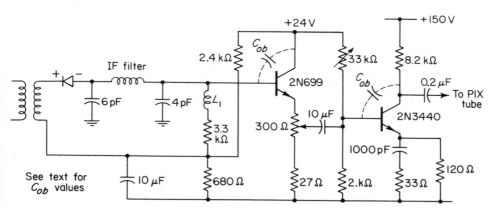

Figure 14.2. TV video amplifier (10 Hz to 4 MHz).

The first stage of the video amplifier is a CC stage with an input impedance of 3 kΩ and an emitter load of about 300 Ω. The 2N699 transistor has a minimum $f_T = 50$ MHz, so the current gain S = 10 implies a gain bandwidth product of about $f_T/S = 5$ MHz. There is no Miller effect in a CC stage, but the collector-to-base capacitance $C_{ob} = 20$ pF does act as a high-frequency shunt which is offset by resonance with the compensating inductance L_1.

The CE output stage of the amplifier provides a voltage gain of about 30 which, with $C_{ob} = 10$ pF, makes the Miller effect capacitance at least 300 pF. This relatively high input capacitance can be tolerated only because the CC stage reduces the equivalent ac source impedance to 50 Ω, or less, depending on the gain setting. Thus, the input capacitance produces a 3 dB loss at 10 MHz or higher. The ac S-factor is so low that the transistor $f_T = 15$ MHz determines the current-gain cutoff frequency. Together the Miller effect and the gain-bandwidth factor reduce the high cutoff frequency to approximately 6 MHz. Since the frequency response is degraded in both stages, the net bandwidth of the two-stage amplifier is about 4 MHz. A more accurate calculation of the frequency response is hardly worth the trouble, since the response of a working amplifier depends to a considerable extent on the capacitance of the components and the stray capacitance between all parts of the circuit. Careful shielding and lay-out are required in wide-band amplifiers.

In a practical TV application the amplifier is used to drive a picture tube cathode, which contributes an additional load capacitance. This additional load and the need to shape the transient response by equalizing components complicate slightly the amplifier circuit. A complete circuit and description of the amplifier as actually used in a TV receiver are described in Ref. 2.

14.3 TUNED AMPLIFIERS (Ref. 2)

In many applications the band-width of frequencies to be amplified is only a small percentage of the center frequency, so a tuned amplifier may be used to select the desired band of frequencies and reject all others. Tuned circuits are commonly used at radio frequencies to eliminate interfering signals or the harmonic distortion of oscillators and Class C amplifiers.

Tuned circuits offer several advantages in high-frequency circuits. The transistor and circuit capacities are made a part of the tuned circuit and the amplifier input and load impedances may be 10 or more times higher than without tuning. Tuned circuits with a tap on the inductor provide a simple means for transforming or matching impedances. In this way the relatively low base-to-collector power gain of a high-frequency transistor is conserved, and the power may be transformed without loss to a following stage or a low-impedance load. Thus, the power gain of a tuned stage is simply the product of the transistor ac current and voltage gains,

$$G_p = SG_v \qquad (14.1)$$

An amplifier tuned for frequencies in the decade above the β-cutoff tends to be unstable or to oscillate unless the voltage gain is limited to about 10 or the feedback is carefully neutralized. Stability problems and tuning interactions between stages generally limit the useful power gain of a broadcast IF amplifier to about 24 dB per stage. For these reasons an IF stage may be constructed with considerable mis-matching between stages or may have resistors connected as gain-reducing loads.

At frequencies near f_T, where the transistor has a low current gain, useful power gain is obtained by making the collector load impedance high enough to produce a base-to-collector voltage gain. Thus, in an iterated stage the inter-stage step-down transformer serves to increase the overall current gain by reducing the voltage gain until both are equal, as required by the TG-IR. If the base-to-collector voltage gain is 10 and the transistor current gain is 1, then with efficient coupling and matching we may expect an iterated power gain of 10 dB. With neutralization, a stage may have a power gain of

12 to 15 dB at the transistor f_T frequency. With a fixed voltage gain, as required for stability, the power gain decreases linearly with increasing frequency because the transistor current gain decreases in the same way. For these reasons, the power gain of tuned RF and UHF stages decrease linearly with increasing frequency and is relatively low at frequencies near the transistor f_T frequency. Curves showing this power gain characteristic of UHF transistors may be found in transistor handbooks.

The circuit of a typical broadcast-receiver tuned IF amplifier is shown in Fig. 14.3. Single-tuned, closely coupled transformers are used between stages with the tuned circuits designed to provide the desired selectivity. The interstage transformers are tuned either by adjustable capacitors or magnetic cores, and one winding usually has so many turns that the collector or the base connects to a tap on the winding. The excess winding is used either to accommodate the tuning slug better or to increase the impedance so that the capacitor is physically small. The collector and base connections may be tapped down on the windings to limit the effect of the internal collector resistance on the Q or to reduce the stage gain by mismatching to avoid the need for neutralization. When only one winding is tuned, the coils should be closely coupled. If greater selectivity is desired, the IF transformers may have both windings tuned and loosely coupled.

When the primary and secondary are both tuned, the circuits may be designed with high Qs and with a high efficiency of power transfer. When the primary and secondary are tuned to the same frequency (synchronous tuning) and are coupled for efficient power transfer at the center frequency, the response falls off rapidly at the ends of the passband and has a high degree of attenuation far removed from resonance. More closely coupled, tuned circuits

Figure 14.3. *IF amplifier—450 kHz.*

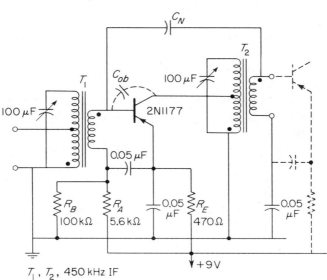

may have a flat response over most of the passband. When the circuits in separate stages are stagger tuned, the flat passband may be extended even farther by accepting slightly reduced gain. The design of tuned coupled circuits is described in Ref. 2.

For small-signal applications a tuned stage is biased class A the same way a stage is biased in an audio-frequency amplifier. As shown in Fig. 14.3 a tuned stage usually has an emitter resistor and a base-circuit voltage divider so that the dc collector current is controlled by the bias resistor and the dc S-factor. In RF applications the emitter resistor is usually bypassed to make the ac current gain as high as possible.

The method by which the tuned amplifier (Fig. 14.3) is biased to make the collector current 0.5 mA is left for study by the reader. The circuit shown in the figure is typical of those used for biasing most transistor radios.

14.4 NEUTRALIZATION

In any high-gain circuit an amplifier oscillates and is useless when the forward gain exceeds the loss from the output back to the input. At frequencies just below the peak frequency of a tuned amplifier, feedback through the collector-to-base capacitance returns a signal in phase with the input signal, and any such stage may be unstable or oscillate. As a practical rule, a stage with tuned-base and tuned-collector circuits oscillates if the forward voltage gain is 30 or more.

The instability produce by feedback in a stage can be offset by providing additional feedback in the opposite phase. If the feedback circuit cancels the effect of all resistance and capacitance feedback, a stage is said to be *unilateralized*. If the circuit cancels the effect of only the reactance feedback, a stage is said to be *neutralized*.

The unilateralization of tuned stages is generally considered impractical because the circuit requires a delicate balance of several feedback effects which makes the adjustments difficult and the amplifier unstable for small changes away from the balance conditions.

For similar reasons the neutralization of a high-gain, multi-stage amplifier is generally time consuming and troublesome because the tuning of a collector circuit affects the tuning of the previous base circuit and may even interact from stage to stage. However, these difficulties may be removed by deliberately reducing the gain per stage sufficiently to ensure a desired degree of stability and ease of tuning. The price paid for the improved performance is lower stage gain, but we obtain ease of design, ease of alignment, and a simpler circuit. When the cost of careful neutralization is important, an amplifier may have a combination of reduced gain and partial neutralization, a practice generally used in the IF amplifiers of transistor radios.

A typical tuned amplifier that uses neutralization is shown in Fig. 14.3.

The neutralization capacitor C_N is connected between the base and the output winding which is phased 180° out of phase with the collector voltage. The neutralized condition is determined by adjusting C_N until the signal voltage at the base does not change when the collector circuit is tuned. With low available circuit gains, where neutralization is most needed, this adjustment is easy and effective. Residual effects that may be observed either require a resistive component of feedback (unilateralization) or indicate additional stray circuit coupling.

14.5 UHF TUNED AMPLIFIER

A UHF tuned amplifier is similar in many respects to an IF amplifier with tuned input and output, except that the circuit must be simplified with as many components grounded on one side as possible. In the UHF spectrum a circuit is usually placed inside a copper enclosure to provide short low-impedance grounds and for shielding.

A typical low-noise, broad-band, 200-MHz amplifier circuit is shown in Fig. 14.4. The amplifier has tuned input and output circuits with neutralization to increase the stability and ease of tuning.

The amplifier uses a CE stage that is biased as if it were a CB stage in order that the input inductor and one tuning capacitor may be grounded. In addition the bias circuit is bypassed by a single feedthrough capacitor so the emitter may have a very short ground connection, which is necessary to prevent regenerative feedback. Similarly, one collector tuning capacitor is grounded and the collector supply is brought in through a feedthrough capacitor. The amplifier has a 20 dB power gain between a 50 Ω source and a 50 Ω load, but a 300 Ω source and load (a 300 Ω line) may be used with similar performance characteristics except for reduced gain.

Each of the tuned circuits has two tuning capacitors which permit an

Figure 14.4. *Broad-band 200 MHz UHF amplifier.*

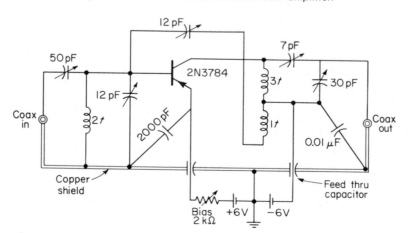

adjustment of both the resonant frequency and the impedance match. The input is connected in series with a tuning capacitor to provide an impedance step-up, while the load is coupled across a capacitance voltage divider which has the advantage of bypassing the harmonic frequencies to ground. When the tuned circuits are designed with a Q of 10, the amplifier bandwidth is 10 per cent of the center frequency, and the 20 MHz frequency band makes this a broadband amplifier.

The transistor may be a high-frequency germanium device which has an f_T exceeding 1 GHz. A germanium transistor generally offers a high f_T or a lower cost than is available with a silicon transistor. Neutralization is required because of the relatively high power gain, but may be omitted by using a transistor with an $f_T = 200$ MHz and by accepting about 15-dB gain. The transistor is operated with a 3-mA emitter current that is adjusted by the variable resistor in the $+6$-V emitter supply. At 200 MHz the transistor current gain is about 5, so 20-dB power gain requires a collector load impedance that is at least 1000 Ω.

14.6 MOS FETs IN RF AND UHF AMPLIFIERS (Ref. 7)

Recent improvements of the insulated-gate, field-effect transistors make the MOS FETs suitable for many VHF and UHF applications. The MOS FETs offer low noise, low distortion, and a low feedback capacitance that permit their use as replacements for vacuum tubes in nearly any high-frequency, low-power amplifier.

The triode types of MOS FETs have drain-to-gate feedback capacities that are typically 0.1 to 0.7 pF with transconductance values of 7500 μmho. MOS FETs with dual gates have a feedback capacitance as low as 0.1 to 0.2 pF with the transconductance exceeding 10,000 μmho. These devices offer unneutralized power gains of 15 to 20 dB at 200 MHz and 10 to 13 dB at 400 MHz. The dual gate reduces the feedback capacitance, and one gate may be used for the input while the other is used for the local oscillator (LO) input of a mixer. The low-value leakage current of the insulated gate eliminates input circuit loading caused by high input signals. Thus, a MOS-FET amplifier tolerates large signals without causing either detuning or Q-reduction.

The circuit of a typical UHF dual-gate, MOS-FET amplifier is illustrated in Fig. 14.5. The input circuit is essentially the same as for a vacuum tube amplifier, while the output circuit has as high a load impedance as the stability permits. Bias is provided to the second gate by the voltage divider R_2 and R_3, or the gate may be returned to an AGC control voltage or a LO signal for mixing. In a single-stage amplifier some MOS FETs may be operated with the gate G_2 connected to the source. Circuits with component values and electrode voltages required for a specific FET may be found in the manufacturer's data sheet.

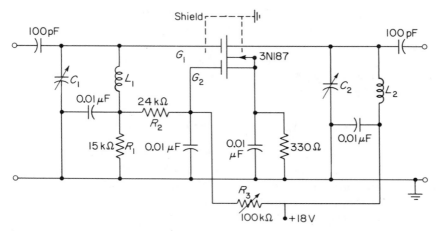

Figure 14.5. *Dual-gate MOS FET RF amplifier—to 100 MHz (frequency depends on values of C and L.)*

The gate-to-drain feedback capacitance of JFETs is typically 1 to 3 pF, which generally makes these devices unsuitable for RF applications. JFETs may, however, be used to replace almost any low-power vacuum tube triode. Of course, such replacements require suitable changes of the bias and reduction of the *B*-voltage to a value suitable for the JFET collector. The replacement of vacuum tubes by MOS FETs requires similar changes, but the tuned circuits and their component values may be used without change.

14.7 RADIO-FREQUENCY POWER AMPLIFIERS (Ref. 5)

When an amplifier is required to supply several watts of power, the transistors are generally operated class-C because the higher efficiency permits more power output from a given transistor. As with class-A power amplifiers, the collector load resistor for a class-C amplifier is determined by the supply voltage and the power output, but the optimum design relations are not easily represented by a formula. Moreover, the characteristics of transistors operated class-C are quite different from their low-power characteristics.

The circuits used for class-C operation are similar to the linear circuits of this chapter except that the resistors for forward biasing the transistors are removed. For example, the amplifier shown in Fig. 14.3 may be operated class-C by removing the emitter bias resistors and connecting the emitter either directly to ground or through a resistor shunted by a large capacitor. The bias is produced by clipping the signal peaks and storing the charge in a capacitor so the bias varies with the signal amplitude. Because the bias tends to suppress most of the signal, the collector current is a series of high-energy pulses at the signal frequency. The tuned collector tank circuit converts these pulses to a sine wave at the signal frequency. If the tank is tuned to a harmonic frequency, the amplifier is called a *frequency multipler*. By frequency multiplication, a crystal frequency standard may be used to regulate the frequency of a UHF transmitter.

14.8 DIODES FOR RF AND UHF APPLICATIONS (Ref. 3)

The semiconductor diode is probably the most versatile and most nearly ideal of all semiconductor devices. The simplicity and quality of the junction diode make it useful in many different ways for LF and RF applications, while at microwave and high frequencies the older point-contact diode remains indispensable. The most common uses of signal diodes are for RF current, voltage, and power measurements; for detection of modulated signals, as in a radio or a TV receiver; and for converting high-frequency signals to lower frequencies, as in a mixer. These applications use the diode as a high-frequency rectifier, and the low-frequency or dc component of the diode current is the desired signal. Other diodes are used as voltage-controlled capacitors for tuning and frequency control, as energy converters for frequency multiplication, and as sources of microwave power or noise.

The characteristics of diodes are determined by the materials and processes used in their manufacture. The alloy and diffused junction diodes that are supplied for low-frequency applications have nearly ideal forward and reverse characteristics, but at radio frequencies they are practically useless. In these diodes conduction charges are stored in the junction during forward conduction and are returned when the diode is reversed biased, so that the diode conducts reverse current. Since the conduction continues for a *reverse recovery time* that is independent of the signal frequency, the efficiency of a junction diode decreases with increasing frequency. The junction diodes designed particularly for RF applications are generally much less efficient that a point-contact diode at frequencies above 100 MHz.

Selected point-contact germanium and silicon diodes have long been the best choice for high-frequency applications. These and the recently developed Schottky diodes are free of stored charge because the current carriers are electrons. However, capacitance across the junction and between the leads becomes a low-reactance shunt that reduces the ON to OFF impedance change and reduces the diode efficiency with increasing frequency. By reducing the circuit impedance, the effect of capacitance may be reduced slightly, but this practice requires an appreciably higher signal power to drive the diode to a low impedance. Thus, at high frequencies there is little choice but to use higher signal levels, and the semiconductor temperature rise becomes an important consideration.

When selected for low-capacitance, the point-contact diodes are preferred for UHF and microwave applications. Germanium diodes are inexpensive and able to rectify low-voltage signals but have a high reverse leakage and are temperature sensitive. Silicon diodes are less sensitive but operate better at high temperatures. The Schottky, or *hot carrier*, diodes are more uniform and more predictable than point-contact devices, but the latter still excel in applications at frequencies above about 1 GHz. With further improvement the Schottky diodes may supplant the point-contact devices.

14.9 DIODE DETECTORS

Detectors use the nonlinear characteristic of diodes to convert the amplitude changes of a high-frequency signal to a corresponding low-frequency or dc signal, which is the information originally modulated on the high-frequency signal. The detectors used at microwave frequencies are small-signal detectors in which the dc output signal is proportional to the square of the high-frequency amplitude. The detected low-frequency signal is easily amplified for presentation as the transmitted information. A linear detector uses the diode to rectify a large signal. The low-frequency output is proportional to the high-frequency amplitude and is more linear than with a square-law detector.

Because tunable amplifiers are expensive and complicated for frequencies above 100 MHz, signals at these frequencies are often detected without prior amplification. Point-contact diodes are used extensively for low-level detection in the UHF and microwave frequency range, but Schottky diodes are more uniform and less subject to burnout by overloads. Both types operate as square-law detectors for input power levels from 0.001 μW, -60 dBm. up to about 1 μW, -30 dBm. Both types require a few microamperes of forward bias current to perform as square-law detectors.

For large-signal detection, Schottky diodes operate as linear, peak-voltage detectors when the peak-to-peak signal exceeds 0.25 V. The hot-carrier diode has the advantage of a decided nonlinearity between forward and reverse voltages of 0.25 V, which are a factor of 2 or 3 lower than for a silicon junction diode. Moreover, hot-carrier diodes have the advantage that their high-reverse-breakdown voltage, at least 70 V, allows the diode to be used at high signal levels and over a wide amplitude variation of the modulated carrier. Thus, these newest members of the diode family offer a distinct improvement for linear detection in many applications.

The circuit for a diode detector is essentially that of any single-diode rectifier with the output driving a low-frequency amplifier. A square-law detector is similar except that the low-frequency output is measured with a sensitive dc meter or is amplified by a high-gain dc or audio amplifier. The circuit of a moderately low-level detector is shown in Fig. 14.6 and the same

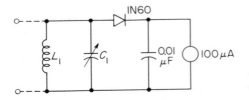

Figure 14.6. *Low-level high-frequency detector.*

circuit may be used as a square-law voltmeter. Examples of high-level linear detectors are illustrated by the video detector that may be found in any TV circuit.

14.10 DIODE VOLTMETERS AT UHF FREQUENCIES

The application of diodes for voltage measurements at high frequencies is illustrated by the circuit in Fig. 14.7. This circuit uses a Schottky diode to charge the 0.01 disc capacitor. The meter is generally calibrated to read rms voltage, but the deflection is actually proportional to the peak-to-peak signal voltage. The variable resistor at the left of the meter is necessary to provide a circuit for dc current and provides a convenient means for calibrating the meter reading. The input capacitor protects the diode from dc voltages, provided the capacitor is not too large.

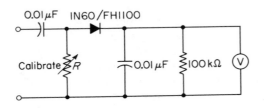

Figure 14.7. *High-frequency ac voltmeter.*

The voltmeter calibration is independent of frequency until the diode capacitance shunts the nonlinear junction and reduces the ac voltage across the diode. Thus, the cutoff frequency varies with the impedance R_S and with the input voltage. When the input signal is sinusoidal and a few volts rms, the rectification efficiency is reduced to 50 per cent when the reactance of the diode capacitance equals the source impedance R_S. For example, the capacitance of the Schottky diode is given as 1.2 pF, so the calculated cutoff frequency of the voltmeter is 14 MHz with a 10 Ω source impedance. The observed cutoff frequency is about 25 MHz, and with a point-contact germanium diode the cutoff frequency is 40 MHz.

The cutoff frequency of the voltmeter may be increased approximately 1 decade for each factor-of-10 reduction of the source impedance. When R_S is 1 kΩ, the cutoff frequency with a germanium diode is about 400 MHz. With lower impedances the rectifier provides useful indications at microwave frequencies. Decreasing the source impedance, forces the diode to rectify at a lower impedance level and reduces the effect of the capacitance, but increases the amount of RF power required to produce a given signal. Probes for high-frequency TV servicing use similar circuits and with germanium diodes give accuracies of ± 2 dB (26 per cent) up to 250 MHz.

The range of input voltages that is handled best by a crystal voltmeter is between 0.1 and 2 V. When the applied voltage exceeds 20 V rms, an attenuator must be inserted to reduce the voltage actually applied to the crystal. Capacitance attenuators are preferred at UHF and GHz frequencies.

Because appreciable power is required to operate high-frequency voltmeters, the diode that converts the ac signal to dc tends to clip the signal on one or both peaks and loads the circuit being tested. The load on the circuit is determined in part by the dc load: hence, the dc meter should have a high sensitivity.

14.11 DIODE MIXERS

A mixer is a circuit used to convert an incoming high-frequency signal to a lower modulated frequency. Conversion is accomplished by driving a diode simultaneously with the high-frequency input signal and a local oscillator (LO). The diode is driven by the sum of the two signals, and the nonlinearity of the diode produces a current that includes sum and difference frequencies. Thus, mixing a 470-MHz UHF TV signal and a 515-MHz local-oscillator signal produces a 45-MHz signal that can be amplified in the IF amplifier of a TV receiver. By this means a UHF tuner is able to convert UHF signals at power levels down to 0.01 μW, depending on the diode conversion efficiency and noise level.

The Schottky diode is a highly efficient mixer. It has less conversion loss and thereby produces lower mixer noise than for any other diode. Its lower impedance level provides better impedance matching, and the high efficiency at high signal levels reduces adjacent channel interference and distortion problems.

The circuit of a UHF TV tuner that employs a diode mixer is shown in Fig. 14.8. The local oscillator is a transistor CB stage tuned by the capacitor C_3. The antenna is connected across the parallel-tuned circuit C_1 and L_1. Because the diode presents a low impedance, the diode loop is a series-resonant circuit C_2 and L_2 inductively coupled to the input circuit. The LO is loosely coupled to the diode, and the power input to the diode is only a few milliwatts. The capacitor C_4 and inductance L_4 are resonant at the 45-MHz IF frequency, so the capacitor has a low impedance at UHF frequencies. The diode may be either a UHF germanium diode or a Schottky diode.

By fabricating the inductors as thin copper strips and enclosing the

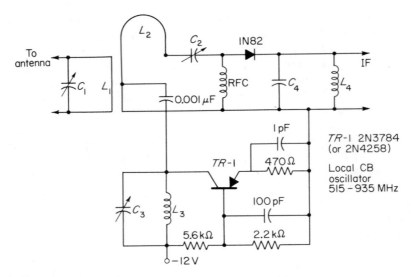

Figure 14.8. *UHF TV tuner (mixer).*

tuned circuits in copper-or silver-plated steel enclosures, the tuned circuits have high no-load Qs which provide a high degree of selectivity and a signal bandwidth of about 10 MHz.

14.12 VARACTOR DIODES

The capacitance of a reversed biased diode may be reduced a factor of 3 or more by increasing the applied voltage. For silicon diodes with a capacitance C_o when the reverse voltage is $V_o = 1$ V, the capacitance for a voltage V_R is given by the equation

$$C = \frac{C_o}{\sqrt{V_R/V_o}} \tag{14.3}$$

which shows that the capacitance varies inversely with the square root of the applied voltage. The control of capacitance by a voltage is particularly attractive at UHF and microwave frequencies, where a variable air capacitor is many times the size of a microcircuit.

Silicon tuning diodes provide a capacitance variation of 3 to 1, and in a few low-frequency, low-voltage types, a capacitance variation of 10 to 1. The voltage-variable capacitors offer advantages in automatic frequency control and for amplifier tuning from 1 MHz through microwave frequencies.

A high-Q capacitor is desirable in most applications. The Q of tuning diodes varies inversely with frequency, and the Q value is generally given at 100 MHz. The Q of a typical diode capacitor varies over the intended frequency range from 20 to 200, and at 100 MHz is between 50 and 500.

The advantages of a diode for tuning include simplified circuits and construction, a higher reactance range, faster response times (vital in frequency sweeping circuits), and the ability to control a multiplicity of devices simultaneously.

14.13 VARACTOR DIODE TUNING

Voltage-variable capacitors, VVC, are commonly used as the voltage-variable tuning element in tuned amplifiers, for automatic frequency control of oscillators, and for frequency modulating an oscillator. In these applications the amplitude of the ac signal across the varactor is relatively small, so the capacitance is only a function of the dc, or low-frequency, control voltage. Tuning varactors are designed to have high Q values and large capacitance variations with voltage.

A circuit used for tuning an L-C resonant circuit is illustrated in Fig. 14.9. The tuned circuit is formed of L_1, which is tuned by C_1 and the diode capacitance C. The diode voltage is varied by the potentiometer R_1. The resistor R_2 isolates the diode from other diodes that may be connected to the potentiometer, and the resistor has a large value to prevent loading of the tuned circuit. There is a negligible current in R_2. The capacitor C_2 is used to block dc from the tuned circuit and may be a small capacitor if the capacitance change in the diode is more than needed for frequency control.

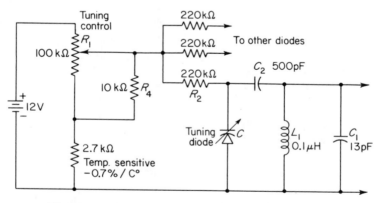

Figure 14.9. *FM varactor tuning circuit, 88–108 MHz.*

14.14 A VOLTAGE-CONTROLLED MICROWAVE OSCILLATOR

Transistor oscillators for service above the transistor f_T rating generally use the transistor in the CB configuration. Since the ac collector voltage is in phase with the emitter voltage, a CB stage tends to be unstable with sufficient collector-emitter feedback capacitance or with an impedance between the base and ground. With tuned emitter and collector circuits the regenerative feedback is generally high enough to make a transistor useful as an oscillator up to about twice the f_T rating. At these relatively high frequencies the stray inductance in the emitter and collector leads may either aid or reduce the regenerative feedback so the physical layout of the components usually has a marked effect on the oscillator performance. Thus, when the parasitic lead inductances and stray capacities become of first order importance, the exact form of a microwave circuit may not be clearly defined.

A high-frequency voltage-controlled oscillator (VCO) is illustrated by the circuit shown in Fig. 14.10. The transistor is connected for CB operation with a $+12$ V bias supply and a -12 V collector supply. The emitter and collector circuits are tuned by the varactors C_1 and C_2. Since the capacitance of a varactor may be changed by a factor of 4, the oscillator frequency may be changed by a factor of 2. Similarly, the oscillator may be frequency-modulated by a signal that varies the voltage applied to one or both varactors.

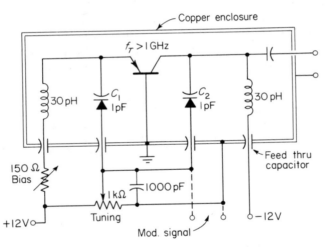

Figure 14.10. VCO or modulated UHF oscillator.

The inductance and capacitance values shown in the oscillator circuit are approximately correct for operating frequencies from 500 to 1000 MHz provided the transistor is biased for an f_T value of at least 1 GHz. At these frequencies the transistor capacitance usually provides adequate feedback, in which case the external capacitor C_f may be omitted. Frequencies a factor of 2 higher may be obtained by interposing an output filter tuned to select the second harmonic. For lower frequencies the CB oscillator operates as a Colpitts oscillator by increasing the collector-to-base feedback capacitance

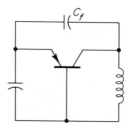

Figure 14.11. *Colpitts oscillator equivalent of a UHF oscillator.*

C_f and retuning the emitter and collector circuits. Tuning the emitter circuit below the operating frequency makes the emitter-to-base impedance a capacitance and tuning the collector circuit above resonance makes the collector impedance an inductance. Thus, the oscillator has the form of a Colpitts oscillator, as shown in Fig. 14.11. The local oscillator used in the TV tuner, Fig. 14.8, is evidently a Colpitts oscillator operating just below the resonant frequency of L_3 and C_3.

14.15 CIRCUITS FOR RF AND MICROWAVE FREQUENCIES

One of the most important considerations at UHF frequencies is the layout of the circuit and the choice of the components. All leads must be as short as possible, grounds must be carefully placed, and each stage and many components must be carefully shielded.

The capacitance to ground or the inductance of a short length of wire may have a considerable effect on the tuning or the impedance of a circuit at UHF frequencies. For example, 1.2 centimeters of wire shaped as a single-turn loop is 10 nH which has an inductive reactance of 30 Ω and resonates with 10 pF at 500 MHz. Similarly, the impedance level of circuits at UHF frequencies must be relatively low because the reactance of 1 pF of stray capacitance is only 320 Ω at 500 MHz.

Component grounding at high frequencies sometimes presents a substantial challenge. A CE stage with too long an emitter ground lead may oscillate, or coupling between leads connected to a common ground may cause feedback problems. Because the inductance of a common ground wire cannot be tolerated, the common side of a circuit is usually a sheet conductor that may be copper or a silver plated brass or steel. When formed to enclose a circuit, the conductor confines the electric and magnetic fields and greatly reduces problems with feedback and interaction between circuits.

At high frequencies bypass capacitors have series inductance and chokes have shunt capacitance, so unexpected resonant effects may occur. Consequently, these elements must be carefully selected, preferably with the aid of a high-frequency bridge. Low-frequency blocking, often observed, is generally a symptom of high-frequency oscillations caused by inadequate bypassing of bias and power circuits. These circuits should be supplied through feedthrough capacitors and bypassed by high-frequency ceramic capacitors connected close to the transistor.

Circuits at UHF frequencies have the advantage of a small physical size that makes then useful in portable equipment. One wave length in free space at 300 MHz is 1 meter, and a quarter-wave antenna is only 25 cm long. Capacitors are easily formed by twisting a pair of wires, and tuning adjustments may be made by simply moving a connecting lead or spacing the turns on a coil.

At microwave frequencies above about 1 GHz, coils, capacitors, and tuned circuits are replaced by (or combined with) coaxial lines and microwave cavities. Physically small circuits, called *thin-film micro-circuits*, are constructed by the photo and etching techniques used in making integrated circuits. For small production lots thick-film circuits are constructed by printing a metal bearing paste on an insulating substrate and heat treating the assembly to produce conductors, resistors, and capacitors. Both types of micro-circuits present difficulties with tuning and circuit adjustments.

Micro-circuits may use either uncased transistors or transistors especially designed for microwave applications. Microwave transistors usually have multiple or wide leads to reduce the effects of lead inductance and shielding to minimize the input and feedback capacitance. The transistor temperature rise is controlled by designing a low internal thermal resistance and providing a contact for an external heat sink.

Throughout the high-frequency spectrum semiconductor devices are still surpassed by vacuum-tube devices. Above 10 GHz traveling wave tubes, magnetrons, and backward-wave oscillators produce kilowatts of power. However, semiconductor devices have important advantages of simplicity, small size, and high reliability.

SUMMARY

The gain and bandwidth of high-frequency amplifiers is limited mainly by the base-emitter capacitance of the transistor and the collector-to-base feedback capacitance. Wide bandwidth, as in a video amplifier, is obtained by designing an amplifier with relatively low values of the current and voltage gains.

The tuning interaction and instability observed in tuned RF amplifiers are caused by feedback through collector-to-base capacitance. Tuning interaction may be reduced by reducing the stage gain, by reducing the feedback capacitance, and by neutralizing the feedback.

Tuned amplifier stages are generally unstable unless the base-to-collector voltage gain is limited to no more than 20 dB. Higher gains may be used when the capacitance feedback is neutralized by an opposing negative feedback.

Above 100 MHz the point-contact and Schottky diodes are used for detection or conversion of low-level signals to lower frequencies. Diode mixers and frequency multipliers are used in microwave applications because of the difficulty of constructing semiconductor amplifiers and oscillators at these frequencies.

The efficiency of a junction diode is impaired at RF and UHF frequencies by stored charge and shunt capacitance. The best diodes for high-frequency mixing and detection are the Schottky hot carrier diodes and the point-contact silicon and germanium devices. Because all diodes and circuit elements have shunt capacitance, the impedance level used in high-frequency circuits is reduced with increasing frequency. The need to use small areas in semiconductors and low impedance levels in circuits forces the designer to use relatively high power levels. Thus, the temperature rise of the semiconductors is usually an important consideration in high-frequency applications.

The capacitance of varactor diodes is reduced by a factor of 5 or 10 by increasing the applied back voltage. These devices have many applications in high-frequency circuits for voltage-controlled tuning, frequency modulation, and frequency changing.

With increasing frequency the dimensions of circuits must be reduced until coils and capacitors are replaced by short lengths of transmission lines, tuned circuits are replaced by microwave cavities, and lumped element circuits become micro-circuits that are made under a magnifying glass or by the photo and etching techniques used in constructing integrated circuits.

The small physical dimensions of the MOS FETs make these devices useful substitutes for vacuum tubes in many RF and UHF applications.

PROBLEMS

14-1. A transistor has a low-frequency current gain of 100 and $f_T = 2$ MHz. (a) What is the current gain at the following frequencies: (a) 500 kHz. (b) 50 kHz. (c) 5 kHz. (d) 20 kHz.

14-2. Assume you built the video amplifier shown in Fig. 14.2 and found the upper cutoff frequency to be only 2 MHz. (a) What defect in the circuit layout should you expect to find? (b) In looking for substitution transistor types, what characteristics would you try to obtain? (c) Explain whether or not a change of the bias might increase the bandwidth. (d) What effect would an incorrect value of the 33 Ω series emitter resistor have on the cutoff frequency?

14-3. (a) Calculate the emitter current to be expected in the stage shown in Fig. 14.3. (b) Explain the use of two 0.05 F capacitors in the bias network. (c) What approximate voltage gain would you expect to find in this stage?

14-4. A tuned CE stage is too unstable to be satisfactorily neutralized. (a) Should the secondary winding of the interstage transformer have more or fewer turns? (b) Explain. (c) What alternate solution would you suggest to make the stage more stable?

14-5. Find the emitter current to be expected in the amplifiers shown in the following figures: (a) Fig. 14.2. (b) Fig. 14.3. (c) Fig. 14.4. (d) Fig. 14.8. (e) 14.10. (f) Why are the current values so different?

14-6. (a) Refer to the RCA data on the 2N187, and explain the Q-point values used in the MOS FET amplifier, Fig. 14.5. (b) What is the transconductance with the indicated Q-point? (c) How high an ambient temperature can you use without a heat sink?

14-7. Examine the video amplifier used in your TV, and explain the need for each component in the circuit.

14-8. Select a transistor suitable for the oscillator shown in Fig. 14.10.

CHAPTER FIFTEEN

Thyristors, Unijunctions, Tunnel and Light-Emitting Diodes

An important group of semiconductors is built as an intertwined pair of transistors, so that the structure has internal positive feedback. These devices respond to a small trigger signal by switching, like a watch escapement, from one stable state to a second stable state. Because the two states are separated by a region in which there is positive feedback and gain in excess of unity, the intermediate region has the characteristics of a negative resistance. Devices that are useful as negative resistances include thyristors, silicon controlled rectifiers, unijunction transistors, and tunnel diodes. Thyristors have extensive applications as switches, especially for the control of ac or dc power. Unijunction transistors are used in low-power applications such as triggered switches for timing, scaling, or pulse-shaping. Tunnel diodes serve as sensitive switches, as oscillators, or as very high frequency amplifiers. This chapter describes briefly some of the applications of negative resistance devices used as pulse-controlled switches or relaxation oscillators.

15.1 THYRISTORS (SHOCKLEY 4-LAYER SEMICONDUCTORS)
(Ref. 12)

Thyristors, an important family of semiconductors, are voltage-controlled devices that operate much like switches. These devices are fabricated as a *pnpn* semiconductor sequence, represented by the stack shown in Fig. 15.1(a). Because each layer interacts with the adjacent layers, three layers form a *pnp* transistor and three form an *npn* transistor. The integrated circuit has positive feedback that makes the device turn ON as a switch. Depending on the manner in which the semiconductor layers are formed and on the terminals provided, thyristors operate as a 2-terminal diode switch, a 3-terminal gate-controlled rectifier or switch, or as a 4-terminal switch and regenerative amplifier.

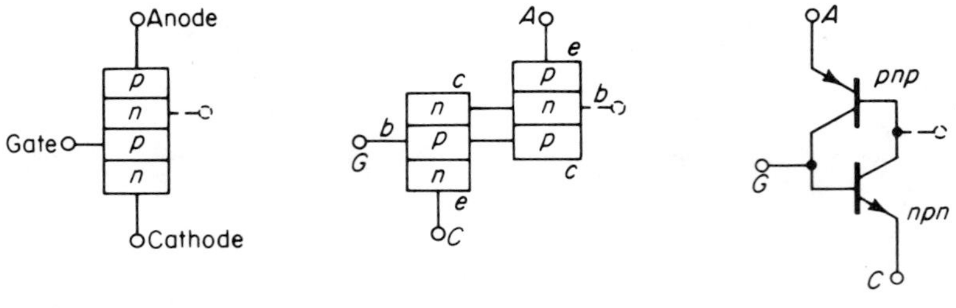

(a) Thyristor   (b and c) Two transistor analogues

Figure 15.1.

The *pnpn* structure is best visualized as consisting of two transistors, a *pnp* and an *npn*, interconnected to form the complementary feedback pair shown in Fig. 15.1(b) and 15.1(c). The collector of each transistor is direct-coupled to the base of the other, thereby making a positive feedback loop. With the cathode connected to the negative of the power supply and the the anode connected to the positive, the center junction is reverse biased and does not conduct unless the transistors have current gain. However, if the current gain taken around the feedback loop exceeds 1, each transistor drives the other into saturation, all junctions become forward biased, and the anode current is limited mainly by the external circuit resistance.

Thyristors are manufactured to provide several types of characteristics. The silicon controlled rectifier (SCR) is a 3-terminal diode that is designed to block both forward and reverse voltage up to 1000 Vdc and to carry up to 500 A turn-on current. A silicon controlled switch (SCS) is a *pnpn* structure which has all four semiconductor regions accessible to allow many different circuit arrangements and applications. The SCS is generally charac-terized as a sensitive SCR designed for anode currents below 0.5 A and for voltage ratings below 60 V. The sensitivity of an SCS is usually specified by closely defined firing currents and voltages. A 4-layer diode has only the anode and cathode terminals accessible and is designed to turn ON within a narrow range of forward breakdown voltages at values specified from 20 V to 200 V. All thyristors block the reverse direction of current flow unless specifically made to do otherwise.

The common symbols for the thyristor family are shown in Fig. 15.2. Two SCR symbols are used interchangeably. A controlled rectifier with a gate at the cathode end is represented in Fig. 15.2(a). The other symbol, Fig. 15.2(b), shows that the device has a *pn* gate at the cathode and a *pn* junction at the anode. Similarly, two symbols are used for the SCS, as shown in Fig. 15.2(c) and 15.2(d). These show that control electrodes are available at both the anode and the cathode. The symbol for the Shockley diode, Fig. 15.2(e), shows that this diode has only two terminals and is not an ordinary diode. SCRs carry 2N numbers and SCSs carry 3N numbers.

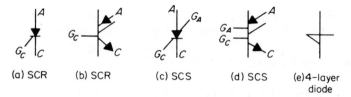

(a) SCR   (b) SCR   (c) SCS   (d) SCS   (e)4-layer
                                                        diode

Figure 15.2. *Thyristor family symbols.*

The important characteristics of thyristors can be easily understood by a study of the 4-terminal transistor circuit shown in Fig. 15.3. The reader is urged to construct the transistor equivalent and to perform the experiments in Sections 15.2 and 15.3. The transistor equivalent makes easier the appreciation of thyristors as 4-layer regenerative switches and as integrated circuits.

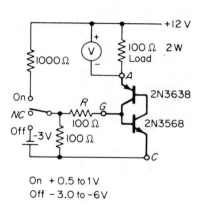

Figure 15.3. *SCR equivalent circuit.*

On + 0.5 to 1 V
Off − 3.0 to −6 V

15.2 SILICON CONTROLLED RECTIFIER (SCR)

The transistor equivalent of an SCR, illustrated in Fig. 15.3, is shown connected to a 100 Ω load and a 12 V supply. The voltmeter across the load is to show when the SCR is turned ON. The gate G is protected from accidental overcurrents by a 100 Ω series resistor, R_1. The shunt resistor is to prevent triggering when the trigger switch is open.

With the anode voltage ON, the SCR is turned ON by applying a gate signal of between +0.5 V and 1.4 V and is turned OFF by connecting a higher negative gate signal, −3 V to −6 V. The turn-on signal of power SCRs is from 1 to 2 V at 30 mA, a value not too different from the gate signal required by the transistor equivalent. Observe that the voltage drop in the SCR is 0.75 V, a little more than the forward voltage drop in one diode. Observe also that the collector-to-base diodes are forward biased when the SCR is ON.

The center OFF-position of the switch is used to illustrate the rate effect which makes an SCR turn ON when the supply voltage is connected. This turn ON cannot be tolerated in some applications and may be avoided by shorting the gate to ground or by making the anode voltage rise slowly by connecting the supply through an RC filter.

15.3 THYRISTOR HOLDING CURRENT AND RATE EFFECT

Because the gate turn-off signal is relatively large, SCRs are usually turned OFF by removing the supply voltage. If the supply is ac, there is no problem except at high frequencies. With dc, however, if the load current does not exceed a certain minimum value called the *holding current*, the SCR opens by itself as soon as the gate signal is removed. The characteristic holding current of the SCR represented by Fig. 15.3 is found by increasing the load resistance to approximately 1 kΩ until the SCR is unable to remain ON when the gate signal is removed. The load current just sufficient to hold the SCR ON is the holding current.

As mentioned earlier, and SCR may turn ON by itself any time the anode supply voltage is raised too rapidly. This turn-on, known as the *rate effect*, is triggered when the anode voltage change reaches the gate by way of capacitors in the 4-layer structure. The rate effect is illustrated by the circuit in Fig. 15.4. The capacitor C_A connected from the anode to cathode is added to eliminate the rate effect normally existing in the transistor equivalent. This

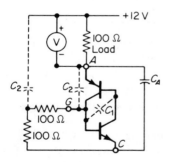

C_A reduces rate-effect (see text)
C_1 or C_2 aggravate rate-effect   *Figure 15.4. Thyristor rate effect.*

capacitor must be small enough to prevent capacitor discharge from damaging the SCR. With the anode capacitor connected as shown, we find that a much smaller capacitor, connected either between the two collectors (C_1) or between the supply and the gate (C_2), makes the transistor equivalent turn ON when the power supply is connected. Because power SCRs have large junction areas, the junction capacitance is large and the rate of change of anode voltage must be held to relatively low values. For a typical medium-power SCR the rate of change of the anode voltage must be maintained below about 300 V per microsecond.

15.4 HALF-WAVE POWER CONTROL WITH SCR

A common application of SCRs is in the control of ac power for motors, lamp dimming, and electric heating. A practical example is illustrated by

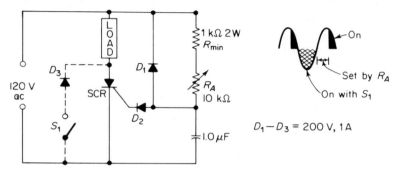

Figure 15.5. *SCR power control.*

the circuit shown in Fig. 15.5, in which a load is supplied half-wave power. When the SCR is ON continuously, it conducts like a plain diode and the load is driven only every other half cycle. The gate control circuit allows the load power to be reduced to a low value by adjusting the resistor R_A, and operation is as follows: During the negative half of the cycle when the SCR is not conducting, the diode D_1 charges the capacitor to the peak negative value of the ac voltage and diode D_2 keeps the capacitor from discharging through the gate to the anode. During the succeeding positive half-cycle, the gate is prevented from conducting until positive current entering the capacitor by way of the resistors R_{min} and R_A has discharged C and raised the voltage to a positive value sufficient to trigger the SCR gate. If the resistor R_A has a high resistance, the SCR is OFF the entire cycle. If R_A has a low resistance, the gate is turned ON for most of the half-cycle. In this manner the load voltage is variable from 0 to 85 V rms. Control from 85 V to 120 V rms is effected by connecting the diode D_3 which conducts load current on the negative half of the cycle. The load voltage with one half power is 85 V rms.

15.5 FULL-WAVE POWER CONTROL WITH A TRIAC (Ref. 7)

Full-wave continuous control of ac power may be provided by using the SCRs as two diodes of a full-wave steering bridge, as illustrated in Fig. 15.6(a). The load is supplied during one half cycle by way of diode D_1 and SCR-B. During the alternate half cycle the load is supplied by D_2 and SCR-A. An advantage of the steering bridge is that full-wave proportional control may be obtained by using a direct-coupled gate circuit. If the load is trans-

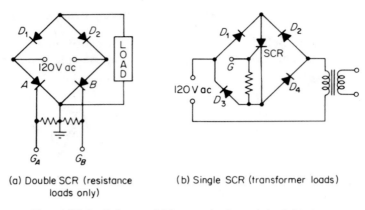

(a) Double SCR (resistance   (b) Single SCR (transformer loads)
    loads only)

Figure 15.6. *Full-wave SCR control using steering bridge.*

former coupled and the alternate halfcycles are not alike, the resulting dc component may saturate the transformer and overload the SCRs. The 4-diode steering bridge shown in Fig. 15.6(b) permits using one SCR and requires only that the ac supply and the gate signal have symmetrical halfcycles.

A Triac is a thyristor that controls full-wave ac power and is like a parallel-connected pair of complementary thyristors with a common gate terminal. Triacs are widely sold in a complete package for the control of 120 V ac lights.

Triacs conduct in both directions with a gate signal of either polarity and are usually connected in the circuit shown in Fig. 15.7. The variable resistor R controls the rate at which charge accumulates in the capacitor C; and the Diac is a bi-directional breakdown device with a negative-resistance

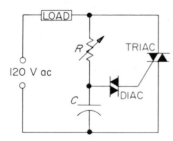

Figure 15.7. *Full-wave power control using a TRIAC.*

characteristic that produces a high gate turn ON current. (Typically, a Diac turns ON at 20 to 30 V and conducts a peak current of 1 A within about 50 μsec.) In the circuit shown a Triac controls power over the high-current part of the cycle and the control package dissipates very little power.

15.6 DC POWER CONTROL

SCRs may be used either as choppers or as switches in an inverter for the control of dc power. As choppers the SCRs interrupt the power source at a high-frequency rate and control the time duration of each pulse of power that is supplied to the load. The average output power is thus regulated by a switch that does not dissipate power. Hence, the power efficiency is high and the SCRs may operate on a small heat sink. The chopper type circuit is known as a *switching regulator*. Similarly, transistors are used as switching regulators.

SCRs are sometimes used as controlled switching elements that replace transistors in a dc-to-ac inverter. These inverter circuits are usually complicated because the SCRs turn OFF slowly (20 μsec) and turn ON more rapidly (2 μsec.). Unless the switching is carefully controlled, both switching elements may be ON at the same time and short-circuit the power supply. Because of this switching problem, SCR inverters are not widely used, except possibly for low-power applications.

15.7 SCR FLASH LAMP CONTROL

A common application of SCRs is found in the control of high-voltage capacitor discharges which are required to fire photo-flash lamps and ignition-timing lamps. The SCR turns ON more rapidly than a mechanical switch, and the switch that triggers the flash is not required to carry the high currents of the capacitor discharge.

The circuit of an automobile timing lamp is illustrated in Fig. 15.8. The flash lamp is a U-shaped tube containing argon gas. Connected in parallel with the lamp is the capacitor C_1, charged to 400 V. When ignited by applying a high-voltage electrostatic field, the lamp discharges the capacitor and emits a brilliant blue-white flash lasting about $\frac{1}{1000}$ of a second. The ignition voltage is obtained by discharging the 0.3 μF capacitor through the primary

Figure 15.8. *Timing flash lamp.*

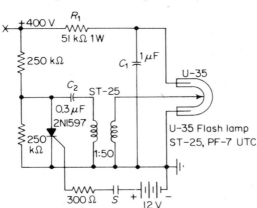

of the air-core transformer ST-25. Because the transformer turns-ratio is 50-to-1 and the capacitor is charged to about 200 V, the ignition voltage may be as high as 10,000 V. As shown in the figure, the capacitor C_2 is discharged by triggering the gate of a 200 V SCR. The gate signal for an ignition-timing system is obtained by connecting the gate through a series resistor to the automobile distributor contacts.

15.8 SILICON CONTROLLED SWITCHES (SCS)

A silicon controlled switch is a 4-layer structure which has all four semiconductor regions accessible. The device is fully equivalent to the transistor circuit shown in Fig. 15.1(c) and may be considered as an integrated circuit which has a *pnp* and an *npn* transistor connected as a positive feedback pair. With all four regions accessible, the positive feedback is easily controlled, and the device may be operated either as a high-gain linear dc amplifier or as a switch. Many applications of silicon controlled switches as pulse generators, voltage-sensitive switches or as SCR replacements are described in the General Electric Transistor Manual (Ref. 12).

15.9 SHOCKLEY 4-LAYER DIODES

Shockley, or 4-layer, diodes are two-terminal devices that are constructed to hold the forward current OFF until a specified breakdown voltage is reached, at which the diode turns ON. The breakdown voltage in the reverse direction is higher than in the forward direction and is not well defined. Shockley diodes are generally low power devices that have forward breakover (switching) voltages from 10 V to 200 V and dissipation ratings from 0.5 W to 5 W.

The operation of a 4-layer breakdown diode is illustrated by the transistor equivalent shown in Fig. 15.9. Suppose that in this circuit the resistor R is made 10 times R_1. If the supply voltage is increased from 0 to 10 V, the base voltage V_B increases from 0 to 0.9 V. As long as the base voltage is less than 0.9 V, the transistor pair is OFF. When the voltage exceeds 0.9 V, the

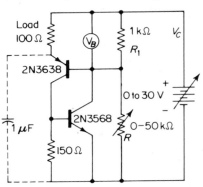

Figure 15.9. *Transistor equivalent of a 4-layer diode.*

Note: Diode conducts when $V_B = 1$ or
$V_C \cong R$ in kilohms

pair turns ON. By adjusting the resistor R, the turn-on voltage may be set as desired. Sometimes the resistor R is replaced by a Zener diode.

15.10 SHOCKLEY DIODE OSCILLATOR

The relaxation oscillator circuit shown in Fig. 15.10 illustrates one application of a low-power 4-layer diode. The diode has a voltage breakdown rating of 10 V. Hence, the capacitor charges through the resistor R until the diode voltage is 10 V. When the diode turns ON, the capacitor is discharged, the diode turns OFF, and the cycle repeats. Because the time required to charge the capacitor to the breakdown voltage depends partly on the amount

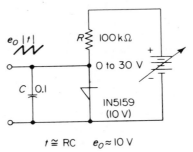

Figure 15.10. *Shockley diode oscillator.* $t \cong RC$ $e_0 \approx 10\,V$

by which the supply voltage exceeds 10 V, the period of the oscillator varies with the supply voltage. An application of this device might be the transmission to a remote location of a signal varying in frequency to indicate the supply voltage. Such a device is called a ***voltage-to-frequency converter.***

15.11 UNIJUNCTION TRANSISTORS (Ref. 12)

The unijunction transistor is a negative resistance device that is, in effect, a filament of silicon to which a rectifying junction has been attached. The circuit symbol of the device represents, as shown in Fig. 15.11, the bar of silicon with the connections base-one and base-two. The rectifying contact, called the ***emitter,*** may be thought of as being located about midway on the bar. When a voltage gradient is set up along the silicon bar, the diode terminal opposite the emitter is at a voltage intermediate (about 60 percent) between the extremes of the interbase voltage. The diode is, therefore, reverse biased up to a "peak-point emitter voltage, V_P." When forward biased, the resis-

Figure 15.11. *Unijunction transistor symbol.*

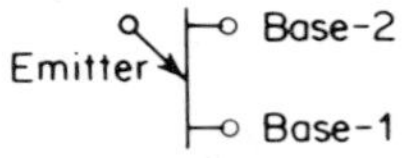

tance from the emitter to base-one drops to a low value. Because the forward
emitter voltage falls as the emitter current increases, the emitter has a negative
resistance characteristic, and a high pulse of current can be transferred from
the emitter circuit to the base-one load.

The negative resistance characteristic and the simplicity of the unijunction transistors make these devices useful as relaxation oscillators, sawtooth
wave generators, and multivibrators. Among the advantages of the unijunction transistor are the inherent stability of the characteristic parameters and a
temperature characteristic that can be controlled by adjusting the externally
connected resistors.

A programmable unijunction transistor (PUT) is a three-terminal
device with characteristics that may be changed by changing the Thevenin
equivalent voltage and resistance of the anode (base-two) supply. By selecting
the characteristics of the anode supply, the designer has somewhat greater
flexibility in circuit design than with the conventional devices.

15.12 UNIJUNCTION EQUIVALENT CIRCUIT

An equivalent circuit of a unijunction transistor is shown in Fig. 15.12.
A typical device has an interbase resistance between base-one and base-two
that is about 10,000 Ω, and the resistance is divided about equally between
R_{B1} and R_{B2}. In operation base-two is made about 20 V positive with reference to base-one, and the bar acts as a simple voltage divider. Because the
cathode side of the emitter diode is at $+10$ V, the emitter conducts when the
emitter voltage exceeds $+10.6$ V. As soon as the emitter conducts, the resistance R_{B1} drops and the device switches rapidly into the conducting mode.
In most circuits the emitter is connected to a capacitor and the rapid decrease
of the emitter voltage causes the capacitor to discharge into a load connected
to base-one. With a typical device an emitter current exceeding 10 mA may
cause the resistance R_{B1} to drop to about 100 Ω. Although the emitter current
lowers the resistance R_{B1} and increases the base-two current, the total interbase resistance remains positive.

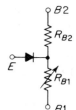

Figure 15.12. *Unijunction transistor equivalent circuit.*

Curves that show the static emitter characteristics of a unijunction
transistor are illustrated in Fig. 15.13. For simplicity, only two curves are presented. The lower curve is for 10 V applied between base-one and base-two,

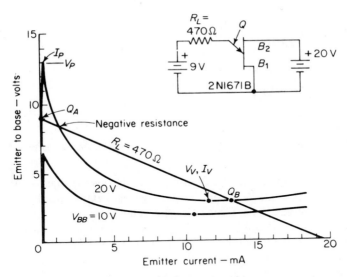

Figure 15.13. *Static characteristics of a unijunction transistor.*

and the upper curve is for 20 V. The vertical portion of the curves at the left shows that the emitter current is negligible when the emitter is below the peak-point voltage V_P. The portion of the curve that slopes downward and to the right represents the conducting emitter characteristic, a negative resistance. If the external emitter circuit resistance, represented by the load line R_L, is low, the emitter Q-point cannot remain on the negative slope and, therefore, moves to the highest possible emitter current. In other words, with a low circuit resistance the static Q-point is either at Q_A at the left of I_P or at Q_B at the right-hand intersection of the load line.

The characteristics of a unijunction transistor are specified by the interbase resistance R_{BB} measured with the emitter open, the current I_P and the voltage V_P of the peak point, the coordinates I_V and V_V of the valley point, and the intrinsic stand-off ratio η. The intrinsic stand-off ratio is actually specified in a way that makes η independent of temperature, but for practical purposes η is the fraction that gives the interbase voltage at the cathode end of the emitter diode as a fractional part of the total interbase voltage. The peak point voltage V_P is approximately 0.6 V greater than the interbase voltage calculated by using η and the total interbase voltage.

15.13 UNIJUNCTION RELAXATION OSCILLATOR (Ref. 12)

As an application of the unijunction transistor consider the relaxation oscillator circuit shown in Fig. 15.14. When the power is turned ON, the

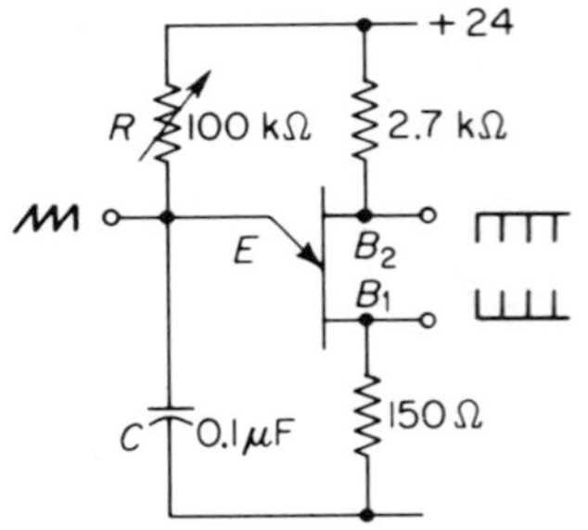

Figure 15.14. *Unijunction RC oscillator.*

capacitor is discharged and the emitter diode is reverse biased. As the capacitor charges through R, the emitter voltage rises exponentially toward the
supply voltage. When the emitter voltage reaches the peak-point voltage, the
capacitor is discharged and the cycle begins anew. The output pulse observed
at base-one is a spike which has a fast rise time and a slower fall time. The
pulse at base-two is similar, except that the polarity is reversed. The voltage
at the emitter has a sawtooth-like waveform.

The unijunction transistor oscillator is noteworthy for its ability to
operate over a wide range of circuit parameters and temperature. Because
the peak-point voltage varies with the supply voltage, the period of the
oscillator is practically independent of the supply voltage and in this respect
is quite different from the 4-layer diode oscillator. The oscillator shown in
Fig. 15.14 will operate up to 100 kHz and can have a frequency stability of
better than ± 1 percent. The oscillator is easily synchronized by application
of a trigger pulse and is useful as a timing device or as a pulse and sawtooth
wave generator.

15.14 UNIJUNCTION MULTIVIBRATOR

The circuit of a unijunction multivibrator, as shown in Fig. 15.15, is
similar to that of the oscillator in Fig. 15.14, except for the addition of diode
D_1 and the resistor R_2. When the capacitor is being charged, the diode is
forward biased and the charging time is determined by R_1 and the capacitor.
However, when the capacitor is discharging, the diode is reverse biased
because the emitter end of the diode is then negative. As long as the capacitor

Figure 15.15. *Unijunction multivibrator.*

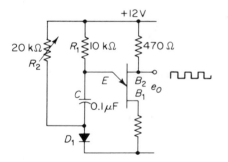

is charged, the discharge current must flow through R_2 with the product R_2C controlling the discharge time. The output of the multivibrator taken at base-two is approximately a square wave, with the ON and OFF intervals separately controlled by R_1 and R_2. In the circuit shown (Fig. 15.15) the ON and OFF times are approximately equal when R_2 is twice R_1.

15.15   TELEMETER TRANSMITTER-RECEIVER

An example of a telemeter signal system is illustrated in Fig. 15.16. The telemeter transmitter is on the left side and the receiver is on the right. A 2500 Ω potentiometer type transducer P is shown as the signal source. The transducer may be an indicator of position, pressure, voltage, or the like. As the transducer changes the voltage applied to the unijunction transistor A, the frequency generated by A varies and is transmitted over the two-wire line x-x.

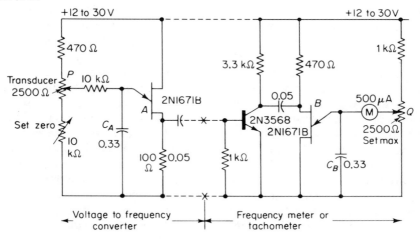

Figure 15.16. *Telemeter transmitter-receiver.*

At the receiving end of the line the pulse signals are amplified by the CE transistor stage and are applied through a capacitor to the base-2 of unijunction transistor B. The meter M reads the average current into the emitter of unijunction transistor B. When there are no incoming pulses the unijunction transistor B does not conduct, and M reads zero. When the base of unijunction transistor B is raised by the incoming pulses the emitter conducts and discharges the capacitor C. The current through the meter is proportional to the number of incoming pulses and, therefore, is a linear indication of the incoming frequency.

The telemeter illustrates an application of the unijunction transistor where the inherent stability of the active element is largely responsible for the reliability and stability of the system. In a similar application the receiver is used as an automobile tachometer. The tachometer operates on the same principle as the telemeter except that the tachometer counts the pulses produced by the automobile distributor. A simple but practical tachometer is described in Ref. 12. Unfortunately, these circuits do not operate satisfactorily if the meter is less sensitive than 500 μA.

15.16 TUNNEL DIODES

A tunnel diode is a small *pn* diode in which the junction is so narrow that the electrical charges can transfer across it by a quantum mechanical effect called **tunneling**. When forward biased, a tunnel diode has a negative resistance characteristic that is useful in amplifiers, pulse generators, and oscillators. The tunnel diode is suitable for logic circuits and memory units in computer applications because of its high switching speed, small size, and low power requirements. The small size and low impedance of a tunnel diode make the device useful at frequencies above 5 GHz. The current-voltage characteristics of tunnel diodes are relatively independent of temperature.

Presently available tunnel diodes are fabricated from either germanium or gallium arsenide. Germanium devices offer high speed and low noise, whereas gallium arsenide diodes provide high power outputs. High-current diodes are used as inverters in circuits which have low-impedance sources (solar cells and thermo-electric generators). Because tunnel diodes operate effectively above 300 MHz, they are particularly suitable for use in microwave amplifiers, oscillators, and mixers. In microwave amplifiers tunnel diodes offer low noise, small size, low power drain, and reliability in severe environments. In oscillator circuits the tunnel diodes provide useful power up to above 5 GHz. Compared with vacuum tube microwave oscillators, the tunnel diode oscillators are less expensive and have simpler circuits. In mixer service the tunnel diodes give high sensitivity and low conversion noise.

15.17 TUNNEL DIODE SYMBOLS

Some of the symbols used to represent tunnel diodes are shown in Fig. 15.17. The symbol designated by *A* is like a diode symbol for a *pn* junction except for the addition of $-g$, which indicates the negative conductance of the device. The symbol *B* is a diode in which the arrow has tunneled through the bar. The symbol *D* has the letter *T* in the structure. All these symbols are in present-day use.

Positive electrode   **Figure 15.17.** *Tunnel diode symbols.*

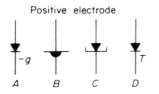

15.18 STATIC CHARACTERISTICS

The first quadrant static characteristic curve of a tunnel diode is illustrated in Fig. 15.18. The important dc parameters that specify the points of inflection are indicated on the curve by the standard letter symbols. For voltages up to about 0.1 V the diode current rises to a peak point having the coordinates V_P, I_P. A further increase of the applied voltage is accompanied by a decrease of the diode current down to the valley-point V_V, I_V. In the region between the peak and the valley-points the diode has a negative resistance because an increase of the diode voltage is accompanied by a decrease of the current.

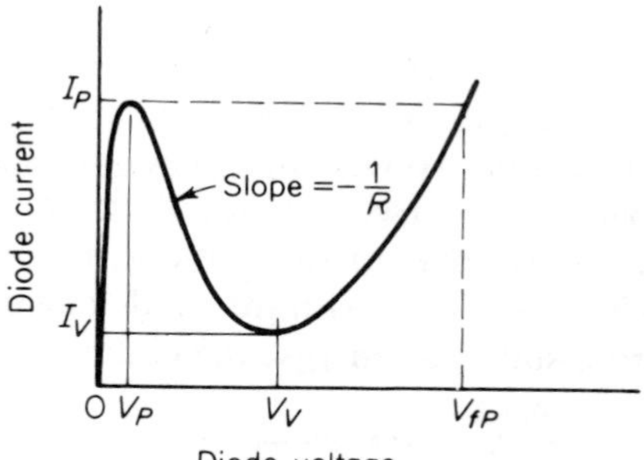

Figure 15.18. *Tunnel diode characteristic curve showing peak and valley coordinates.*

Tunnel diodes are characterized by the peak-point current and are available with peak current values from about 0.1 mA to 100 mA. The static current-voltage curve of a 4.7 mA tunnel diode is shown in Fig. 15.19. Most low-power tunnel diodes have peak-point currents that are 6 to 8 times the valley-point current, and the peak and valley voltages are about 0.06 V and 0.5 V, respectively, as shown in Fig. 15.19. For the tunnel diode represented by the curve the negative resistance is about 50 Ω. This resistance may be estimated from the peak and valley coordinates. For example, for the diode represented by Fig. 15.19 the peak voltage is 0.06 V and the valley voltage is 0.5 V, while the peak current is 4.6 mA and the valley current is 0.6 mA.

Figure 15.19. *Typical tunnel diode characteristic current-voltage curve.*

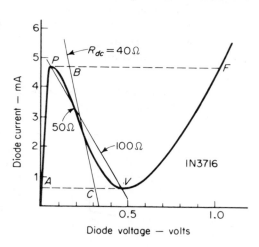

From the differences, respectively, an increase of 0.44 V is accompanied by a decrease of 4 mA. The large signal incremental resistance is, therefore, $\frac{0.44}{-0.004}$, which is $-110\,\Omega$. However, the slope over the central portion of these curves tends to be about 2.2 times the slope calculated from the peak and valley coordinates. The greater slope of the curve implies a lower small-signal negative resistance, or about $50\,\Omega$ for this diode. When required, a different value of the negative resistance is obtained by selecting a diode which has a different peak-point current.

15.19 TUNNEL DIODE RELAXATION OSCILLATOR

A tunnel diode relaxation oscillator is made by driving a tunnel diode from a low-voltage supply in a series R-L circuit, as shown in Fig. 15.20. The supply voltage and the dc resistance must be of such value that the dc load line intersects the diode curve at only one point and on the negative resistance part of the curve as in Fig. 15.19. If the dc load line is vertical when compared with the diode curve, and the inductance is large enough to store sufficient energy, the circuit operates as a *relaxation oscillator*.

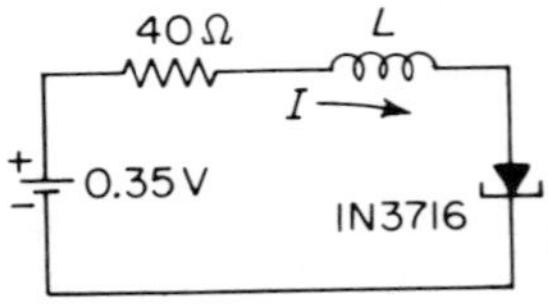

Figure 15.20. Tunnel diode relaxation oscillator circuit.

When the supply is turned ON, the current rises and energy is stored in the inductor. At the peak point (P in Fig. 15.19) the current is increasing because there is still not enough voltage drop across the tunnel diode and the resistor to equal the applied voltage. At the peak, the tunnel diode switches toward an open circuit at F, the voltage across the inductor reverses, and the current begins to fall. At the valley point (V in Fig. 15.19), the voltage across the tunnel diode still exceeds the supply voltage, so the current is decreasing. However, the tunnel diode now switches to an effective short at A, the current begins to rise, and the cycle repeats.

The waveform shown in Fig. 15.21 for a relaxation oscillator is interesting because the tunnel diode switching voltages appear at the points of dis-

Figure 15.21. Tunnel diode relaxation oscillator waveform.

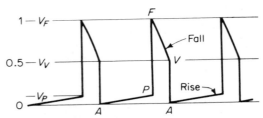

continuity on the wave. The signal wave shows from A to P the slow rise of
the tunnel diode voltage caused by the rise of current through the inductor,
the fast switching of the tunnel diode from P to F, the slow fall of the diode
voltage from F to V, and, finally, the rapid switching of the tunnel diode
from V to A. The signal output voltage equals the difference between the
voltages at F and A.

15.20 TUNNEL DIODE SINUSOIDAL OSCILLATOR

A tunnel diode relaxation oscillator may be converted to a sinusoidal
oscillator by adding a capacitor, so that energy is stored and exchanged
between the inductor and the capacitor. Moreover, by increasing the circuit
resistance so that the dynamic load line is parallel to the diode curve, the
voltage excursions are limited, and the signal is nearly sinusoidal. High
operating frequencies are obtained by reducing the inductance and using the
junction capacitance of the diode either with or instead of a tuning capacitor.
The circuit of a typical high-frequency oscillator is shown in Fig. 15.22.

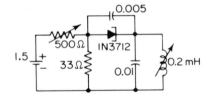

Figure 15.22. Typical tunnel diode oscillator, 100 kc.

15.21 TUNNEL DIODE SWITCHES

Tunnel diodes are used as voltage-sensitive switches that operate from
monostable or bistable Q-points. A bistable tunnel diode switching circuit is
shown in Fig. 15.23 with the static curve and the dc load line. The circuit
employs a variable low-voltage supply and a 300 Ω series resistor. The load
line AB represents the conditions that exist when the supply voltage is 1.0 V
as given by the intersection of the line with the abscissa. Because the load
line passes through the negative resistance region and intersects both positive
slopes of the curve, there are two stable Q-points, A and B. If the supply
voltage is raised from 0 V to 0.6 V, there is only one stable Q-point, as at A'.
If the supply is increased to 1.4 V, the Q-point moves from A' through A
to the peak point P and then switches to the only stable Q-point at B'. When
the supply voltage is lowered to 0.6 V, the Q-point moves from B' through B
to the valley point V and switches to A'. As indicated, the positive slopes of
the curve may be traced by changing the supply voltage, but a Q-point on the
negative slope represents an unstable condition, and the circuit current
changes until a stable Q-point is reached.

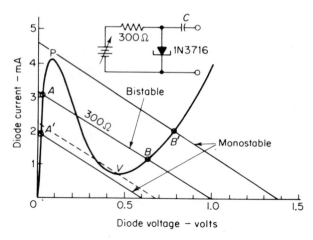

Figure 15.23. *Tunnel diode switching circuit and load lines.*

A more common way of switching the bistable circuit is to leave the supply voltage fixed, say at 1.0 V, and to insert a signal pulse that momentarily increases the tunnel-diode current. This signal may be inserted by way of the capacitor *C* shown in Fig. 15.23. The net effect is the same as that produced by a sudden increase of the supply voltage. A convenient way of introducing the signal is to connect the capacitor momentarily to the battery. Similarly, the diode may be switched from *B* to *A* by grounding the capacitor. The fact that the charge on a very small capacitor (*C* = 20 pF) is able to effect switching is evidence that a tunnel diode is capable of switching from one state to another in times that may be measured in fractions of a microsecond.

Many simple and practical switching circuits are possible when a tunnel diode is direct-coupled to a transistor or an SCR. Many of these hybrid circuits are described in the literature. Single-stage tunnel diode amplifiers that provide 30 db of gain at VHF frequencies are described also.

15.22 BACK DIODES (TUNNEL RECTIFIERS)

Tunnel diodes may be manufactured especially for service as low-voltage rectifiers. In a tunnel rectifier a substantial reverse current flows at very low voltages while the forward current is relatively small. Consequently, tunnel rectifiers can provide rectification at smaller signal voltages than conventional rectifiers where the source and load impedances have suitably low values. Because the tunnel rectifier is made with a high reverse current and a low forward current, the tunnel rectifier is sometimes called a ***back diode***. The high-speed capabilities of tunnel rectifiers are useful in computer applications for level sensors or logic switches.

A back diode is represented by the standard symbol for a diode in order to designate correct use of the device, although the junction is reversed. The bar of the symbol represents the cathode but is connected to the *p*-type material because the direction of easy current flow is opposite to that of a conventional diode.

The characteristic current-voltage curve of a tunnel rectifier is shown in Fig. 15.24. Also shown in this figure is the familiar curve of an ordinary germanium rectifier. Although the tunnel rectifier is connected backward, the curves in the first quadrant represent the high-conduction characteristic of both diodes. The curves in the third quadrant represent the low-conduction characteristic of back diodes. As long as the device is used inside the rectangle where the impressed signal is less than 0.4 V, the back diode is useful as a rectifier and the regular germanium diode is not useful. If the peak signal level is more than a few tenths of a volt, the tunnel diode is not a satisfactory rectifier. Back diodes are used in low-voltage microwave mixing and detecting applications where the diodes offer the advantages of low noise, high sensitivity, and reliability. Back diodes are also generally useful for providing higher negative resistances than are available in tunnel diodes. When forward biased, the diodes provide a very stable voltage reference that is independent of temperature over a wide temperature range.

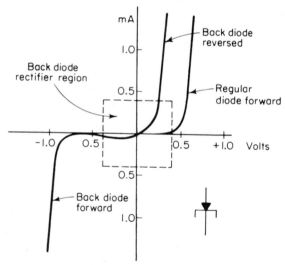

Figure 15.24. *Back diode characteristic curve.*

15.23 LIGHT EMITTING DIODES

Light emitting diodes (LEDs) are semiconductor diodes that emit light when forward biased. They are used in indicator lamps called *solid state lamps* (SSLs) or are built into systems as intrusion alarms, as small intense light sources in instruments, or as a light source for isolating one circuit from another. These devices have a high resistance to shock and vibration, high reliability, and lifetimes estimated to exceed 100,000 hours. LEDs may be designed with turn-on times from 1 μs to as short as 1 ns, which makes them useful for data transmission and in computers and numerical control equipment.

The most frequently used LEDs emit a red light (660 nM) and are usually operated with 20 to 150 mA at 1 to 2 V. The infrared devices emit at about 900 nM with similar input power requirements. Amber, yellow, green, and blue light sources are being used but they all require higher input power at 2 to 4 V and heat sinks. An important advantage of LEDs and SSLs is that their current and voltage requirements are compatible with transistor and integrated circuits, while their small size and reliability make them attractive optical devices for today's electronic applications. A familiar application of LEDs is in their use as light sources for the numerical indicators of pocket computers.

SUMMARY

A *thyristor* is a series of semiconductor junctions which have internal positive feedback. Most thyristors are used as regenerative switches that are controlled by low-power trigger signals. A silicon controlled rectifier blocks forward current flow until triggered by a positive gate signal. A silicon controlled switch has two gate terminals which permit external control of the feedback, thereby providing considerable flexibility in applications and circuit design. A 4-layer diode is a two-terminal switch that conducts when a specified voltage is exceeded.

A *unijunction transistor* is a silicon bar which has an emitter (control) diode near the midpoint. When the diode conducts, the emitter circuit has a negative resistance characteristic that makes the unijunction transistor useful as a simple oscillator, pulse generator, or multivibrator.

A *tunnel diode* has a negative resistance characteristic that is useful from dc up to microwave frequencies. A tunnel diode operates as a monostable or bistable switch when the dc load line intersects the positive slope of the current-voltage curve. An oscillator, either sine wave or relaxation, is formed when the circuit resistance is less than the negative resistance of the tunnel diode and the dc load line intersects only the negative slope of the current-voltage curve. Because the dc load line must intersect the negative slope only, a tunnel diode oscillator requires a low resistance, low voltage supply.

The equivalent negative resistance of a tunnel diode is approximately inversely proportional to the peak current rating of the tunnel diode. Back diodes have low peak current values and are used as demodulators or switches that operate at voltage levels as low as 0.1 V.

PROBLEMS

15-1. (a) Examine the transistor equivalent of an SCR and explain which junctions are reverse biased when voltage is applied overall and the gate signal is OFF. (b) Which junction in the 4-layer equivalent is required to withstand the overall anode-to-cathode voltage?

15-2. (a) Show by a sample calculation how much the frequency of a 4-layer diode oscillator changes with the supply voltage. (b) Explain the reason the frequency of a unijunction oscillator is independent of the supply voltage.

15-3. An SCR has a back resistance of 4 MΩ and a forward voltage drop of 0.75 V at 10 A. (a) Calculate the power dissipated when the SCR is OFF and the anode dc voltage is 400 V. (b) What is the power dissipated when the SCR is ON? (c) What percentage of the load power is lost in the SCR when the device is ON?

15-4. (a) List the inflection point parameter values one expects to find for a 1 mA tunnel diode. (b) What is the expected value of the dynamic resistance?

15-5. (a) Sketch an approximate curve for a 10 mA tunnel diode and specify the load line resistance in ohms that would make the device operate as a bistable switch from a 10 Vdc supply. (b) Repeat for a monostable switch to be operated with negative-going signals.

15-6. (a) Refer to Figs. 15.19, 15.20, and find the voltage across the inductor just before the diode opens at the peak point. (b) Repeat for the valley point. (c) Explain the statement in the text that the current in the inductance is increasing up to the moment at which the diode switches.

15-7. Describe the circuit characteristics of each of the following: (a) thyristor; (b) tunnel diode; (c) unijunction transistor. (d) Which of these devices have a useful negative resistance characteristic? (e) How are back diodes used?

CHAPTER SIXTEEN

Laboratory Instruments and Techniques

The practical application of transistors requires experience with electronic measurements and with laboratory methods. This chapter discusses the tools and experimental procedures most commonly used in laboratory studies of transistor equipment and helps the reader acquire enough confidence and understanding to enjoy the challenge of practical problems. It is shown that most problems can be resolved quickly by the effective use of simple, readily available instruments.

Skill in finding and correcting application difficulties is a valuable asset which can be promoted by analysis and study of practical laboratory methods. However, technical skills are acquired primarily by observing other persons at work, by laboratory experience, and by a critical appraisal of one's own methods.

16.1 LABORATORY INSTRUMENTS

That most day-to-day application problems can be resolved quickly and easily with a volt-ohmmeter (VOM) is too often forgotten. Because of this instrument's broad usefulness two volt-ohmmeters should always be at hand, one to be used as a voltmeter and the other as an ohmmeter or milliammeter. Employing two instruments helps avoid damage to the milliammeter by preventing its accidental use in place of a voltmeter. Effective use of these instruments requires at least a basic familiarity with the internal resistance of the lowest meter ranges. The voltmeter resistance is given on the face of the meter, and the ammeter resistance can be obtained by referring to the manufacturer's service manual. One should certainly know the ohmmeter circuit and memorize the open-circuit voltage and the short-circuit current of the low ranges. The short-circuit current should be known so that high-frequency transistors are checked using currents within the transistor rating. Similarly, the emitter diode of most transistors should not be subjected to reverse voltages as high as are produced on the high resistance range of some ohmmeters. The internal resistance of an ohmmeter is the resistance value read at the mid-scale corresponding to the range in use.

The ohmmeter is commonly used to make rough checks of the forward and reverse characteristics of diodes. Good low current silicon diodes usually show a reverse resistance of more than 20 MΩ. The forward resistance of diodes measured with an ohmmeter is variable and meaningless. Because the diode voltage drop tends to remain close to 0.6 V regardless of the current, an ohmmeter with a 1.5 V battery tends to read six-tenths of the way towards full scale (zero ohms), as if the battery voltage were reduced to 0.9 V.

Zener diodes check as ordinary diodes when the ohmmeter voltage is below the breakdown voltage. The double anode Zener diodes are reverse biased with either polarity and should show as open below the breakdown rating.

The polarity of a diode can be checked with an ohmmeter if we know which way the ohmmeter leads are polarized. Ohmmeters are made with either polarity, so one should check his instrument, using a marked diode. The diode is an open circuit when the positive side of the ohmmeter is connected to the positive side of the diode. Remember that a diode can be connected to a voltage supply without a limiting resistor if the positive end is connected to the positive side of the circuit.

For testing and performance studies of low-power amplifiers we need a regulated dc power supply that can be adjusted to operate from 0 to 30 V at 0.5 A. A supply which has an adjustable current limit is helpful in protecting transistors from damage at the hands of inexperienced and careless experimenters.

A laboratory oscillator for low-frequency studies should supply 10 V rms to a 500 Ω load and have a 2 MHz upper frequency limit. Amplifier studies are simplified by using a voltage divider that acts as a low impedance source. A simple voltage divider can be constructed on a terminal board and mounted by wire leads on the oscillator output terminals. The resistor network shown in Fig. 16.1 supplies a series of 10-to-1 signal steps from 1 mV

Figure 16.1. *AC voltage divider (low output impedance).*

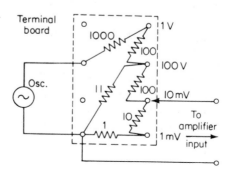

up to 10 V. Except at the two highest output levels the equivalent generator impedance is low enough to be treated as a constant voltage source.

For voltage gain measurements the oscillator output should be maintained at 10 V peak-to-peak. If the oscilloscope is set on 10 V peak-to-peak, full scale, then when the amplifier input signal is 10 mV, full scale represents a voltage gain of 1000. In other words, any time the full scale voltage calibration is numerically the same as the input mV, the gain is 1000. With this understanding we can easily interpret the corresponding meaning of other settings of the input signal or of the oscilloscope. A good practice is to favor

Plate 2. *A typical experimental set-up that can be used to adjust the Q-point and measure the gain of a three-stage amplifier like the one shown in Fig. 5.8 (photo courtesy of Warren B. Davis, Houston, Texas).*

signal frequencies and capacitance sizes that are multiples of 4 because the reactances are multiples of 10. For example, at 40 Hz the reactance of 400 μF is 10 Ω and 0.4 μF is 10 kΩ. The substitution of 500 μF or 0.5 μF introduces only a minor error.

Almost any late model calibrated oscilloscope having a dc amplifier is adequate for transistor circuit studies. As a matter of practical convenience and to ensure reliability, it is best to avoid an oscilloscope that offers controls or performance characteristics that are not regularly needed. A second beam is useful for monitoring the oscillator output or for displaying a reference signal to help locate intermittent and transient disturbances in a complex system. A differential input is useful for observing off-ground signals, such as the base-to-emitter voltage in an emitter feedback stage, or for observing signals that have a large common mode interfering signal.

For biasing and amplifier studies we need four or more resistance substitution boxes, two capacitor substitution boxes, and a good supply of $\frac{1}{2}$ W resistors. The experimenter needs a small selection of ceramic and mylar capacitors and a dozen high capacitance 25 V electrolytic capacitors. It is convenient to equip some of the commonly used capacitors with clips for insertion in circuits without soldering. A few silicon diodes and a source of Zener diodes are needed. Three 4.5 V C batteries tapped at 1.5 V intervals and taped together make a conveniently adjustable auxiliary 13.5 V power supply.

16.2 BREAD BOARDS

Experimental circuits are assembled in many different ways. Over the long run it is better to assemble circuits on terminal boards, using soldered connections. Place components that are less likely to be changed near the board and attach capacitors and transistors without cutting the leads. The collector load resistors should extend off the board, as shown in Fig. 16.2, and be brought together where they connect to the power supply or to a Zener regulator. The terminal boards may be supported by clamping the mounting studs between lengths of $\frac{1}{2}$ in. $\times \frac{1}{2}$ in. or 1 cm. $\times$ 1 cm. iron bars. The iron bars give weight to the assembly and can be clamped together if they are drilled and tapped for two screws. The bars should be insulated by wrapping with tape, and sometimes they should be grounded.

A supply of clip leads is needed for circuit connections, and it is helpful to color code these to indicate a variety of lengths or types of clips. To avoid capacitance feedback and hum pickup, the leads should not be longer than needed. A wood top workbench is generally preferred for electronic work,

Figure 16.2. Bread board assembly.

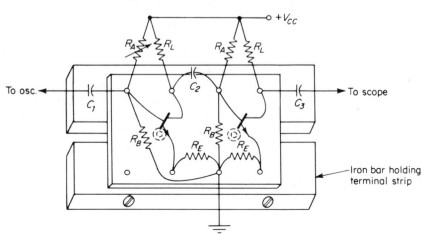

but the underside of the bench top should be covered with a grounded screen wire. The screen acts as a ground plane and reduces 60 Hz pickup and amplifier feedback. Most troubles with radio frequency pickup can be eliminated by reducing the bandwidth of the oscilloscope, and switches for this purpose are usually provided inside the scope.

16.3 BIASING ADJUSTMENTS

Most circuits in this book should operate satisfactorily after a minor adjustment of the bias. A VOM measurement of the collector-to-emitter voltage is the quickest way to determine whether a transistor is over-biased or under-biased. An over-biased condition may indicate that the power supply is connected with reversed polarity or that the transistor has suffered punch-through. High-gain amplifiers are sometimes difficult to bias. In difficult cases it helps to begin with a high input signal and to set the bias so that the output waveform is approximately a square wave. With this as a starting bias, the input signal should be reduced and the bias adjusted until the signal peaks are clipped the same on both sides. Generally, this bias gives the best room-temperature Q-point. If both peaks cannot be made to clip symmetrically, the amplifier should be examined for an error that causes overload to occur in one of the low-level stages. If it seems impossible to bias an amplifier except at full-on and full-off, we should suspect that the amplifier has so much positive feedback that it behaves as a flipflop. In this case we should locate and remove the positive feedback or reduce the number of stages.

16.4 FREQUENCY RESPONSE

Amplifiers should be checked for adequate frequency response. An amplifier may oscillate merely because the input and output leads, used for testing the amplifier, are placed too close together. Often it is desirable to limit the high-frequency cutoff by inserting a shunt capacitor across a high impedance point about midway through the amplifier. At other times it may be necessary to reconstruct an amplifier or to shield components to reduce capacitance feedback to a tolerable level. Low-frequency oscillations are usually caused by inadequate filtering or by coupling in the collector supply. A good rule is to connect only two or three stages to the same filter capacitor or Zener diode. In a 3-stage amplifier it is sometimes sufficient to filter or regulate the input stage bias supply separately from the remaining stages.

Insufficient high-frequency response is probably caused either by the Miller effect in high voltage-gain stages or by the transistor input capacitance in high current-gain and power stages. Further information concerning the cause and correction of high-frequency cutoff is given in the author's *Analysis and Design of Transistor Circuits* (Ref. 1).

Insufficient low-frequency response probably means that one or more coupling and emitter capacitors are too small. The capacitance values given in the circuits in this book should ensure that the 3 dB cutoff (one-half power) is below 40 Hz. A convenient way of measuring the input impedance of an amplifier is to lower the signal frequency until the input base signal is 3 dB below the signal on the generator side of the input capacitor. At this frequency the reactance of the capacitor equals the input impedance looking into the base. This measurement is easily made if the generator has a low impedance. The output impedance of a CB or CE resistance-coupled stage is the resistance of the collector resistor.

16.5 PERFORMANCE TESTS

We must recognize the importance of carefully checking electronic equipment, despite the urgency of delivery time schedules. The most important check is the simplest and most neglected—visual examination of every electrical connection for good soldering and examination of every component for its indicated value. The majority of all service difficulties can be traced to a failure to make careful visual checks. Many service faults can be located by carefully searching for component failures or wiring defects. This examination of a circuit must be performed by a person who is convinced there are errors. Therefore, the man who assembled the equipment is usually the least qualified to find impending trouble.

The intended application of a circuit usually suggests the obvious electrical tests that are needed to check equipment. Ordinarily, amplifiers would be checked for correct frequency response, gain, and the overload characteristics. A high-gain amplifier should be tested to detect any tendency to oscillate or to exhibit abnormal noise. For some applications we must measure the time required by an amplifier to recover from an overload. High-gain direct-coupled amplifiers usually require a well regulated power supply. Good practice requires checking a high-gain amplifier to know it will operate at high or low supply voltages that may exist at times. Any deviation from the normal operating voltage range should be interpreted as indicating either poor circuit design or a faulty component. Square wave tests should be used to examine the transient performance of an amplifier, especially when an accurate reproduction of the waveform is required.

16.6 THERMAL TESTS

Careful thermal tests of transistor equipment are needed to ensure thermal stability and reliable operation at extreme ambient temperatures. The simplest thermal test is to check all transistors and resistors for overheating by touching the components with the little finger. This finger is able to withstand continuously a temperature of 50°C and to withstand 60°C for a second at most. Within this range temperature can be estimated fairly well, and 60°C is a good conservative operating temperature for germanium devices. Silicon transistors in low-gain circuits are usually operative up to 100°C, the temperature at which water sizzles on the transistor case. A soldering iron brought carefully into contact with a silicon transistor to raise the case temperature may show that more elaborate temperature tests are unnecessary. This simple test often reveals either that a circuit is much too sensitive to a small temperature rise or that it is surprisingly insensitive to high temperatures. Silicon power transistors can be expected to operate at case temperatures above 100°C. High-gain dc amplifiers designed to operate over a wide temperature range must be carefully tested while exposed to the expected ambient temperatures.

16.7 TRANSISTOR TESTING

Well designed circuits should operate satisfactorily if the transistors are neither open nor shorted and have a current gain of at least 50. With an ohmmeter, we can check most transistors easily and safely to establish that they pass these tests. The simplest test is to check that the transistor is not short-circuited. A transistor is a pair of back-to-back (coupled) diodes, having the base as the common side of both diodes. An ohmmeter which has one side connected to the base should show a low resistance when the other side is connected to either the emitter or the collector. With the ohmmeter connections interchanged, the meter should indicate the higher resistance of the reverse biased diodes. Generally, the diodes do not check alike and the emitter diode tends to have the lowest forward and reverse resistance. Neither diode should be expected to show anything like the resistance ratio of a plain diode. When power transistors are checked, a suspected transistor should be compared with a new one of the same or similar type before a transistor is rejected as defective.

Punch-through causes a short-circuit from the collector to the emitter and is found by measuring the collector-to-emitter resistance. Good power transistors may show a reverse biased collector-to-emitter resistance as low as 1 kΩ, and the resistance usually changes by a factor of at least 5 when the ohmmeter is reversed. With punch-through, the collector-to-emitter resistance is only a few ohms and does not change with the polarity. Transistors that behave as good back-to-back diodes and do not have punch-through are usually operative.

The emitter current and voltage ratings of high-frequency transistors should be considered before transistors or transistor circuits are checked by using an ohmmeter.

16.8 CURRENT GAIN MEASUREMENTS

The current gain of a transistor can be measured with acceptable accuracy by using an ordinary ohmmeter and a resistor. This method uses the ohmmeter circuit to supply current; but, because β is proportional to the current, it is read on a convenient current or voltage scale. The meter is calibrated to read the correct current gain by selecting the resistor that supplies the transistor base current.

For power transistors the ohmmeter should be set on the lowest resistance range, and for low-power transistors the ohmmeter should be set on a range that reads 50 to 500 Ω at mid-scale. The high resistance ranges are unsatisfactory and may damage small, high-frequency transistors. Now, disregarding the fact that the meter is set on ohms, as shown in Fig. 16.3, a convenient meter scale, either for voltage or current, is chosen to represent the current gain and a number is selected to represent a current gain of 100. As shown in Fig. 16.3, the meter must be read at the low-current end of the scale where the voltage across the ohmmeter terminals is about 90 percent of the open circuit voltage.

We calibrate the meter to read current gain by selecting the base resistor, as shown in Fig. 16.3. For example, the calibrating base resistor is 100 times the resistance at the scale number representing a current gain of 100. If the transistor has a current gain of 100, the meter will respond as if the collector

Figure 16.3. Ohmmeter beta test.

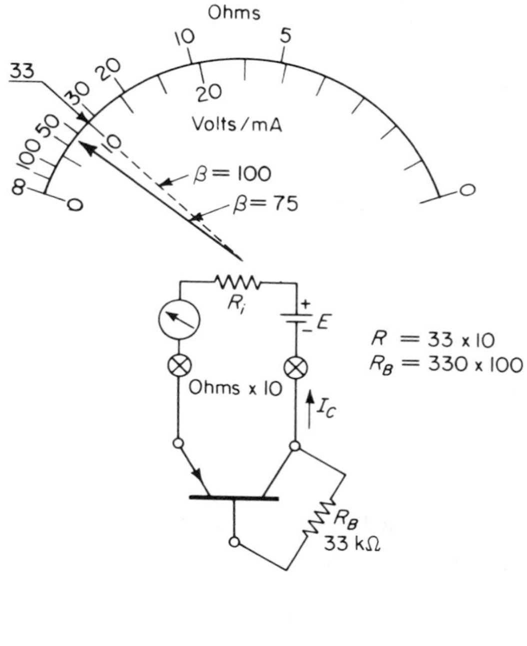

$R = 33 \times 10$
$R_B = 330 \times 100$

circuit resistance is $\frac{1}{100}$ the base resistor. If the transistor has a current gain of only 50, the meter current will be one-half and read 50. *The transistor effectively reduces a fixed base resistance by the current gain, but the current is read as current gain rather than as a resistance.*

It is meaningless for practical purposes to require even a 10 percent accuracy in a current gain measurement because current gain varies so much with the emitter current. This variation with the current can be observed by measuring the current gain on two different ohmmeter ranges. A measurement of β should always be made either to know that β is high compared with the amplifier S-factor, or to ensure that the transistor is reasonably good. There is no value in saying a transistor has a particular current gain unless the emitter current is given also.

Power transistors can be tested at higher currents than are attainable with an ohmmeter by using the circuit shown in Fig. 16.4. This circuit requires only a 10 V or 12 V power supply, a 100 mA meter, and a 10 kΩ resistor. Only the 10 kΩ resistor is needed if a suitable power supply is available. Many laboratory power supplies have a meter that will read 100 mA, and the 10 Ω that protects the meter can be omitted if the power supply is current limited. A transistor is tested by connecting the power supply from collector to emitter, using the correct polarity. The collector current is read when the 10 kΩ base resistor is connected to the collector. (Actually, the base resistor can be connected to whichever side gives a reading.) Because the base current is 1 mA, the current gain is numerically the value of the collector current in mA. The transistor has to dissipate about 1 W during this test, and a heat sink is not usually required. Most small transistors that are rated to carry 100 mA will withstand 1 W power long enough to permit a meter reading,

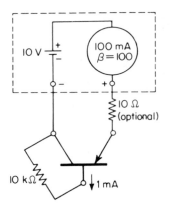

Figure 16.4. *Power transistor beta test.*

$$R_B = \frac{V}{I} \times \beta = \frac{10}{0.1} \times 100 = 10^4 \,\Omega$$

For *npn* reverse connection to power supply

and the transistor will not be damaged if it can be held between the fingers during the test.

The high range of a sensitive ohmmeter will show the equivalent collector leakage resistance of any good transistor. In a β measurement this resistance is negligible if the meter reading is large compared with the reading shown with the base resistor removed. The leakage current is most noticeable in testing power transistors.

That the transistor current gain is given by the meter current can be easily proved. The total resistance in the ohmmeter circuit is the sum of the internal resistance R_i and the external resistance R read on the ohmmeter scale. By Ohm's law, the current I equals the battery voltage E divided by the total resistance; therefore:

$$I = \frac{E}{R + R_i} \tag{16.1}$$

By using the ohmmeter at the low-current end of the scale where the scale readings are over ten times the mid-scale value, the internal resistance R_i may be neglected. Calling the calibration current I_{100} and the ohmmeter reading R_{100}, we have from Eq. (16.1):

$$I_{100} = \frac{E}{R_{100}} \tag{16.2}$$

Because the ohmmeter voltage across R_{100} is approximately the voltage across the base resistor, the base current is:

$$I_B = \frac{E}{100 R_{100}} \tag{16.3}$$

and the collector current is:

$$I = \beta \frac{E}{100 R_{100}} \tag{16.4}$$

Combining equations (16.2) and (16.4) and solving for β, we find:

$$\beta = 100 \frac{I}{I_{100}} \tag{16.5}$$

Equation (15.5) states that β is proportional to the meter current. The principal error in the measurement comes from the low voltage available at the collector and the fact that the base-emitter voltage reduces the voltage across the base resistor. However, as stated above, an accurate measurement of β is not of practical importance unless the emitter current is measured.

16.9   CIRCUIT TROUBLESHOOTING

Troubleshooting in simple transistor circuits may be difficult, even when the normal performance of the circuit is understood. The first requirement for troubleshooting is to have as much knowledge as possible concerning the circuit and its intended purpose. Transistor circuits are complex combinations of amplifiers, switches, regulators, and gates. Therefore, it is important to know which transistors are intended to operate as switches and which are intended to be amplifiers.

The measurement of all dc voltages is usually a good way to locate troubles. The easiest first step in troubleshooting is to find, by measuring collector-to-emitter voltages, which transistors are turned full-ON and which are OFF. Finding a transistor ON or OFF in a linear amplifier will often direct attention to the real difficulty. Sometimes it is helpful to measure the base-emitter voltages to know that the transistors are forward biased. The base-emitter voltage cannot give a clear indication that a transistor is operative, but the voltage will sometimes indicate a circuit condition that makes it impossible for a transistor to operate.

The current gain of a transistor or of a series of direct-coupled stages can be measured by opening a base connection or a bias resistor. The gain test is performed by inserting a measured or an estimated input current and observing the change of the collector current. Very often a gain test can be made by adding a resistor in parallel with the bias resistor or the base resistor. We can determine with a gain test whether the amplifier responds to a large signal or follows only small signals. The gain test is useful for observing the ability of a transistor to respond to a switching signal.

Table 2 gives a practical procedure that is helpful for troubleshooting many amplifier and switching circuits. In brief, the procedure is as follows: (1) The collector dc Q-point voltages are observed under operating conditions. These may indicate which transistors operate as switches and which as amplifiers. Also, the Q-point check may show which stage is defective. (2) With the power off, dc resistance measurements are made to determine whether the affected transistor is *pnp* or *npn* and whether there are simple shorts or open circuits. (3) The circuit voltages are measured to determine whether the transistor may be switched ON and OFF when the bias is changed. This test may indicate that the circuit has either a defective transistor or an improper bias.

Experience in troubleshooting may be secured by simulating a circuit problem and following the tests outlined in Table 2. A low β transistor is simulated by shunting the base-emitter diode with a low resistance; punchthrough, by a collector-to-emitter short; and an open, by cutting a lead. The experimenter is urged to acquire troubleshooting experience by simulating defects and practicing the test procedure.

It is helpful to have a planned procedure for locating troubles. With a working circuit at hand we are in a better position to measure operating voltages and plan a series of tests. Wherever possible, the tests should be arranged to establish by halves which half of a circuit is operative and which half is inoperative. The bad half can be examined by halves, etc. Unlike vacuum tube circuits, a fault in a transistor circuit, particularly in direct-

Table 2. *In-Circuit Tests of Transistor Circuits*

Measure	Indicates
1. DC collector Q-point voltages	Whether transistors operate as switches or amplifiers; may reveal questionable stages.
2. Base-emitter diode resistance*	Whether *pnp* or *npn*, and whether a good diode or short.
3. Collector-emitter resistance*	Punch-through, or collector diode short.
4. Emitter and collector resistors*	Approximate value of resistors or wiring error.
5. Base-emitter voltage drop	Whether forward or reverse bias and amount of bias.
6. Collector-emitter voltage drop	Whether collector current agrees with indicated bias.
7. Emitter and collector voltages	Whether emitter and collector currents are nearly equal.
8. Collector voltage with emitter diode shorted	That transistor can be turned OFF, low value of β, or improper bias.
9. Collector voltage with bias resistor shunted by $R_A/3$ or $R_A/4$	That transistor can be turned ON, low β, or improper bias.
10. If in-circuit tests fail, carefully check the circuit with one or more transistors removed and test these transistors.	Look for incorrectly installed semiconductors, off-value resistors, defective capacitors, open wiring or solder connections.

*Use ohmmeter with power off. For all other tests use voltmeter with power on.

coupled amplifiers, may show an adverse effect one or two stages ahead of a fault. For example, in a stage having emitter feedback, an open collector reduces the input impedance, a condition that may be interpreted as indicating a low voltage-gain in the driver stage.

Feedback circuits which have more than one stage in the feedback loop may present very difficult troubleshooting problems. A closed feedback loop causes all transistors in the loop to respond as a unit, with the result that it may be difficult to know which stage is at fault. Finding a circuit difficulty with the feedback loop open is easier than with the loop closed, but it is sometimes nearly impossible to open the loop without disturbing the circuit Q-points. At times a feedback amplifier has to be examined one or two stages at a time in locating troubles.

Troubleshooting requires skill, patience, ingenuity, and a detailed understanding of circuits. However, a close examination for physical defects will often reveal a faulty component or connection. We should take pride in locating obvious defects by simple methods and use complex tests and instruments only when their use is expedient.

SUMMARY

An ohmmeter, perhaps the most useful of all laboratory instruments, deserves study and practice in its use until the instrument's characteristics are clearly appreciated and understood. Besides its use in troubleshooting an ohmmeter is a convenient and reliable instrument for measuring transistor current gain and for checking diodes. The simple visual examination of a defective circuit frequently reveals a source of trouble or gives a valuable clue.

The careful testing and study of electronic equipment must be encouraged for its value in revealing potential service failures and unsatisfactory operation. Such testing is valuable also for training service personnel and for developing a familiarity with new equipment. The skills to find and correct service difficulties require as much study and analysis as are required to understand the underlying theory. These practical problems are challenging and are of more real monetary value than is generally appreciated.

The day-to-day experiences in an electronics laboratory (college or industrial) present continuing opportunities to improve one's knowledge, to acquire new skills, and to increase one's value to an employer. These opportunities deserve frequent and critical appraisal of work habits. Technical competence is maintained today only by continually studying new techniques and seeking new knowledge. Happily, the pursuit of knowledge can be one of the most rewarding pleasures of a lifetime.

Annotated Bibliography

Every student needs a small library of carefully selected reference books to supplement the texts required in courses. Because each author presents a different point of view, a difficult concept may be understood more easily by consulting several writers. A book that brings only one or two good chapters into a library is well worth the cost. Good books are the least expensive source of reliable information, and books that are not at hand are practically useless. One should acquire good books regularly and share their treasures frequently.

The books annotated below are listed alphabetically by the author. Those books designated by an asterisk (*) tend to support the material of this text; the others are more specialized.

*1. Cowles, L. G., *Analysis and Design of Transistor Circuits*. New York: Van Nostrand Reinhold Co., 1966. A practically oriented engineering introduction to transistor circuits and design. Practical circuits are described with components and gain-impedance values indicated. The book includes single-stage amplifiers, transistor pairs, direct-coupled amplifiers, power amplifiers, noise, thermal problems, medium-frequency amplifiers, and FETs.

*2. Cowles, L. G., *Transistor Circuit Design*. Englewood Cliffs, New Jersey: Prentice-Hall, Inc., 1972. A design manual emphasizing feedback as a design tool, with simpler methods for predicting amplifier stability. Covers circuits from dc to microwaves, audio and RF power amplifiers, and integrated circuits. Includes new material on transistor and FET circuit design, UHF and video amplifiers, and microwave devices. Describes microwave diodes used as oscillators, mixers, detectors, varactors, and frequency multipliers. The Appendix has design data and charts.

3. Millman, Jacob, *Vacuum-Tube and Semiconductor Electronics*. New York, N. Y.: McGraw-Hill Book Company, 1958. A clearly written college text that considers vacuum tubes and semiconductors together. Recommended for students and teachers interested in the physical theory of semiconductor parameters, electronics, gas tubes, oscillators, and feedback amplifiers.

*4. Mitchell, Ferdinand H., *Fundamentals of Electronics*, 2nd ed. Reading, Mass.: Addison-Wesley Publishing Co., 1959. An excellent, clearly written introduction to vacuum tube electronics and applications. This book is recommended for the beginner or as review material for anyone wishing to improve his understanding of basic electronics and circuit theory.

*5. *Radio Amateur's Handbook*. West Hartford 7, Conn.: The American Radio Relay League, 50th ed., 1973. A well known handbook containing a wealth of practical radio circuits and how-to-build information. The material of general interest includes electrical laws and circuits, vacuum tubes, radio circuits, measurements, and miscellaneous data. The handbook is readily available and inexpensive. No library is complete without a copy.

6. *RCA Receiving Tube Manual*. Harrison, New Jersey: Radio Corporation of America, 1971. A concise description of vacuum tube circuits and applications with vacuum tube data, RC amplifier design charts, and working circuits.

*7. *RCA Transistor, Thyristor & Diode Manual*. Somerville, New Jersey: Radio Corporation of American, 1971. Contains an outline of transistor circuit theory, practical applications, and recommended circuits. The manual includes data describing most of the RCA semiconductor devices.

8. *Reference Data for Radio Engineers*, 4th ed. New York, New York: International Telephone and Telegraph Corporation, 1956. A handbook of data concerning components, networks, bridges, transformers, and electron tube circuits. *Reference Data* provides a particularly useful source of mathematical formulas, tables, and radio engineering data.

9. Slurzberg, Morris, and Osterheld, William, *Essentials of Radio*, 2nd ed. New York, N. Y.: McGraw-Hill Book Company, 1959. An easy introduction to circuit analysis, vacuum tubes, amplifiers, and radio circuits.

10. Terman, F. E., *Electronic and Radio Engineering*, 4th ed. New York, N. Y.: McGraw-Hill Book Company, 1955. An excellent radio engineering text, primarily concerned with vacuum tube circuits. The book is a valuable reference source of information about components, amplifiers, modulators, noise, feedback, and electronics.

11. Thomas, Harry E., *Handbook for Electronic Engineers and Technicians*. Englewood Cliffs, New Jersey: Prentice-Hall, Inc., 1965. A practical man's handbook covering electronic construction, components, and instruments. Describes testing procedures for radio receivers, transmitters, radar, and servo systems. Includes shop and mathematical tables.

*12. *Transistor Manual*, 7th ed. Electronics Park, Syracuse, New York: General Electric Co., 1964. The GE Transistor Manual has many practical circuits that illustrate applications of transistors, silicon-controlled switches, unijunction transistors, and tunnel diodes. The circuits include audio amplifiers, radio receivers, digital computer elements, servos, and experimenter projects. The volume has a tabulation of transistor specifications and is the most useful transistor handbook.

13. U. S. Navy, *Basic Electronics*, 1964 ed. Washington, D. C.: U. S. Government Printing Office, NAVPERS 10087-A. A carefully prepared Navy training manual on vacuum tubes and their applications in radio and radar systems. Includes material on power supplies, tuned circuits, modulation, and test equipment.

14. *Semiconductor Data Book*. Phoenix, Arizona: Motorola, Inc. A useful cross reference for the registered semiconductors with data given for many types and including typical UHF and microwave circuits.

Laboratory Experiments

LABORATORY EXPERIENCE

The laboratory should be welcomed as the testing ground for ideas and as an opportunity to gain practical experience. We soon find that there is much to be learned in the laboratory besides equations and circuit theory. Experiments with transistors and circuits reveal many unexpected rewards for a questioning mind, and there is pleasure in finding that the rewards are often greatest when the simplest circuits are closely examined. The laboratory should be approached with a spirit of discovery and not merely to confirm what is presumed to be known.

A laboratory experiment is never quite as simple as expected, and the beginner should be encouraged to understand thoroughly whatever is observed before proceeding to the next part of an experiment. Elementary experiments usually lead to valuable insights, and much is lost by complicating the circuit and by using elaborate test equipment. More is gained by repeating the experiment using different component values or supply voltage to discover the limitations of a theory or a device.

Based on long experience in the laboratory, the author feels that with rare exceptions all measurements should be made with an oscilloscope and a volt-ohmmeter. The oscilloscope has the advantage of convenience and a wide frequency response, and it gives a continuous display of the signal waveform. The volt-ohmmeter is convenient, portable, and usually accurate. By restricting measurements to these two instruments, we become fully aware of their limitations, we become experienced in their use, and we save time and acquire accuracy. We should understand the uses and advantages of other instruments, but they should be considered only when they are really needed. An instrument should not be chosen to obtain more accuracy than is useful or because it is slightly more convenient.

The experiments in this section have been prepared to show that much can be learned by simple measurements. The procedures can be used as guides, but it is always better for the experimenter to develop his own techniques and chart his own route to understanding. Experience in the laboratory should aim to show the student how to work with practical problems and devices, how to apply theories, and how to extend his knowledge and understanding. The laboratory should provide many rewarding experiences and much enjoyment if entered with an open mind and a desire to learn.

LABORATORY RECORDS AND REPORTS

A clear and complete record of all laboratory experiments should be kept in a bound notebook. The record should include a brief statement of the purpose of the experiment, a circuit diagram, and a list of the instruments with serial numbers. Numerical data should be recorded as read in neat columns with headings. If in error, original data should never be erased. Draw a line through defective data and record the new data below. Be certain that the record shows the correct units at the head of each column.

Carefully check data in the laboratory for errors and determine that the calculated results are reasonable. Review the adequacy of data to make certain that the circuits and the experimental procedure will be understood at a later date. One can hardly be too careful in recording minute details and explanations. A procedure and data that seem quite obvious today are soon forgotten and may be useless a few weeks later. When possible, ask another person to see whether the recorded data are clear and complete.

Most industrial laboratories require that each page of data is signed and dated. The practice of dating and numbering each page is a worthwhile habit that all should endeavor to acquire.

A technical report is intended to show another person what has been accomplished. The report is an opportunity to exhibit the value and quality of your work. A short, carefully prepared description of your work tells the reader more than a long recounting of elementary and obvious details. A laboratory report should include the following sections:

1. *Introduction or Abstract.* Clearly and immediately give the purpose of the experiment, and state what was actually accomplished.
2. *Procedure.* As concisely as possible, outline the procedure or method of investigation used so that another person could repeat the experiment.
3. *Data.* Submit a copy of the original data and the calculated data with sample calculations.
4. *Curves.* Curves exhibit the results of an experiment and should be carefully and neatly prepared. Because ratios of numbers are usually more important than differences, the scales used for curves should be logarithmic.

Linear scales generally distort the physical meaning of data, although a linear scale is sometimes used to conform with custom (e.g. the exponential discharge of a capacitor). The points in a series of measurements may be selected in an approximately geometric series by consulting a dB table (*See* the Appendix).

5. *Conclusions.* Seek out the meaning of the observed data and decide just what the data and curves really show. Present a brief statement of your conclusions and a summary of the findings of the study. Include a description of any unusual or unexpected results. Date and sign the report.

EXPERIMENTS

EXPERIMENT NO. 1. CHARACTERISTICS OF DIODES

This experiment demonstrates the most common characteristics of diodes.

1. Using a power supply and a resistance substitution box, measure the dc input resistance of your oscilloscope, checking all sensitivities from 0.1 V to 50 V full-scale. The oscilloscope is to be used as the dc voltmeter wherever indicated in the figures for this experiment.

2. Obtain and plot the reverse current-voltage curve for the collector diode, using the circuit in Fig. E-1.1. Calculate the voltage across the diode from the supply voltage and the oscilloscope dc deflection. Calculate the reverse current, using the oscilloscope reading and the oscilloscope dc resistance. For comparison repeat the data, using a 500 mA silicon diode.

3. Obtain data and plot the forward current-voltage curve for the collector diode, using the circuit in Fig. E-1.2. Calculate the current from values of the supply voltage and the series resistor, but neglect the diode voltage drop in the calculation.

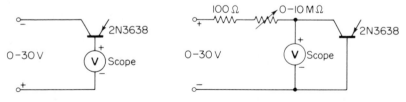

Figure E-1.1. *Reverse diode.*   **Figure E-1.2.** *Forward diode.*

Figure E-1.3. *Zener diode.*

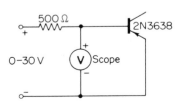

4. Using the circuit in Fig. E-1.3, obtain data and plot the Zener breakdown curve for the emitter diode of a silicon transistor. The 500 Ω resistor protects the transistor from overheating and should not be a lower resistance than indicated. Calculate the Zener current, using the supply voltage and the diode voltage to obtain the voltage drop across the resistor.

5. Using an ohmmeter, check several unknown transistors to observe their characteristics as back-to-back diodes and to determine whether they are *pnp* or *npn*. Find the ohmmeter polarity, using a marked diode. Using several TO-3 or TO-36 power transistors, find whether they are *npn* or *pnp* and identify the base connection.

6. Replace the power supply in Fig. E-1.1 by an oscillator, and observe the characteristics of a diode as a simple rectifier. Observe particularly that the diode acts as a resistor at signal levels below about 0.2 V. Make an approximate check of Eq. (7.1) for the percent ripple, remembering that the half-wave ripple is about twice the full-wave ripple.

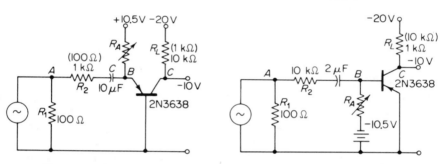

Figure E-2.1. *CB amplifier.*   **Figure E-2.2.** *CE amplifier.*

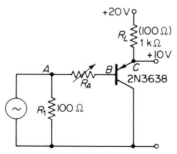

Figure E-2.3. *CC amplifier.*

EXPERIMENT NO. 2. BASIC AMPLIFIER CHARACTERISTICS

Each of the basic amplifiers has distinct characteristics of its own. This experiment illustrates the distinguishing characteristics of the CB, CE, and the CC amplifiers (see Figs. E-2.1, -2.2, and -2.3) and shows how the current gain and the input impedance can be measured in an operating stage.

The amplifier in each figure is arranged so that, by adjusting the bias resistor to make the Q-point voltage 10 V, the current gain can be directly read as the resistance ratio R_A/R_L. The voltage gain of each amplifier is the ratio of the ac voltage at C to the ac voltage at B, and the input impedance is calculated by comparing the ac voltage at B to the generator voltage at A. The input impedance of the CB and the CE amplifiers is low compared with the bias resistor, and the resistor can be neglected.

1. For each amplifier in turn adjust the bias so that the voltage drop in the collector resistor is 10 V. This places the Q-point close to the optimum for large signal outputs. Find h for the transistor, using Shockley's relation. From the resistor ratio R_A/R_L find the current gain β. Compare the measured voltage gain with the value predicted by the TG-IR. Compare the measured input impedance with the value calculated from h and β (or α).

2. For a fixed value of the bias resistor R_A, substitute several transistors in the CE amplifier and observe the variations of the Q-point and of the large signal clipping. Change the load resistor to 100 kΩ and observe how β changes with the collector current.

3. Tabulate for comparison the results obtained for each amplifier when using both values suggested for the load resistor.

4. Using an X-Y recorder, obtain collector-current vs. collector-voltage plots for several transistors of the same type number. It is interesting to show how much these characteristics vary with the temperature of the transistor and with the collector current.

5. For each amplifier connect a load resistor, using a 2000 μF series capacitor. By varying the load resistor, measure the internal output impedance of each amplifier. Find how the input impedances vary with the load resistor and check with the TG-IR.

EXPERIMENT NO. 3. TRANSISTOR PARAMETERS AND GAIN

An amplifier that does not have feedback presents opportunities for the study of basic transistor relations. Because low emitter currents are used in this experiment, the transistor should be a low-current planar silicon transistor. The ac input resistance of your oscilloscope should be measured with an oscillator and resistor box by the method given in Part 1 of Experiment 1.

1. Examine each amplifier in Fig. E-3.1, using the component values given in the table. Adjust the resistor R_A for a maximum signal just below overload. The resistor used for R_A should provide both a coarse and a fine adjustment. For each amplifier measure the dc base and collector voltages, using the dc scope. Measure the ac signal voltages where indicated in the figure. Correct the data for the effect of the oscilloscope on the measured values, and record all data. Estimate the temperature sensitivity of

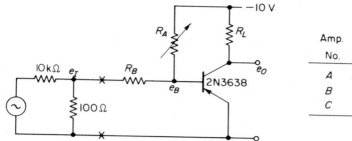

Figure E-3.1. Transistor parameter.

Amp. No.	R_B (ohms)	R_L (ohms)
A	10^4	10^4
B	10^5	10^6
C	10^3	10^2

each amplifier first by holding the transistor in your fingers and then while warming it carefully with a soldering iron.

2. Measure the transistor current gain by opening R_B, readjusting the bias, and observing the change in collector voltage when R_A is changed about 30 percent from the bias point.

3. From the data calculate the collector current, the input impedance R_I, and the transconductance g_m. Find h, using Shockley's relation. Tabulate the data and compare g_m with $1/h$. From the measured input impedance and h find an approximate value for β. Draw a SC equivalent circuit for the transistor in use.

4. It is interesting to repeat the experiment, using a germanium transistor and a medium-power transistor.

5. Calculate the rms input power, output power, and the power gain. Estimate the maximum useful temperature range for each amplifier.

EXPERIMENT NO. 4. EMITTER FEEDBACK BIASING

A study of the emitter feedback stage shows how easily the gain and the impedance characteristics of an amplifier can be found by inspecting the circuit. The experiment demonstrates also that h of the transistor is not important except when the emitter resistor is bypassed to increase the gain. The components that are to be used in each amplifier are indicated in the table adjacent to Fig. E-4.1.

Figure E-4.1. Emitter feedback amplifier.

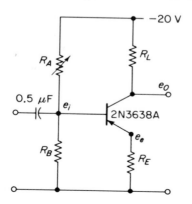

Amp. No.	R_B (kilohms)	R_E (kilohms)	R_L (kilohms)
A	51.0	5.1	68.0
B	51.0	0.51	68.0
C	5.1	0.51	6.8

1. For amplifier A adjust the bias to give the best waveform at the maximum output signal, and measure the peak-to-peak output voltage. With the input signal reduced, measure all ac and dc voltages and find the input impedance. Tabulate for comparison R_I with R_B, G_v with R_L/R_E, h with R_E, and show the voltage gain to the emitter, as an emitter follower. Tabulate the calculated no-load dc base voltage, and the emitter voltage.

2. Repeat step 1 for amplifier B and for C.

3. Examine the voltage gain and the low-frequency cutoff characteristics that are obtained for amplifier B when using a 1.0 μF emitter bypass capacitor. From the cutoff characteristics obtain a measured value for h and for β.

4. Examine the temperature characteristics of one of the amplifiers, and for each temperature measure the base-to-emitter voltage drop. With silicon transistors the temperature drift is caused partly by the V_{BE} drift and partly by the change of β. Determine for this example which effect predominates.

5. Study the performance characteristics of the emitter-coupled amplifier in Fig. 3.12 and the differential amplifier in Fig. 3.13 as examples of emitter feedback stages.

EXPERIMENT NO. 5. Combined Collector and
Emitter Feedback Biasing

This experiment illustrates the simplicity and advantages of the combined feedback stage.

1. Examine the Q-point and gain characteristics of the combined feedback stage shown in Fig. 3.10. Show that the stage has significant feedback, and compare the measured and calculated voltage gains.

2. Examine the effect on the Q-point made by connecting a bias resistor between the base and the ground. Set the bias for a maximum output signal amplitude and compare the voltage gain and the input impedance with that obtained without the bias resistor.

3. Obtain a family of low-frequency response curves similar to those shown for the feedback amplifier in Fig. 10.7. Observe how the characteristics change when the generator impedance is reduced to 100 Ω.

4. Examine the temperature characteristics of the stage without the bias resistor and determine the amount of drift with temperature and its cause.

5. Remove the capacitor C_f and the emitter resistor R_E and observe the characteristics of the collector feedback stage.

EXPERIMENT NO. 6. FET Amplifiers

This study of FET and MOS FET amplifiers is to show how the voltage gain of a stage depends on the Q-point conditions and on the figure of merit of the FET. A stage is used to illustrate the Miller feedback effects.

1. Examine zero bias operation of several types of FETs, using circuits of the form shown in Fig. 4.16. Using a 500 Ω load resistor, measure the stage voltage gains and calculate g_o and V_P. Show that the attainable voltage gains conform to Eq. (4.14).

2. Bias an FET and show that the effect of bias is to change g_m in agreement with Eq. (4.10). Show that voltage gain in excess of the value given by Eq. (4.14) can be obtained.

3. Examine the characteristics of stages constructed as in Fig. 4.17, where an FET is used as a constant current load.

4. Examine the feedback characteristics of several FET stages having source feedback. Show whether or not each stage has significant feedback.

5. Using an FET amplifier having a stage gain of about 10, connect an 0.01 μF capacitor from drain to gate. Drive the amplifier, using a 1000 Ω generator, and measure the high cutoff frequency. Show that the input capacitance is given by Eq. (4.25).

6. Study a MOS amplifier of the type shown in Fig. 4.9. Show that the input impedance is given by a resistance form of the Miller effect equation. Study the operation of a direct-coupled MOS amplifier.

7. Using a variable drain resistor and a variable gate bias voltage, show that the voltage gain attainable in an enhancement-mode stage is greatest when the FET is biased just above cutoff.

EXPERIMENT NO. 7. Direct-Coupled 2-Stage Amplifiers

The 2-stage amplifier shown in Fig. E-7.1 offers an interesting example for the study of a direct-coupled amplifier and for the application of the TG-IR.

Figure E-7.1. *Two-stage feedback amplifier.*

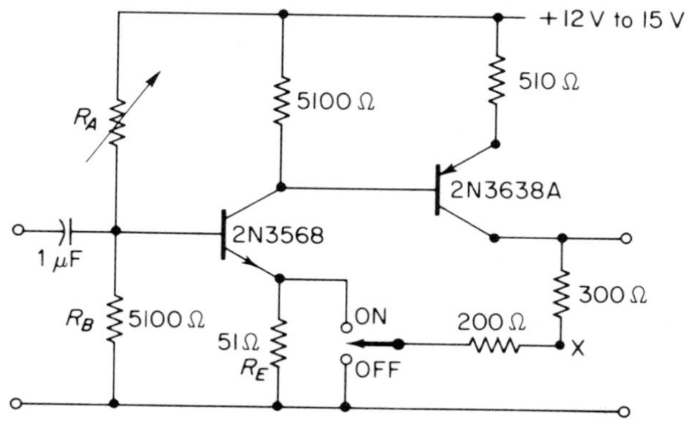

1. Connect the amplifier with the load grounded (switch to OFF) so that there is no feedback. Adjust the bias resistor, and then measure the peak-to-peak sine wave output voltage, the overall voltage gain, and the input impedance. Compare the measured voltage gain with the gain predicted by the TG-IR.

2. Connect the load resistor to the emitter (switch to ON), readjust the bias, and repeat the measurements of step 1. The overall current gain with feedback is R_B/R_E. Compare the measured voltage gain with that given by the TG-IR. Measure the amount of feedback (see Sec. 4.10).

3. Short the second-stage emitter resistor, readjust the bias resistor, and repeat steps 1 and 2. Compare the measure gains with those given by the TG-IR.

4. Find what determines the size of the capacitor that must be connected to point X to bypass the ac feedback.

5. Measure the output impedance of the amplifier by shunting the output terminal with a capacitor. Observe the effect on the output impedance caused by reinserting the second-stage emitter resistor.

6. The amplifier in Fig. 5.8 is easily adjusted to provide a voltage gain of 10. A comparison of the amplifier in Fig. 5.8 with the amplifier in Fig. E-7.1 makes an interesting study for the experimenter. A study of the sensitivity of both amplifiers to changes in the supply voltage and temperature is a profitable project.

7. The Darlington compounds of Figs. 5.5 and 5.6 furnish interesting examples for comparing measured and calculated gain-impedance relations.

EXPERIMENT NO. 8. The Equivalent Circuit of a Transformer

Most transformers are easily tested to determine their characteristics and suitability for a particular application. This experiment demonstrates the principal limitations of an audio transformer.

1. Connect a low-power, low-inpedance audio transformer (Stancor TA-59) to a source and a load having the rated winding impedances. Vary the signal frequency from 20 Hz to 200 kHz, and observe the frequency characteristic between the 3 db cutoff points. Using the low cutoff frequency, calculate the equivalent primary inductance. Measure the ac turns ratio and the dc resistances. Draw an equivalent circuit representing the transformer.

2. Apply an input signal at a convenient mid-band frequency (440 Hz). Measure the generator voltage, the input voltage, and the output voltage when the transformer is operated, using the nominal impedances. From these measurements calculate the transformer efficiency.

3. Decrease the source and the load impedances by a factor of 10 and observe the effect of this change on the frequency response and on the efficiency. Repeat with the terminations a factor of 10 larger than the nominal impedances.

4. Measure the equivalent primary and secondary inductances with a bridge and compare with the inductance value obtained from the frequency response measurements. Short the secondary, and measure the equivalent primary leakage inductance. Compare with the equivalent circuit.

5. Output transformers used in push-pull amplifiers have a center tap on the primary winding. Terminate the secondary winding of a push-pull transformer in the rated impedance. (a) Find by measurement or calculation the impedance, looking into one half of the primary when the other half is open. (b) Find the power delivered to the load when the full primary is driven by a 2 V source that has the rated primary impedance. (c) Find the power delivered when one side of the primary is driven by a 1 V source with one-half the rated primary impedance. The latter condition corresponds to one transistor of a push-pull pair being open.

EXPERIMENT NO. 9. PUSH-PULL POWER AMPLIFIERS

This experiment illustrates the characteristics of a push-pull power amplifier and the adjustment of biasing for both Class A and Class B operation.

1. Connect the amplifier, as shown in Fig. 6.3, using a pair of 2N2869 transistors mounted on a 8 cm. × 15 cm. heat sink. Adjust the bias so that the total collector power input is about 7 W. For a geometric series of load resistor values adjust the input signal so that the output signal shows about 5 percent distortion. For each load, measure the output voltage, calculate the output power, and plot the power to find the optimum load resistance.

2. Using the optimum load resistance, measure the power gain of the amplifier, and study the effect on the gain and the output power of operating at a frequency just below the 3 db cutoff frequency.

3. Apply a 400 Hz input signal reduced to $\frac{1}{5}$ the drive required for maximum output power, and measure the internal output impedance by changing the load resistance. Using the load resistance that approximately matches the internal impedance, examine the performance characteristics of the amplifier.

4. Adjust the bias for Class B operation and measure the power output, the efficiency, and the no-signal standby power. Carefully observe the differences between Class A and Class B operation, particularly at maximum power output.

5. Measure the turns ratio of the output transformer and the winding dc resistances. Calculate the power loss in the copper at full output and, assuming the core loss equals the copper loss, calculate the indicated efficiency.

6. Replace the transistors with a pair of silicon 2N3638's, and avoid over-heating the transistors by sustained signals. Explain why the crossover bias needs readjusting and note the additional bias power that is required. Using the optimum load resistance found in step 1, measure the maximum output power. Then find what level of output power can be used continuously without overheating the transistors.

7. Replace the transistors with any 150 mW germanium transistors. Lower the collector voltage to 8 V and operate the transistors with a sustained signal. At 5 minute intervals increase the collector voltage 1 V at a time until the transistors overheat and "run away." Observe that the voltage from collector to emitter under runaway conditions is abnormally low and that the collector input power is abnormally high. If the experiment is performed carefully, the transistors should recover and operate normally after cooling.

EXPERIMENT NO. 10. SWITCHING CIRCUITS

1. Construct the switching amplifier shown in Fig. 8.4. Measure the base-emitter and base-collector voltages in both stages and verify that the stages operate as switches. Increase the base resistor of the input stage until the collector-to-emitter voltage of each stage, in turn, is 1 V. Decide whether the base currents in the original circuit are at least $\frac{1}{10}$ the collector currents and adequate for reliable switching.

2. Construct a 1-stage inverter, using the circuit in Fig. 8.6. Verify that saturated switching is obtained by showing that the collector diode is forward biased. Find the range of input voltage over which the transistor remains saturated. Reduce the load resistance until the transistor is pulled out of saturation and observe that it dissipates more heat.

3. Construct the saturated flipflop shown in Fig. 8.7. Trigger one of the stages by grounding a collector and verify that the stages switch into saturated Q-points. Connect the collectors to each other and find the active region Q-point. Disconnect the collector tie. Drive the emitter, using a square wave or pulse generator and observe the collector, base, and emitter waveforms.

4. Construct a multivibrator, using the circuit in Fig. 8.10. By varying both base resistors, verify that the switching frequency depends on the value of the base resistors and the coupling capacitor. Select capacitors over a 30-to-1 range and plot the frequencies to obtain a relation between frequency and the RC product.

5. Construct the Schmitt trigger shown in Fig. 8.12, using 10 kΩ in the first collector and 1 kΩ in the second. By varying the voltage at the input terminal, observe the operation of the trigger and measure the hysteresis voltage. Plot the input base voltage as a function of the input current and verify that the peak point of the curve is not retraced when the input voltage is decreased. Connect a 3 kΩ resistor between the base and +3 V on a bias battery. Construct the corresponding load line on the *V-I* chart and verify that the base *Q*-point can be made to rest at two points on the load line.

6. Construct the monostable flipflop shown in Fig. 8.11. By grounding the input base, observe that the output stage can be switched into saturation. If a pulse source is available, study the performance of the mono as the input pulse amplitude and repetition frequency are varied. Connect a low-frequency sine wave source through a bias battery to the input base and adjust the bias so that the output signal is a square wave. As the input frequency is increased, observe the operation as a frequency divider. Disconnect the capacitor that couples the output back to the input and observe the performance of the circuit as a squaring amplifier. Using a feedback capacitor of several hundred microfarads, observe the operation of the mono as a way of producing a time delay.

EXPERIMENT NO. 11. DIODE SWITCHES AND LOGIC CIRCUITS

A purpose of this experiment is to illustrate the characteristics of NOR and NAND logic and to show how logic gates are utilized in practical circuitry. A synchronous detector is examined to illustrate phase-sensitive signal detection and the narrow-band characteristics of signal correlators.

1. Set up the NOR and NAND logic circuits shown in Fig. 9.17 and Fig. 9.18 with manual switches at the inputs and a meter at the output so that each student becomes familiar with the characteristics of logic operations. The logic circuit in Fig. 9.14 and a voltage doubler might be set up as interesting demonstration projects.

2. Construct the FET synchronous detector shown in Fig. 9.11 and use the lowest range voltmeter of a volt-ohmmeter as the output indicator. Observe the characteristics of the system as a phase-sensitive detector and measure the effective bandwidth. If two oscillators are available, increase the input and switching frequencies and observe the increased *Q* of the system.

Make the input frequency successively 2, 3, 4, and 5 times the switching frequency and observe the output signal. Repeat by making the switching frequency integral multiples of the input frequency. What bearing does the observed characteristic have on the use of a phase-sensitive detector?

3. If a fast square wave generator is available, the switching times of a saturated transistor switch make an interesting study. Long switching times
can be obtained by using a power transistor and reducing the circuit
resistances. Observe the turn-ON, turn-OFF, and the delay time.

EXPERIMENT NO. 12. Active Filters and Tuned Amplifiers

1. Construct the amplifier shown in Fig. 10.6 and make C_2 equal $2\,\mu F$.
Drive the amplifier from a low-impedance source and find the input
resistance of the stage by observing the frequency at which the gain falls
3 dB. Drive the amplifier from a 20 kΩ source, connect a capacitor from
the collector to base, and adjust the capacitor so that the 3 dB high-frequency cutoff falls at about 3 kHz. Show that the cutoff frequency is
correctly explained by the Miller effect.

2. Drive the amplifier shown in Fig. 10.6 by a low-impedance source and
study the performance characteristics illustrated in Fig. 10.7.

3. Study the characteristics of the tuned amplifier shown in Fig. 10.14. Plot
the frequency characteristics for two values of the Q control and find
the circuit Q. Remove the signal generator and examine the characteristics
of the circuit as an oscillator. A medium value of β gives the best waveform. Measure the voltage gain from the emitter to the base while the
circuit is oscillating.

4. Construct one or more of the active filters and obtain frequency characteristics.

5. Construct a simple integrating amplifier which has a time constant $R_1 C_0$
of about 10 seconds. Examine the characteristics of the integrator when
the input signal is, in turn, a voltage step, a sine wave, a square wave, and
an impulse.

6. Construct for class demonstration the analog computer shown in Fig.
10.29.

EXPERIMENT NO. 13. Oscillators

Experimental confirmation of the Barkhausen conditions is probably
the most meaningful objective of any study of oscillators. When the feedback
network is opened, an impedance network must be added to duplicate the
closed loop conditions, and a simple way of terminating the open loop is
not always evident. The Franklin oscillator is studied first because the open
loop is easily terminated by resistors.

1. Set up a Franklin oscillator, adjusting the bias for a compromise between
good waveform and amplitude stability when the supply voltage is

changed ± 10 percent. Select a point where the feedback loop can be opened without disturbing the circuit impedance, or terminate the open ends of the loop in an equivalent load. Drive the input end of the loop at the signal level observed with the loop closed. Vary the frequency over a small range and record the frequency and the loop gain when the input and output signals are in phase. Explain why the observed conditions may not confirm the theory exactly.

2. Set up an oscillator that appears interesting, adjust the bias for stable operation, and measure the frequency range over which satisfactory operation is obtained when one or more elements are varied as a frequency control. Measure the frequency stability for a ± 5 percent change of the supply voltage and compare with the stability specifications of a manufactured audio oscillator. Measure or estimate the percent distortion of the output waveform.

3. An open-loop study of a transistor phase-shift oscillator provides an interesting exercise in finding the best way of terminating the open ends of the loop. Be sure to consider the loading effects of instruments used in the experiment.

4. Construct the inverter shown in Fig. 11.13 and plan and execute a series of performance tests which include the full-load efficiency, frequency stability and percent regulation.

EXPERIMENT NO. 14. WAVE-SHAPING

An understanding of wave shapes and of the causes of waveform distortion provides clues that are needed to solve many everyday application problems. The following experiments present interesting examples of clipping, chopping, distortion, and square wave testing.

1. Construct several clippers, making a record of the circuits used, and sketch the observed waveforms with the clipping levels indicated.

2. Examine the performance of a self-biased FET amplifier and sketch a load diagram that explains both the small-signal and the large-signal performance.

3. Examine the FET chopper described in the text and explain the difficulty observed at large signal inputs. If two oscillators are not available, use a filament transformer and a 500 Ω potentiometer to supply the 60 Hz input signal.

4. Design a resistance network by which a signal at line frequency is superimposed on an oscillator signal, and reproduce the waveforms illustrated in Fig. 13.22.

5. Examine the square wave response of the high-pass filter shown in Fig.

13.20. Repeat, using the low-pass filter shown in Fig. 13.21. Use a variable frequency square wave generator, if available, or make a square wave using the diode limiter shown in Fig. 13.4. Demonstrate examples of wave-shaping by integration and examples by differentiation.

6. Examine the square wave response of an audio amplifier. A high performance hi-fi amplifier is difficult to test. Therefore, an inexpensive amplifier should be examined first.

EXPERIMENT NO. 15. THYRISTORS

This experiment presents a study of the SCR as a power-control device and a study of the 4-layer diode as a relaxation oscillator.

1. A study of thyristors should begin with an experimental examination of the transistor feedback pair, following the outline given in Chap. 15. The transistor equivalent represents 4-layer devices accurately and gives an intuitive understanding of regenerative switching. Proceed to the remainder of the experiment only after the characteristics of a transistor regenerative pair are thoroughly explored.

2. Set up the SCR circuit shown in Fig. E-15.1, using any SCR capable of blocking 200 V or more and capable of carrying 1 A average. Examine the waveform of the load voltage and explain the observed wave shape.

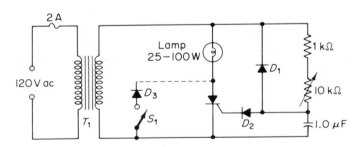

SCR = 200 V, 1 A (2N1597)
D_1, D_2, D_3 = 200 V, 1 A
T_1 = 100 W, 120 V isolation transformer

Figure E-15.1. *SCR phase control of power.*

3. Construct a 4-layer diode oscillator using the circuit shown in Fig. E-15.2. Measure the frequency range over which the oscillator operates and determine to what extent the frequency varies with the supply voltage. Explain what you have observed concerning the frequency variation.

4. Set up a voltage-variable, Triac, 150 W lamp dimmer (General Electric Co. S100F1 Sub-Assembly), and conduct a temperature rise test of the heat sink.

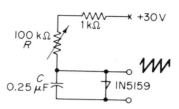

Figure E-15.2. *Four-layer diode oscillator.*

EXPERIMENT NO. 16. UNIJUNCTION TRANSISTORS

The unijunction transistor has a stable negative resistance characteristic that is easily observed experimentally. Most applications of the unijunction transistor use the device as a relaxation oscillator or as a monostable trigger circuit. Both circuits are examined in this experiment.

1. Assemble a unijunction transistor oscillator, using the circuit shown in Fig. E-16.1. Observe the operation of the circuit as a variable frequency oscillator, being careful not to short the base-one load when a large emitter capacitor is in use. Observe the waveforms and signal magnitudes available at the emitter, base-one, and base-two. Observe how the frequency varies with the supply voltage. Reduce the capacitor and measure the upper frequency limit.

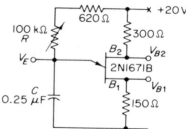

Figure E-16.1. *Unijunction transistor oscillator.*

2. Disconnect the capacitor C and vary R from 10 MΩ down to the 620 ohm minimum resistance shown in Fig. E-16.2. As R is varied, measure the emitter voltage E and measure or calculate the emitter current I. Plot the emitter voltage-current characteristics and obtain values for V_P, I_P, V_V, I_V, and η.

Figure E-16.2. *Unijunction transistor test circuit.*

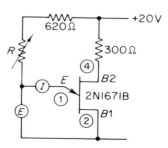

3. Connect the unijunction transistor, as shown in Fig. E-16.3 and show that the transistor has two stable Q-points that can be predicted by the curve obtained in step 2 above. Be sure to use the current-limiting resistors shown in series with B_1 and B_2.

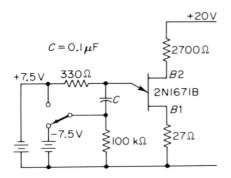

Figure E-16.3. Bistable unijunction switch.

4. Construct the multivibrator or the telemeter transmitter-receiver that is described in Chap. 15.

5. Construct the staircase wave generator or the tachometer described in Projects 10 and 12, following Experiment No. 17.

EXPERIMENT NO. 17. TUNNEL DIODE APPLICATIONS

The simplicity of the tunnel diode and the inherent speed of the device have opened many new applications areas. This experiment demonstrates the characteristics of tunnel diodes and shows how the circuit dc load line determines whether the device performs as a switch, an oscillator, or an amplifier. *CAUTION: Do not connect an ohmmeter to a tunnel diode.*

1. The experimental circuit shown in Fig. E-17.1 uses a 30 V laboratory power supply with a voltage divider that reduces the tunnel diode supply to $\frac{1}{10}$ the input voltage and provides a 300 Ω Thevenin equivalent source resistance. The tunnel diode should have a peak point current between 2 and 5 mA. When these conditions are met, the circuit is made to operate as a bistable or a monostable switch by varying the supply voltage. The

Figure E-17.1. Tunnel diode switching circuit and Thevenin equivalent.

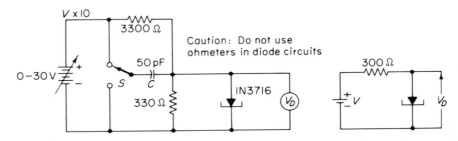

small capacitor and switch provide a fast pulse for triggering the tunnel diode from one state to the other.

Set up the circuit and measure the voltages V_D at which the tunnel diode switches as the supply voltage is raised and lowered. Make a rough plot of the tunnel diode current-voltage characteristic curve and estimate the dynamic resistance from the peak and valley coordinates. Estimate the time duration of the pulse produced by capacitor C as an estimate of the diode switching speed.

2. Set up the relaxation oscillator shown in Fig. E-17.2, but omit the capacitor. For L use an inductor which has 2 to 20 mH inductance with less than 20 Ω resistance, or use one winding of a line-to-line transformer. Adjust R to make the total resistance about 20 Ω. Examine the performance of the relaxation oscillator as the supply voltage is varied. Measure the inflection point voltages by use of the diode signal break points.

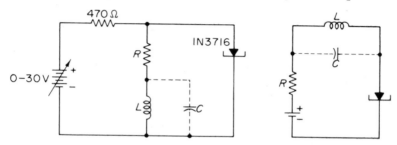

Figure E-17.2. Tunnel diode oscillator and Thevenin equivalent.

3. Select a capacitor that resonates with the inductance when the inductive reactance equals the negative resistance of the diode. Vary the capacitor over a 10-to-1 range and find the frequency range that produces a reasonable sinusoidal output. Compare the peak-to-peak output voltage with the linear portion of the diode curve. Examine the waveform of the signal across the tuned circuit and compare with the signal across the tunnel diode.

4. Construct a tunnel diode high-frequency amplifier, using a circuit given in the General Electric *Tunnel Diode Manual.*

ADDITIONAL PROJECTS FOR STUDY OR DEMONSTRATIONS

1. Transistor Tester   Ref. 12, p. 387, Ref. 7, p. 237
2. Demonstration of Shockley's Relation   Text, Sec. 2.6–2.9 incl.
3. 5 Watt Complementary Symmetry   Ref. 1, p. 190
   Amplifier

Appendix

SEMICONDUCTORS USED IN THIS BOOK

Transistors

JEDEC No.	Type	β min. max.	V_{CBO} (volts)	Power (watts)	Use	Source
2N277	Ge-pnp	25–70	−40	25	High power	Delco
2N699	Si-npn	40–120	120	0.6	$f_T > 50$ MHz	Motorola
2N1177	Ge-pnp	100-up	−30	0.08	RF amp.	RCA
2N1304	Ge-npn	40–200	25	0.15	Amp./Sw.	Tex. Instr.
2N1305	Ge-pnp	40–200	−30	0.15	Amp./Sw.	Tex. Instr.
2N1502	Ge-pnp	25–100	−40	10	Med. power	Solitron
2N1711	Si-npn	100–300	75	0.8	Amplifier	Fairchild
2N2484	Si-npn	100–500	60	0.3	Low noise	Fairchild
2N2869	Ge-pnp	50–160	−60	15	Med. power	RCA
2N2870	Ge-pnp	50–160	−80	10	Med. power	RCA
2N3053	Si-npn	50–250	60	3	Low power	RCA
2N3214	Ge-pnp	25–100	−60	6	Med. power	Delco
2N3215	Ge-pnp	25–100	−40	6	Med. power	Delco
2N3440	Si-npn	40–160	250	10	Med. power	Fairchild
2N3568	Si-npn	40–120	80	0.3	Amp./Sw.	Fairchild
2N3569	Si-npn	100–300	40	0.3	Amp./Sw.	Fairchild
2N3638	Si-pnp	20–130	−25	0.3	Amp./Sw.	Fairchild
2N3638A	Si-pnp	100-up	−25	0.3	Amp./Sw.	Fairchild
2N3784	Ge-pnp	20–200	−30	0.15	$f_T > 700$ MHz	Motorola
2N5191	Si-npn	25–100	60	20	Low power	Motorola
2N5194	Si-pnp	25–100	−60	20	Low power	Motorola

Field-Effect Devices

JEDEC No.	Channel Type	g_m approx. (μmho)	I_{DSS} (mA) min. max.	V_{DS} (volts)	Use	Source
2N2386	p	1000	1 −15	−20	Amplifier	Tex. Instr.
2N3086	n	800	0.8– 3.0	40	//	Crystalonics
2N3370	n	1500	0.1– 0.6	40	//	Un. Carbide
2N3687	n	1000	0.1– 0.5	50	//	Un. Carbide
2N3696	p	1000	0.5– 1.5	−30	//	Un. Carbide
2N4220	n	2500	0.5– 3.0	30	Switch	Motorola
2N4360	p	4000	4 −12	−20	//	Fairchild
2N4351	n-MOS	1000	1 −10	25	Amplifier	Motorola
2N4352	p-MOS	1000	1 −10	−25	//	//
3N187	n-MOS	7000	5 −30	20	Dual-gate	RCA

Miscellaneous Devices

JEDEC No.	Type	Use	Source
1N60	Ge — 30 V	UHF signal	Sylvania
1N82	Si — 5 V	UHF mixer	Sylvania
(Diodes)	Silicon	Rectifiers, 1 A, 400 V	Any available
1N5159	4-layer diode	Switch, 10 V	Motorola
2N1597	SCR	Controlled rectifier	Gen. Electric
2N1671B	Unijunction	Oscillator	Gen. Electric
1N3716	Tunnel diode	Oscillator/Switch, 4.7 mA	Gen. Electric
μA741C	IC OP amp.	Compensated amp.	Fairchild
FH1100	Schottky	UHF Diode	Fairchild

SEMICONDUCTOR INTERCHANGEABILITY

Most low-power transistor circuits operate satisfactorily when similar devices with equivalent current gains are substituted, provided the bias is adjusted. Circuits operating at high frequencies or switching at high speeds require transistors that meet the frequency and speed requirements indicated in the data sheets by the cutoff frequency f_β or the switching times. The demands on power transistors are more exacting, and many devices are inferior even though more expensive. The substitution of power transistors is not recommended because the recommended types have been carefully selected. For equipment repairs, a better device should be substituted, if possible, and the bias should always be adjusted.

Field-effect transistors are improving rapidly, and substitutions can be made providing the g_m and the I_{DSS} ratings are similar. New types are priced reasonably, while the prices of the early types remain high. Amplifier applications require high g_m at low I_{DSS} values and the most inexpensive devices generally prove unsatisfactory.

Silicon diode types are not specified in the circuits because almost any 400 V, or 200 V, 1 A diode should be satisfactory. For experimenter purposes the voltage rating of diodes should be 3 to 4 times the rms voltage input to the diode. The current rating should be 3 to 5 times the diode dc current. Excellent high-current high-voltage diodes may be obtained at reasonable prices. Diodes produced in small quantities may be surprisingly expensive.

TRANSISTOR BASING DIAGRAMS

The diagrams are given only as guides because there are many exceptions. All diagrams are viewed towards leads (bottom view).

Small metal case or round plastic

Plastic, round or flat

Power, diamond or Delco

Silicon controlled rectifier

Unijunction

FETs: No standard basing. Identify gate with ohmmeter

Diodes: The cathode, or plus, is the marked end.

ZENER DIODE VOLTAGES

(Standard Tolerance ±5%, ±10%, ±20%)

2.4	4.3	7.5	13	24	43	75	130
2.7	4.7	8.2	15	27	47	82	150
3.0	5.1	9.1	16	30	51	91	160
3.3	5.6	10.0	18	33	56	100	180
3.6	6.2	11.0	20	36	62	110	200
3.9	6.8	12.0	22	39	68	120	

TYPICAL BASE-EMITTER VOLTAGE CHARACTERISTICS

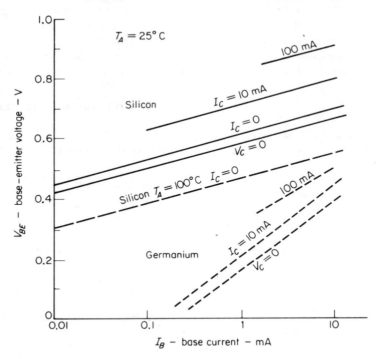

Base-emitter characteristics for typical low power silicon planar transistors and germanium transistors, dotted. Note: All curves move down 2 mV/°C increase of junction temperature.

TYPICAL COMMON BASE COLLECTOR CHARACTERISTICS

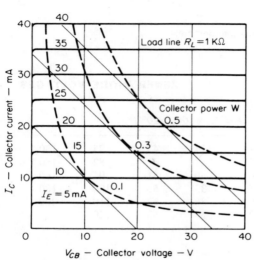

Common base collector characteristics for a typical low power silicon planar transistor; dotted durves show collector power.

TYPICAL COMMON EMITTER CURRENT GAIN CHARACTERISTICS

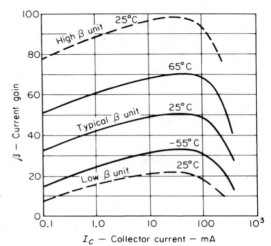

Current gain characteristics of a typical low power silicon planar transistor. Note variation with temperature and with transistor.

TYPICAL COMMON EMITTER COLLECTOR CHARACTERISTICS

Common emitter collector characteristics for a typical low power silicon planar transistor. Note that base current changes with β.

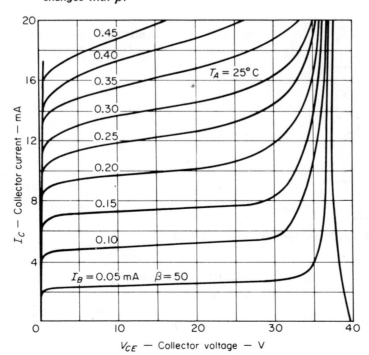

TRANSISTOR AMPLIFIER APPROXIMATIONS

For: $R_E + \beta' R_B \gg h = 26/I_E$ (ohms, mA)

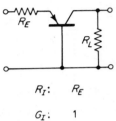   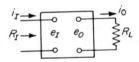   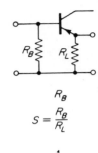

R_I:	R_E	R_B	R_B
G_I:	1	$S = \dfrac{R_B}{R_E}$	$S = \dfrac{R_B}{R_L}$
G_V:	$\dfrac{R_L}{R_E}$	$\dfrac{R_L}{R_E}$	1

TRANSISTOR GAIN-IMPEDANCE RELATION:

$$G_v = G_i \frac{R_L}{R_I} \quad \text{or} \quad G_i = G_v \frac{R_I}{R_L}$$

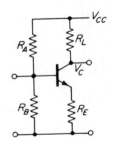

SINGLE-STAGE EMITTER FEEDBACK DESIGN

1. Select: $\qquad\qquad\qquad R_1, R_L, V_{CC}$

For iterative stages make: $\quad R_I = R_L$

2. Select: $\qquad\qquad\quad S = 20 \quad$ for *high gain*

$\qquad\qquad\qquad\qquad\quad S = 10 \quad$ for *stability*

$\qquad\qquad\qquad\qquad\quad S \leqq 5 \quad$ for *power*

then: $\qquad$ $G_v = S\dfrac{R_L}{R_I}$ $\qquad$ $G_i = S$

3. Make: $\qquad$ $R_B = R_I$ $\quad$ and $\quad$ $R_E = \dfrac{R_B}{S}$

4. Bias to make: $\qquad$ $V_C \geqq \dfrac{V_{CC}}{2}$

Try: $\qquad$ $R_A \simeq R_B \dfrac{R_L}{R_E}$

SINGLE-STAGE COMBINED FEEDBACK DESIGN

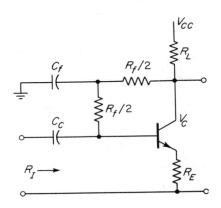

1. Select: $\qquad$ $R_S,\ R_I,\ R_L,$ and V_{CC}

For iterated stages make: $\qquad$ $R_S = R_L$

For significant feedback make:
$$R_I = 3R_S$$

2. Select: $\qquad$ $S \leqq \dfrac{\beta}{3}$ $\quad$ for Q-point stability

and $V_C = \dfrac{S}{\beta} V_{CC}$

3. Make: $\qquad$ $R_E = \dfrac{R_I}{\beta},$ $\quad$ and $\quad$ $R_f = SR_L$

Then: $\qquad$ $G_v \cong \dfrac{R_L}{R_E}$

4. For capacitors at f_1

try: $\qquad$ $X_{Cf} = \dfrac{R_f}{20},$ $\quad$ and $\quad$ $X_{Cc} = R_I$

FORMULAS

Ohm's law: **Power:**

$$\text{dc:} \quad I = \frac{E}{R}, \qquad \text{ac:} \quad I = \frac{E}{Z} \qquad W = EI = \frac{E^2}{R} = I^2 R$$

Reactance:

$$X_L = 2\pi f L = \omega L \qquad X_C = \frac{10^6}{2\pi f C} = \frac{10^6}{\omega C} \qquad (C \text{ in } \mu\text{F})$$

Resonant Frequency: **Rise Time:**

$$f_0 = \frac{1}{2\pi\sqrt{LC}} \qquad\qquad t_r \cong \frac{1}{3f_h}$$

Energy Stored:

$$W = \frac{LI^2}{2} \qquad W = \frac{CV^2}{2} \qquad \text{(watt-sec)}$$

Q-factor:

$$Q = \frac{\omega L}{R_s} \quad (R_s \text{ in series}) \qquad Q = \frac{R_p}{\omega L} \quad (R_p \text{ in parallel})$$

$$Q = \frac{1}{\omega C R_s} \qquad\qquad\qquad Q = \omega C R_p$$

Dissipation Factor:

$$D = \frac{1}{Q} = \omega C R_s$$

Admittance: **Impedance:**

$$Y = \frac{1}{Z} = G_p + jB_p \qquad\qquad Z = R_S + jX_S$$

Conductance: **Susceptance:**

$$G = \frac{1}{R_P} = \frac{R_S}{R_S^2 + X_S^2} \qquad B = \frac{1}{X_P} = \frac{-jX_S}{R_S^2 + X_S^2}$$

Power Transfer:

$$P_O = \frac{e_I^2 R_L}{(R_G + R_L)^2} \qquad P_{\max} = \frac{e_I^2}{4R_G} \qquad R_L = R_G$$

Voltage Loss:

$$\frac{e_O}{e_I} = \frac{Z_L}{Z_G + Z_L}$$

Voltage Gain:

$$G_v = \frac{g_m R_L}{(R_L/R_O) + 1} \cong g_m R_L \qquad R_O \gg R_L$$

Transformer Equation:

$$e = 4.4 f N A_C B_m \times 10^{-8}, \qquad A_C B_m \cong \text{lines}$$

ALTERNATING CURRENT RELATIONS

Ohm's law (ac):

$$I = \frac{E}{Z}$$

$$P = EI \cdot P.F.$$

$$P.F. = \cos\theta = \frac{R}{Z}$$

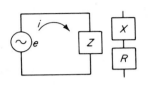
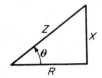

Series R and L:

$$|Z| = \sqrt{R^2 + X_L^2}$$

$$\theta = \tan^{-1}\frac{X_L}{R}$$

$$Q = \frac{X_L}{R}$$

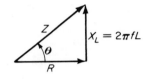

$X_L = 2\pi fL$

Series R and C:

$$|Z| = \sqrt{R^2 + X_C^2}$$

$$\theta = -\tan^{-1}\frac{X_C}{R}$$

$$Q = \frac{X_C}{R}$$

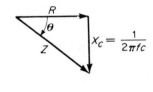

$X_C = \frac{1}{2\pi fc}$

Series R, L, and C:

$$|Z| = \sqrt{R^2 + (X_L - X_C)^2}$$

$$\theta = \tan^{-1}\frac{X_L - X_C}{R}$$

$$Q = \frac{X_L}{R} \qquad \omega_0 \equiv \frac{1}{\sqrt{LC}}$$

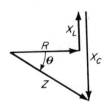

Series Equivalent of Parallel Impedances:

$$Z = \frac{Z_1 Z_2}{Z_1 + Z_2} = R_E + jX_E$$

The series impedance is equivalent at all
frequencies; hence, the three quantities
θ (which is equal to $\tan^{-1} X_E/R_E$), X_E,
and R_E all vary with frequency.

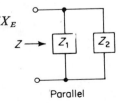

Parallel

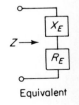

Equivalent

Parallel R and L:

$$R_E = \frac{RX_L^2}{R^2 + X_L^2}$$

$$X_E = \frac{R^2 X_L}{R^2 + X_L^2}$$

$$|Z| = \frac{RX_L}{\sqrt{R^2 + X_L^2}}$$

$$\theta = \tan^{-1} \frac{X_L}{R} \qquad Q = \frac{R}{X_L}$$

Parallel R and C:

$$R_E = \frac{RX_C^2}{R^2 + X_C^2}$$

$$X_E = \frac{-R^2 X_C}{R^2 + X_C^2}$$

$$|Z| = \frac{RX_C}{\sqrt{R^2 + X_C^2}}$$

$$\theta = -\tan\frac{X_C}{R}$$

$$Q = \frac{R}{X_C}$$

Parallel R and X for $Q > 3$:

$$R_S R_P = X^2$$

$$X_P = X_S$$

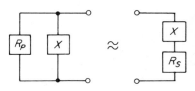

Parallel L and C with series R:

$$Z \cong \frac{Z_L Z_C}{Z_L + R_S + Z_C}$$

$$|Z| \cong \frac{(\omega_0 L)^2}{\sqrt{R_S^2 + (X_L - X_C)^2}}$$

$$|Z| \cong Q\omega_0 L \qquad \text{at resonance } (X_C = X_L = \omega_0 L)$$

For parallel R_p add:

$$R'_s = \frac{(\omega_0 L)^2}{R_p}$$

TRANSFER CHARACTERISTICS

Impedance Loss:

$$\frac{I_2}{E_1} = \frac{1}{Z_1 + Z_2} \qquad \frac{E_2}{E_1} = \frac{Z_2}{Z_1 + Z_2}$$

$$\frac{I_2}{I} = \frac{Z_1}{Z_1 + Z_2} \qquad \frac{E_2}{I} = \frac{Z_1 Z_2}{Z_1 + Z_2}$$

$$\frac{I_3}{I} = \frac{Z_1}{Z_1 + Z_2 + Z_3} \qquad \frac{E_3}{I} = \frac{Z_1 Z_3}{Z_1 + Z_2 + Z_3}$$

$$\frac{I_3}{E_1} = \frac{Z_2}{Z_1 Z_2 + Z_2 Z_3 + Z_1 Z_3}$$

$$\frac{E_3}{E_1} = \frac{Z_2 Z_3}{Z_1 Z_2 + Z_2 Z_3 + Z_1 Z_3}$$

$$\frac{E_4}{E_1} = \frac{Z_2 Z_4}{Z_1 Z_2 + (Z_1 + Z_2)(Z_3 + Z_4)}$$

$$\frac{I_O}{I} = \frac{Z_1 Z_2}{Z_1 Z_2 + Z_2 Z_3 + Z_1 Z_3}$$

$$\frac{E_O}{I} = \frac{Z_1 Z_2 Z_3}{Z_1 Z_2 + Z_2 Z_3 + Z_1 Z_3}$$

Wheatstone Bridge:

For $e_O = 0$,

$$R_1 R_4 = R_2 R_3 \qquad \text{or} \qquad R_4 = \frac{R_2 R_3}{R_1}$$

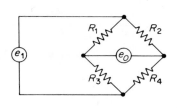

For a small off-balance,

$$e_O = \frac{e_1 R_3}{(R_3 + R_4)^2} \Delta R_4$$

Reactance Loss:

Low pass, phase lag:

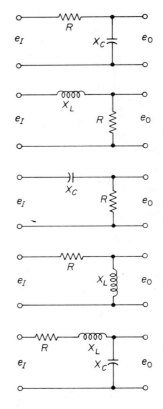

$$\left|\frac{e_o}{e_I}\right| = \frac{X_C}{\sqrt{R^2 + X_C^2}} \qquad \theta = -\tan^{-1}\frac{X_C}{R}$$

$$\left|\frac{e_o}{e_I}\right| = \frac{R}{\sqrt{R^2 + X_L^2}} \qquad \theta = -\tan^{-1}\frac{X_L}{R}$$

High pass, phase lead:

$$\left|\frac{e_o}{e_I}\right| = \frac{R}{\sqrt{R^2 + X_C^2}} \qquad \theta = \tan^{-1}\frac{X_C}{R}$$

$$\left|\frac{e_o}{e_I}\right| = \frac{X_L}{\sqrt{R^2 + X_L^2}} \qquad \theta = \tan^{-1}\frac{R}{X_L}$$

Series resonant circuit:

$$\left|\frac{e_o}{e_I}\right| = \frac{X_C}{\sqrt{R^2 + (X_L - X_C)^2}}$$

$$\theta = \tan^{-1}\frac{R}{X_L - X_C}$$

Q meter application: when $X_L = X_C$, max e_o,

$$\left|\frac{e_o}{e_I}\right| = \frac{X_L}{R} = Q$$

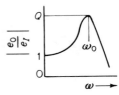

Shunt Tuned Circuit:

$$\left|\frac{e_o}{e_I}\right| = \frac{L/C}{\sqrt{(L/C)^2 + R^2(X_L - X_C)^2}}$$

$$\theta = \tan^{-1} R\frac{X_L - X_C}{L/C}$$

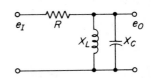

FOURIER SERIES OF COMMON WAVEFORMS

$$y(t) = \frac{4A}{\pi}\left(\sin\theta + \frac{1}{3}\sin 3\theta + \frac{1}{5}\sin 5\theta + \cdots\right) \qquad \theta = 2\pi\frac{t}{T}$$

Square wave

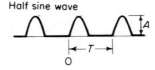

$$y(t) = \frac{2A}{\pi}\left(\frac{1}{2} + \sin\theta + \frac{1}{3}\sin 2\theta - \frac{1}{15}\sin 4\theta + \cdots\right)$$

Half sine wave

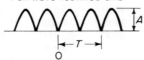

$$y(t) = \frac{4A}{\pi}\left(\frac{1}{2} + \frac{1}{3}\sin 2\theta - \frac{1}{15}\sin 4\theta + \frac{1}{30}\sin 6\theta + \cdots\right)$$

Full wave rectified sine

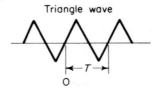

$$y(t) = \frac{8A}{\pi^2}\left(\sin\theta - \frac{1}{9}\sin 3\theta - \frac{1}{25}\sin 5\theta - \frac{1}{49}\sin 7\theta - \cdots\right)$$

Triangle wave

$$y(t) = \frac{A}{\pi}\left(\frac{\pi}{2} - \sin\theta - \frac{1}{2}\sin 2\theta - \frac{1}{3}\sin 3\theta - \cdots\right)$$

Sawtooth wave

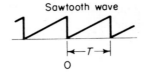

COMPOSITION RESISTOR VALUES

Standard EIA Resistance Values. Figures in bold type are 10% EIA values. All values listed are available in 5% tolerance.

Ohms	Ohms	Ohms	Ohms	Ohms	Ohms	Megs.	Megs.
1.0	9.1	**82**	**680**	**5600**	**47000**	0.36	3.0
1.1	**10**	91	750	6200	51000	**0.39**	**3.3**
1.2	11	**100**	**820**	**6800**	**56000**	0.43	3.6
1.3	**12**	110	910	7500	62000	**0.47**	**3.9**
1.5	13	**120**	**1000**	**8200**	**68000**	0.51	4.3
1.6	**15**	130	1100	9100	75000	**0.56**	**4.7**
1.8	16	**150**	**1200**	**10000**	**82000**	0.62	5.1
2.0	**18**	160	1300	11000	91000	**0.68**	**5.6**
2.2	20	**180**	**1500**	**12000**		0.75	6.2
2.4	**22**	200	1600	13000	Megs.	**0.82**	**6.8**
2.7	24	**220**	**1800**	**15000**	**0.1**	0.91	7.5
3.0	**27**	240	2000	16000	0.11	**1.0**	**8.2**
3.3	30	**270**	**2200**	**18000**	**0.12**	1.1	9.1
3.6	**33**	300	2400	20000	0.13	**1.2**	**10.0**
3.9	36	**330**	**2700**	**22000**	**0.15**	1.3	11.0
4.3	**39**	360	3000	24000	0.16	**1.5**	**12.0**
4.7	43	**390**	**3300**	**27000**	**0.18**	1.6	13.0
5.1	**47**	430	3600	30000	0.20	**1.8**	**15.0**
5.6	51	**470**	**3900**	**33000**	**0.22**	2.0	16.0
6.2	**56**	510	4300	36000	0.24	**2.2**	**18.0**
6.8	62	**560**	**4700**	**39000**	**0.27**	2.4	20.0
7.5	**68**	620	5100	43000	0.30	**2.7**	**22.0**
8.2	75				**0.33**		

RMA RESISTOR COLOR CODE

Black — 0 Green — 5
Brown — 1 Blue — 6
Red — 2 Purple — 7
Orange — 3 Gray — 8
Yellow — 4 White — 9

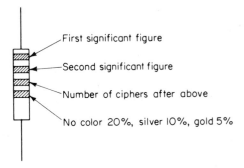

First significant figure
Second significant figure
Number of ciphers after above
No color 20%, silver 10%, gold 5%

The first 4 colors are used most frequently. The cipher color is useful for evaluating the order of magnitude of a resistor. A gold band means $\times \frac{1}{10}$ and silver means $\times \frac{1}{100}$.

PARALLEL RESISTANCE NOMOGRAM

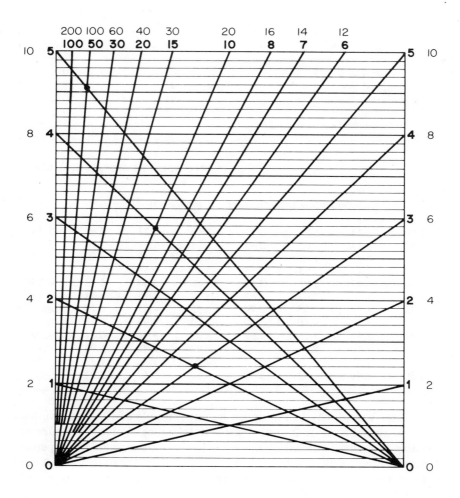

Use all inside numbers or all outside numbers. Use left side for the lower resistance.

Examples: $2\ k\Omega \parallel 3\ k\Omega = 1.2\ k\Omega$

$800\ \Omega \parallel 2\ k\Omega = 570\ \Omega$

$100\ k\Omega \parallel 10\ k\Omega = 9.1\ k\Omega$

RMA COLOR CODE EXAMPLES

Examples

Colors	Ohms	Tolerance
Brown, black, red	1000	20%
Red, red, red	2200	20%
Red, red, black, silver	22	10%
Red, red, gold	2.2	20%
Green, brown, green, gold	5.1 Meg	5%

MIL-BELL RESISTANCE VALUES

Table for Determining MIL-BELL Standard Resistance Values

1.00	**1.33**	**1.78**	**2.37**	**3.16**	**4.22**	**5.62**	**7.50**
1.02	1.37	1.82	2.43	3.24	4.32	5.76	7.68
1.05	*1.40*	*1.87*	*2.49*	*3.32*	*4.42*	*5.90*	*7.87*
1.07	1.43	1.91	2.55	3.40	4.53	6.04	8.06
1.10	**1.47**	**1.96**	**2.61**	**3.48**	**4.64**	**6.19**	**8.25**
1.13	1.50	2.67	2.67	3.57	4.75	6.34	8.45
1.15	*1.54*	*2.05*	*2.74*	*3.65*	*4.87*	*6.49*	*8.66*
1.18	1.58	2.10	2.80	3.74	4.99	6.65	8.87
1.21	**1.62**	**2.15**	**2.87**	**3.83**	**5.11**	**6.81**	**9.09**
1.24	1.65	2.21	2.94	3.92	5.23	6.98	9.31
1.27	*1.69*	*2.26*	*3.01*	*4.02*	*5.36*	*7.15*	*9.53*
1.30	1.74	2.32	3.09	4.12	5.49	7.32	9.76

Note: Values may be multiplied by any multiple of 10. Values in **bold** type are available in 1, 2, and 5 percent tolerance. Values in *italics* are available in 1 and 2 percent tolerance. All values are available in 1 percent tolerance. Frequently, resistances from 100 ohms up are shown as a 4 digit number. The first 3 digits indicate the significant value; the last digit indicates the number of zeros to follow. Thus: 1-0-0-0 is 100 Ω, 2-1-5-2 is 21,500 Ω, and 1-4-7-4 is 1.47 MΩ. If a letter follows a number: *F* means 1 percent tolerance, *G* means 2 percent, and *J* means 5 percent.

DECIBEL FORMULAS AND TABLE

By definition the decibel is a logarithmic measure of power ratios:

$$dB \equiv 10 \log_{10} \frac{P_2}{P_1}$$

Ratios greater than 1 represent a power gain, +(plus) dB. For ratios less than 1, representing a loss, the dB value is found using the reciprocal of the ratio and the dB value is negative, —(minus) dB.

For convenience, the dB is used as a measure of voltage ratios, where:

$$dB = 20 \log_{10} \frac{E_2}{E_1}$$

However, voltage ratios expressed in dB units do not represent power ratios unless the voltages are measured across equal resistances.

Also, 0 dBm is used as a 1 mW power reference level.

A Practical Decibel Table

db	Voltage ratio	Power ratio	dB	Voltage ratio	Power ratio
0	1.00	1.00	14	5.0	25
1	1.12	1.26	16	6.3	40
2	1.26	1.59	17	7.1	50
3	1.41	2.00	20	10.0	100
4	1.58	2.51	25	17.8	316
5	1.78	3.16	30	31.6	10^3
6	2.00	4.00	40	10^2	10^4
8	2.51	6.31	60	10^3	10^6
10	3.16	10.0	80	10^4	10^8
12	4.0	15.9	100	10^5	10^{10}

STANDARD PREFIXES FOR UNITS

Multiple	Prefix	Symbol	Multiple	Prefix	Symbol
10^{12}	tera	T	10^{-1}	deci	d
10^9	giga	G	10^{-2}	centi	c
10^6	mega	M	10^{-3}	milli	m
10^3	kilo	k	10^{-6}	micro	μ
10^2	hecto	h	10^{-9}	nano	n
10	deka	dk	10^{-12}	pico	p

REACTANCE-FREQUENCY CHART

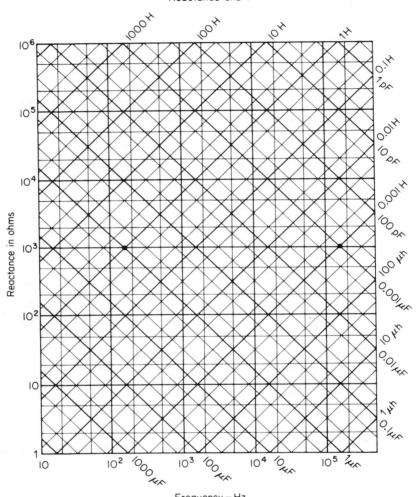

Reactance chart

Frequency – Hz

Note: All intermediate lines are at 2 and 5. For frequencies multiplied by 10, 100 or 1000, use μF and H reduced by 10, 100 or 1000.

Useful values:

$$1\ \mu\text{F} \approx 1\ \text{H} \approx 10^3\ \Omega \qquad \text{at} \qquad 160\ \text{Hz}$$

$$1\ \text{pF} \approx 1\ \mu\text{H} \approx 10^3\ \Omega \qquad \text{at} \qquad 160\ \text{MHz}$$

$$1\ \text{nF} \approx 1\ \text{mH} \approx 10^3\ \Omega \qquad \text{at} \qquad 160\ \text{kHz}$$

Index

Index

G

H

I

L

M